ALEXANDER REIN

THE BLUE STREAK

A HACKER'S GUIDE TO

SPECIAL RELATIVITY

TRAFFORD PUBLISHING

To *Kosta*, my brother

who first made me aware of the grand system of natural laws
governing just about every aspect of the extrenal world

National Library of Canada Cataloguing in Publication

Rein, Alexander
    The blue streak : a hacker's guide to special relativity
/ Alexander Rein.
Includes bibliographical references and index.

ISBN 1-4120-0153-6
    1. Special relativity (Physics) I. Title.
II. Title: Hacker's guide to special relativity.

QC173.65.R45 2003   530.11   C2003-901743-5

This book was published *on-demand* in cooperation with Trafford Publishing.
On-demand publishing is a unique process and service of making a book available for retail sale to the public taking advantage of on-demand manufacturing and Internet marketing. **On-demand publishing** includes promotions, retail sales, manufacturing, order fulfilment, accounting and collecting royalties on behalf of the author.

Suite 6E, 2333 Government St., Victoria, B.C. V8T 4P4, CANADA

Phone	250-383-6864	Toll-free	1-888-232-4444 (Canada & US)
Fax	250-383-6804	E-mail	sales@trafford.com
Web site	www.trafford.com	TRAFFORD PUBLISHING IS A DIVISION OF TRAFFORD HOLDINGS LTD.	
Trafford Catalogue #03-0521		www.trafford.com/robots/03-0521.html	

10   9   8   7   6

CONTENTS

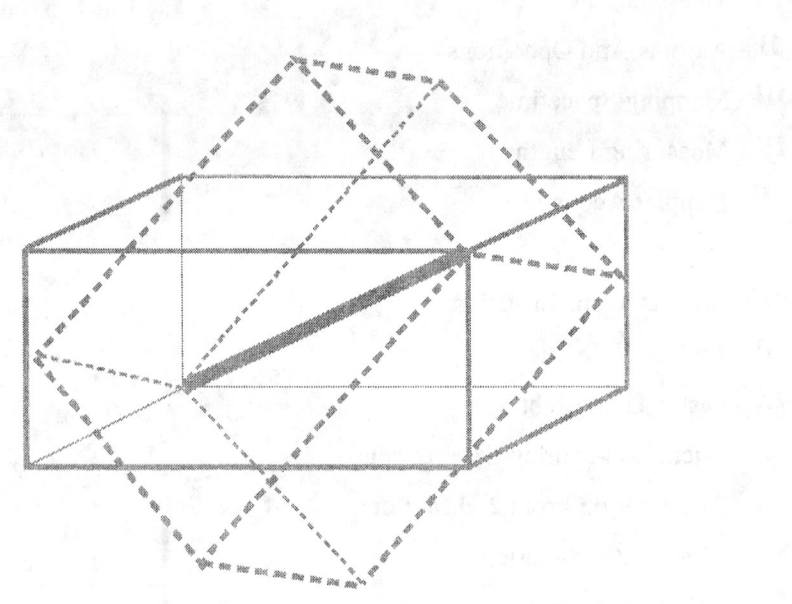

A - PREFACE

To whom it may concern

The hacker's approach to Special Relativity grew out of an attempt *to de-mystify the puzzling features of the theory to intelligent but intuition-blocked lay persons* by a strategy tailored to their handicap. The results of this attempt are contained in this book and you, too, are invited to become *a hacker*. The *invitation* is extended to all those *eager* to dig into Special Relativity but *lack confidence* in their math skills and are *intimidated* by the complexities of the theory. Potential hackers come at all ages, in all walks of life and with varying backgrounds in the sciences. And there are those high school and college *students* who were exposed to Relativity in the usual, superficial manner but nevertheless became intrigued by its mind-bending features. Perhaps *physics instructors* can also be joined as honorary hackers, helped to recall their own initial puzzlement and made to quit talking over the heads of their students.

Why Special Relativity?

The *Quantum Theory*, one of the two grand pillars of the 20th century physics, is basic to all sciences but is inaccessible to those with *limited knowledge* of the subatomic realm and *totally illiterate* in its complex mathematical language. With the *Special theory of Relativity*, the foundation of the other pillar, our prospects are much better. *For a fairly good grasp of its fundamentals, only a primitive, commonsense understanding of Space, Time and Motion is enough to get started.* But contrary to expectations, this theory is not pertinent to our everyday life because *it explains nothing we can see, hear and touch.* Still, there are enough reasons to study Relativity: (**1**) *technical* and *academic* career considerations, (**2**) just to gain *general insight into the most puzzling nature of Space and Time*, (**3**) to develop *a broader view of the overall structure of the universe*, (**4**) *to remain aware of the current scientific advances of our times*, but more often than not, there may simply be (**5**) *a nagging curiosity*, a desire to understand at least some of its enigmatic features.

This *curiosity* about Relativity ranges from being attracted to something *inscrutably mysterious* to a simple, straightforward wish to understand some of its *concrete particulars*. It is those fuzzy details of Relativity that have been an endless source of fascination. Popular accounts in various publications have certainly played their part keeping interest in Relativity alive by never failing to point out *its extraordinary consequences, often accompanied by impressive but dangerous-looking equations*. Discretely omitting their derivations may well have reduced math anxiety and prevented sudden development of reader block but still allowing access to some worthwhile wisdom from the more user-friendly text and pictures

Double trouble: math and intuition

(**1**) *A lack of math skills is the more obvious problem.* For instance: The first equation in Chapter I *(about the 4 dimensions)* displays the Theorem of Pythagoras in a deceptively simple notation: $a^2 + b^2 = c^2$. The letters *(a, b, c)*, numbers *(here only the number 2)* and operational signs *(+, -, =)* are familiar but what the equation actually says remains unclear to most readers. The more complex the equations the more mystifying their impact. The equations, however, become less alarming once their derivations are understood.

It's the derivations that are more important than the equations themselves because *not only do they explain the equations but they also lead, logically and mathematically, from the underlying Principles to the theory and from the theory to its consequences.* Mathematics is indispensable for understanding the theory but as the language of Special Relativity, it has always been a serious obstacle to the mathematically challenged.

(2) *Intuition blocks are less obvious but more obstinate than the challenges posed by mathematics.* Everyone has them to some extent. We are accustomed to the routine of slow speeds *(compared to the Velocity of Light)* and our minds are set on automatic pilot, locked in on the usual, familiar appearances. No wonder *our intuition is balking at things too strange and too abstract.* We are at ease in a world still primitive enough to be understood in terms of nothing more complicated than the ancient, *Aristotelian system of natural laws.* Much of what we know of the sciences was learned casually, without more than a passing thought as to what makes things work. As a result, we handle all kinds of objects without awareness of Newton's laws of motion, ride motorized vehicles without reflecting upon the thermodynamics that propels them, watch television without understanding its complicated circuitry and use all kinds of electronic gear that would not exist without the industrial application of Quantum Theory. The obedient response of our mechanized world to mere turning of switches is just another form of *magic* created by clever inventors good at converting arcane abstractions into practical applications. It is not at all surprising that *our initial reaction is often disbelief when first confronted with the seemingly unreal effects of Relativity.* Lacking simple mental models for the abstractions within the theory, we have *difficulty grasping key concepts* such as the *Equivalence of Linear Motion at uniform Velocity with Non-motion,* the *Invariant,* the *Four-dimensionality of Spacetime, Relativity of Simultaneity* and *the same, constant velocity of Light to all Observers.*

And the solution is.....

What you'll find here is an *intuition-oriented, geometrically conceptualized and mathematically developed exposition* of Special Relativity. It is *built up from scratch around First Principles* based on our *familiar, common-sense, primitive notions of Space, Time and Motion* which, in turn, are then continually *updated* in the exposition process. Throughout the book, a serious attempt is made to demonstrate that *Special Relativity is not anti-intuitive* as it is claimed to be. More specifically:

(1) *The insufficiently understood key concepts are first explained at length* to overcome the initial, ground-level obstacles to intuition.

(2) The *four-dimensional cornerstone of the theory, the Invariant, is presented as a geometric structure analogous to the Diagonal of a two-dimensional rectangle and of a three-dimensional box-like Frame of Reference to which Time can be added* without serious conceptual difficulty *as the Fourth Dimension.*

(3) The *tools for investigating and the terminology for describing Motion* are then provided by actually exploring Motion in a step-by-step fashion, always visualizing the crucial steps by suitable illustrations.

(4) The Spacetime *"terrain"* *(curved but not spherical)* and its various *"domains"* *(Past, Present and Future from Big Bang to Big Crunch or Chill)* are then defined and mapped.

(5) The *standard topics* of the theory are finally elaborated by the already introduced step-by-step method and its mathematical results, the *equations*, are not just displayed in print as customarily done in undergraduate textbooks but are *actually derived* from specific Spacetime situations using only elementary, highschool algebra, the simplest possible math for the task. These derivations are additionally *written out in longhand* for the benefit of those still inexperienced or whose math has become rusty. A minimal introduction to complex numbers is included.

(6) The theory of Special Relativity is covered in sufficient detail to make the book an optional supplement in a *college-level physics course.* It can also serve as a source of information and insight in *adult and high-school science clubs* but, above all, it was meant to be *a self-study manual*, a classroom at home, a do-it-yourself teaching program.

(7) Besides standard fare, *two speculative topics* are included: (a) a "Faster than Light" chapter deals with its main reputed consequence, the reversal of Time Arrow, at speeds that have "crashed" the "Light Barrier" and (b) a tentative answer is given to the question as to how a very-very fast-moving, familiar object would appear not only to our wide-open eyes but also to a super-fast, yet-to-be-invented picosecond shutter-speed camera.

Models and sources of inspiration

In my search for a personal tutorial touch, I was inspired and guided by a number of excellent works available for lay readership, most importantly by Hermann Bondi's "Relativity and Common Sense, A New Approach to Einstein." Bondi's remarkable ability to make a complex topic accessible set an example to be imitated. That mathematical derivations in *Sam Lilley's "Discovering Relativity for Yourself"* could be made transparently clear also came as a surprise so his way of setting up equations just had to be adopted. The descriptive terms, *Home Frame* and *Other Frame* along with many helpful pointers to Spacetime Map construction were taken from *Thomas Moore's "A Traveler's Guide to Spacetime."* Numerous other works listed under "Books Recommended For Independent Study" were combed for features useful in preparing the initial draft but at the end, it was impossible to recall all publications and accidental sources of specific ideas that helped to organize all the accumulated information and finally put everything together. A determined effort was made to *improve* on all explanations, line drawings and mathematics that in one form or other have become *standard teaching tools*. New *illustrations* were drawn whenever concepts and their ramifications could be visually amplified. To convey various *degrees of emphasis*, bold, bold-italic, all-capital lettering and parentheses were used. To award *special status to the key elements of the theory such as Space, Time, Motion, Light, Event, Moment, Observer, Signal, Length, Distance, Lighttimes, etc.,* those words in the text were routinely *capitalized*. Only *Minkowskian Spacetime Geometry* was used in preference to other versions like those by Brehme, Loedel and Gudermann. The somewhat variable *terminology* and *visual representations* in the literature necessitated a selection of sorts for the sake of uniformity throughout the text. On rare occasion, *new terms and abbreviations* were invented to bring the concepts conveyed into prominence or provide economy of expression. In all this, my own work was figuratively nothing more than *re-inventing wheels*. And following a time-honored custom, *I claim all mistakes* both big and small and those of commission and omission. These are all mine.

Intended omissions

After describing what is contained in this volume, it is only fair to point out *what you don't find.* These omissions are not so much about the theory itself but deal with the *general history of science, the borderline areas of physics leading to Special Relativity* and *the work done by individual contributors in the development of Special Relativity* at the turn of the century. These topics and the sources where you can find them are listed next:

(1) The *basic perceptions about nature* first put into words in a distant era shrouded in the mists of antiquity were soon followed by formulation of *ideas, categories of meaning and rules of reasoning.* Nascent harbingers of *scientific concepts* emerged from a *mix of mythology and inspired speculation* eventually evolving into *"natural philosophy,"* the true precursor of the theoretical-observational science known to us today. This topic is less relevant to our study and only one source is listed: D-6. Brief introductions to Relativity are given in A-1 through 5.

(2) A history of trail-blazing *discoveries, conflicts within theories and radically new developments often necessitated conceptual re-orientations in science.* These are described in a major work by a contemporary historian, T.S. Kuhn: D-7.

(3) A narrative, *non-mathematical overview of classical mechanics and electro-magnetism* at the border of old and new physics by Einstein and Infeld: D-1. Also see D-4.

(4) A description of the turn-of-the-century *experimental work with Light, radio waves and electrons (making up the new-fangled cathode rays)* taking place *at the cutting edge of science directly leading to Relativity*, by A. Pais: D-5.

(5) *Biographical detail about Einstein* as a person and a scientist in relation to his work, also by A. Pais: D-3.

Books recommended for independent study

(A) INTRODUCTORY AND LESS MATHEMATICAL. General orientation for those without previous exposure to Relativity.

(1) Nigel Calder: *Einstein's Universe.* The Viking Press, 1979.

(2) D.E. Mook, T. Vargish: **Inside Relativity.** Princeton University Press, 1987.

(3) Albert Einstein: *Relativity.* Original edition for popular audience *(translated)*, 1916.

(4) A. Einstein: *Relativity, The Special and the General Theory*. 1916, 15th Edition 1952. Three Rivers Press.

(5) Gary F. Moring: *The Complete Idiot's Guide to Understanding Eintein,* alpha books, 2000.

(B) MORE MATHEMATICAL BUT STILL QUITE ACCESSIBLE. The most important features of the first three sources in this group have been incorporated in this book. But *H. Bondi'* s book also contains an excellent introduction to Classical Mechanics connecting nicely with his exposition of Relativity. *T. H. Moore's* book covers relativistic mechanics intuitively and also features a short chapter on relativistic particle physics. *S. Lilley's* book also takes up General Relativity in considerable detail but still without the complex math tools like tensors and non-Euclidian geometry, both unforgiving obstacles for the mathematically challenged. The other books listed here have a number of strong points making them attractive enough to those able and willing to go boldly beyond the limits set in this volume:

(1) Hermann Bondi: *Relativity and Common Sense, A New Approach to Einstein.* Doubleday & Co., 1964. Dover Publications, Inc., 1080.

(2) Sam Lilley: *Discovering Relativity for Yourself.* Cambridge University Press, 1981

(3) Thomas A. Moore: *A Traveler's Guide to Spacetime, An Introduction to the Special Theory of Relativity.* McGraw-Hill, Inc., 1995.

(4) Lewis C. Epstein: *Relativity Visualized.* Insight Press, 1982.

(5) James H. Smith: *Introduction to Special Relativity.* Dover Publications, Inc., 1965, 1993.

(6) George F.R. Ellis and Ruth M. Williams: *Flat and Curved Space-Times.* Clarendon, 1981.

(7) Nick Herbert: *Faster than Light.* New American Library, 1988.

(8) Lawrence Sklar: *Space, Time, and Spacetime.* University of California Press, 1974.

(9) Edwin F. Taylor and John A. Wheeler: *Spacetime Physics, Introduction to Special Relativity,* 2nd Ed., W. H. Freeman and Co., 1992.

(10) Banesh Hoffman: *Relativity And Its Roots,* W. H.. Freeman and Company, 1983.

(C) MORE SUBSTANTIAL and considerably more technical, requiring greater expertise in physics and mathematics:

(1) Ray Skinner: *Relativity for Scientists and Engineers.* Dover Publications, Inc., 1982.

(2) E. A. Robinson: *Einstein's Relativity in Metaphor and Mathematics.* Prentice Hall, 1990.

(3) Arthur I. Miller: *Albert Einstein's Special Theory of Relativity*. Addison-Wesley, 1981.

(4) H.A. Lorentz, A. Einstein, H. Minkowski, H. Weyl: *The Principle of Relativity.* Original history-making articles on Special and General Relativity. Dover Publications, 1952.

(5) W. L. Burke: ***Spacetime, Geometry, Cosmology,*** University Science Books, Mill Valley, 1980.
(6) Max Born: ***Einsteins's Theory of Relativity,*** 1924 completely revised and reissued in 1962, Dover Publications, Inc., 1962.

(D) WITH MORE HISTORICAL PERSPECTIVE, very good for gaining a better understanding of why and how Special Relativity came into being. The references listed in this group make up for the brevity of general remarks in the Introduction and at the beginnings of most of the chapters:

(1) A. Einstein, L. Infeld: ***The Evolution of Physics.*** Simon and Schuster, 1966.
(2) Stanley Goldberg: ***Understanding Relativity.*** Birkhäuser, 1984.
(3) Abraham Pais: ***The Science and the Life of Albert Einstein.*** Clarendon Press, 1982.
(4) G. F. Moring: ***The Complete Idiot's Guide to Understanding Einstein.*** Alpha Books, 2000. This surprisingly good volume for the general public should not be missed.
(5) Abraham Pais: ***Subtle is the Lord.*** Oxford University Press, 1982.
(6) ***The Pre-Socratics,*** Edited by A.P.D. Mourelatos. Princeton University Press, 1993
(7) Thomas S. Kuhn: ***The Structure of Scientific Revolutions.*** Chicago Univ. Press, 1970.

(E) SOURCES CROSSING OVER TO OTHER AREAS OF PHYSICS is a very brief list of the many books that can be highly recommended for their pertinent cosmological content:

(1) Paul Davies: ***About Time.*** Touchstone *(Simon & Schuster)*, 1995.
(2) David Wick: ***The Infamous Boundary.*** Copernicus *(Springer-Verlag)*, 1996.
(3) Alan H. Guth: ***The Inflationary Universe.*** Addison Wesley, 1997.
(4) Craig J. Hogan: ***The Little Book of the Big Bang***. Springer-Verlag New York, Inc., 1998.
(5) Mario Livio: ***The Accelerating Universe***, John Wiley & Sons. 2000.
(6) J. Richard Gott: ***Time Travel In Einstein's Universe,*** Houghton Mifflin Company. 2001.

Respectability of hacking

In this book we'll ***construct the theory of Special Relativity from scratch by applying First Principles (explained further in Introduction)*** to our primitive notions of Space, Time and Motion. Each chapter begins with a brief general statement pertinent to the topic. From there on, it's all ***"nuts and bolts"*** in the genuine hacker style. It is worth pointing out that there is an interesting ***historical parallel*** to our chosen assignment. The earliest scientific investigators were complete amateurs consumed by intense curiosity about nature, actually ***"hackers"*** in the best sense before the word was even coined and before anyone could earn a living as a scientist. While we are not in a position to discover anything previously unknown, ours is still ***a quest*** of sorts, a serious attempt at re-discovery, to find out for ourselves what until recent years may well have been the exclusive domain of scientists, specialists, academics and engineers in full command of the basics of Relativity but a subject not yet brought to our attention in a sufficiently sensible, detailed and intuition-oriented way. Not until recently, perhaps.

What is our workshop?

Special Relativity deals with 4-dimensional Spacetime. But you should be forewarned that ***you'll never get a chance to step into all these four dimensions after the first chapter.*** In this book, almost all our studies are conducted in an environment that has absolutely no semblance to our ***three-dimensional SPACELAND*** *(also known as Sphereland)* or, with just a couple of exceptions, to the somewhat alien ***two-dimensional FLATLAND*** *(also known as*

Surfaceland) you may have heard about. After the first chapter, you get immediately "beamed up" to a strictly *one-dimensional LINELAND* for practically all the exercises. But studying fast Motion in the 1-D setting is *dangerous.* It is *like standing in the middle of a busy one-lane highway featuring two-way traffic.* Viewed a little differently, just imagine standing in pitch-black darkness equipped with a powerful, narrow-beam laser flashlight and a one-track mind intent on examining mechanical behavior of objects moving at high Velocities in the direct line of sight either towards you or away from you.

The benefits gained from the 1-dimensional arrangement are limited to keeping the equations simple and math anxiety under control, certainly a fair trade-off. But wait*!* There is no need to risk your valuable life for brief moments of wisdom. There is a way to cheat Fate. Haven't you heard of *"virtual reality?"* Well, this is our method of madness here, everything virtual, that is imaginary like virtual objects, virtual Motion, virtual Observers, all action staged as *"virtual experiments,"* also known as "thought experiments" *(named so by Einstein).* And it becomes so elementary*!* First, write down the specifics of the set-up and map it, then develop its consequences and, finally, figure out the math. Instead of using a computer program to visualize the virtual experiment at hand or do the inevitable equation crunching, everything is accomplished with our own *computer "greyware,"* our brains. It may be primitive, Stone Age computing but for resourcefulness and instant availability, it sure is hard to beat.

About the Index

A number of *key words, technical terms and abbreviations* are found throughout the text. These are usually introduced informally, at times only mentioned or used without formal definitions but always accompanied by explanations as to their meaning. The page references show where these are first used, where descriptions or explanations are located or where you find the context wherein these can be understood. If only a chapter number is given, the entire chapter is dedicated to that particular term, concept or topic.

Of the numerous *scientists* whose work significantly advanced the development of Special Relativity, only a few are mentioned. *The lack of emphasis on history is reflected by the absence of their full names and of all biographical detail.* It is not meant to minimize their importance in science or their contributions to the development, application or exposition of the theory. Those mentioned in this book are: Galileo, Newton, Lorentz, Poincare, Einstein, Weyl, Hilbert, Minkowski, Michaelson, Morley and Terrell. Others important in philosophy or in the physical sciences but only peripherally related to Relativity are: Aristotle, Pythagoras, Ptolemy, Kepler, Copernicus, Roemer, Bradley, Cherenkov, Feynman, Guth, Bohr and Kuhn.

About the Title

By the way, the distinctive expression in the title comes from an *old phrase about things moving real fast and proverbially called "Blue Streaks."* For a book about very fast Motion in Space and Time, I could not have found a more appropriate name. I hope that everything in this volume is interesting, stimulating, illuminating, perplexing, tantalizing, frustrating, etc. but totally rewarding of your best efforts.

A warning

Before you continue reading, it is only fair to point out that there is *a hidden danger* lurking in Chapter I and beyond. Once you master *(or re-master) elementary algebra (and a little trig),* fail to panic at the thought of *imaginary numbers,* become able to navigate around *intuition blocks* and manage marching in step to *step-by-step instructions,* you may find yourself *mentally trapped for good in Spacetime.* And there is *no turning back !!!*

B - INTRODUCTION

Benchmarks of the 20ᵀᴴ century

Toward the end of the 19th century, the physical sciences were placed firmly upon the laws of Motion set forth by Isaac Newton in his monumental *Principia.* The remarkable success of these laws in all areas of *Classical Mechanics* also ordained that the fundamental approach to other natural phenomena was to be based on mechanics, even when no obvious Motion of purely physical matter was involved as it was with Light, electricity and magnetism, the science of heat *(thermodynamics)* and the newly discovered radioactivity of certain minerals. At the turn of the century, further developments in these areas led to *Relativity* and *Quantum Mechanics,* both benchmarks of the 20th century.

Two theories of Relativity

The complete Theory of Relativity was formulated by Einstein in two creative spurts. The *SPECIAL THEORY* saw publication in 1905 and described the consequences of the simplest and the most basic kind of Motion, one that is linear in direction and uniform in Velocity. The mass-energy equivalence *(E = mc²)* came three months after the original paper. The *GENERAL THEORY* was an extension of the Special Theory and was wrought in a single-minded, thought-intensive, one-man project over ten additional years resulting in a reformulation of Newton's laws of gravitation in terms of non-Euclidian geometry. Special Theory, the simpler of the two, is invariably the entry point in all discussions of Relativity and a prerequisite to understanding General Theory. All controversies about Relativity have long been settled and it has found wide application in many areas of physics notably in electromagnetism, quantum theory and celestial mechanics. The absence of complete harmony between General Relativity and Quantum Theory is not entirely relevant to Special Relativity and is not taken up in this book. Mathematically, Relativity is self-consistent and elegant. Esthetically, it is beautiful and exhilarating.

Three roads to Special Relativity

Historically, there were three roads to Special Relativity: (1) experiments with cathode rays, (2) the Michaelson-Morley experiment, and (3) a re-evaluation of the theoretical foundations of Classical Mechanics.

Cathode ray experiments led the way. After the discovery that cathode rays were made up of streams of electrons "boiled" out of the heated cathode and propelled by high voltage toward the anode in the vacuum tube, the race was on to measure their chief properties such as charge, momentum and mass. Along with progress, a problem emerged. *Differences in measurements were far greater than could be blamed on experimental errors.* Scientists were groping for explanations. *Konrad LORENTZ*, a Dutch mathematician and the leading "armchair" *(theoretical)* physicist derived a *"fudge factor"* from electron velocity data to correct for the observed discrepancies. This is generally known as an "ad hoc" type of a solution. It led him to a cautious but *prescient conclusion that electrons in the cathode ray tube followed a variable "local time" in contrast to the constant "universal time"* believed to be progressing evenly, at the same pace, in all parts of the universe. In addition, Lorentz devised a versatile formula to convert, or "transform," measurements in one Inertial System into measurements in another System thereby creating the framework for an emerging Theory of Measurement applicable also to situations other than electrons in cathode ray tubes. Experimental science began to march toward *a revision of Classical (Newtonian) Mechanics.*

The next road to Relativity was paved by the unexpected *failure of the Michelson-Morley experiment in 1987 to prove the existence of "ether," the hypothetical medium for transmission of Light.* This medium was believed necessary for Maxwell's theory of electromagnetism and the experiment was designed to detect its influence upon the Velocity of Light in a moving system. The initial results were equivocal but over aperiod of years, the instrumental setup was gradually refined and the experiment was finally repeated under well-controlled conditions but with the same negative result. The Ether Theory became a casualty to this lack of verification. Other developments soon followed.

The third road consisted of *a rethinking in depth about the fundamental assumptions of classical (Newtonian) mechanics.* Because experiments in the 19th century had already generated a great deal of data that did not fit comfortably into existing theoretical mold, several heads at the cutting edge of new thinking were working out equations that described much of the newly discovered empirical relationships, essentially piecing together what started to look like *a novel physical theory. Lorentz, Poincare, Weyl* and *Hilbert* were in the forefront of this *new theoretical movement.* Relativity *(in this developmental phase),* still lacking a name, was described by later historians as already being felt "*in the air,*" so to speak. It would have been completed sooner or later, perhaps in a few more years. But it was *Einstein*, then an obscure clerk in the Swiss patent office who put the various loose ends together and showed it to be a consequence of the *EQUIVALENCE PRINCIPLE*, also known as the *PRINCIPLE of RELATIVITY* , contained in the writings of Galileo and subsequently expanded by Newton with his Laws of Motion. Three years after Einstein's famous 1905 paper, *a geometrical, four-dimensional interpretation* of the Theory was presented by Einstein's one-time teacher, *MINKOWSKI.* It immediately became the preferred way to visualize and understand Relativity. Minkowski also created much of the Spacetime terminology now in use and indispensable for describing it.

It is *important* to point out that unlike the later General Theory of Relativity, *Special Relativity was NOT a brainstorm of a single scientist. The initial ideas were formulated clearly during renaissance by Galileo and were later extended by Newton with his laws of Motion. Considerable theoretical progress was driven by difficulties with cathode ray experiments and everything already discovered was generalized and spectacularly reduced to FIRST PRINCIPLES by Einstein. The 4-dimensional conceptualization by Minkowski further prepared the way for the General Theory. The latter, in turn, provided gravitation a better explanation than given by Newton and supplied the broad field of cosmology the first crucial theoretical tool that changed it from idle speculation into a respectable scientific discipline.*

The First Principle, unbelievable my dear Galileo

Before the "nuts and bolts" of Special Relativity, we must take a good look at the *PRINCIPLE OF EQUIVALENCE or of RELATIVITY.* Note, please, that the Theory is a mathematical-geometrical elaboration of the Principle from which it was developed. Here in the Introduction, we do not start with the Theory, not yet. We'll just explain the Principle that is absolutely the most important statement about Motion and a result of an astute observation *(in the early 16th century)* by *GALILEO* who persuasively argued for it in his writings. He noticed that *the mechanical behavior of objects in a coach moving along on a smooth road was exactly the same as in a coach parked at the roadside.* That there should be no difference is obvious today. But what Galileo came up with amounted to an *ORIGINAL CONJECTURE* not only contrary to what was then held as true but what also violated the laws of Motion set down by an authority no less than *Aristotle.* A case in point: if you release an object directly above your toes, it does, of course, fall down upon them. It was then believed that if you repeated the same test in exactly the same way in a moving vehicle, your toes would move forward along with the vehicle and the object would fall behind your toes missing them altogether *(by the Distance traveled by the coach during the object's fall).* Similarly, if in a moving vehicle you would throw an object up vertically, your hand, held in the same position, could not possibly catch it because, again, the hand would move forward while the object was in flight. Since Galileo, it has become firmly established that *in uniformly moving systems, such as a fully enclosed vehicle traveling smoothly and without acceleration, all the known laws of mechanics described by Newton are completely valid and operate the same way as they are known to operate in a non-moving system.* For this reason, *a State of Uniform Motion cannot be distinguished from a State of Rest by any test based on the laws of mechanics.* Both of these states can, therefore, be considered *EQUIVALENT* if not entirely identical. This single principle is key to the Special theory of Relativity. In this form, it is referred to as the *GALILEAN PRINCIPLE of RELATIVITY. Special Relativity is what you get if you start out with the Principle of Relativity and proceed to its mathematical consequences.* And here it is, our first obstacle to intuition. Are we conquering it *?*

I can't believe I am moving

The smooth flight of a large commercial airliner flying at night with all the external visual clues as to Motion obliterated by darkness pretty well duplicates Galileo's experience in a smoothly riding coach. You can walk up and down the isle, have liquids served in a cup, even play ping pong or engage in complex mechanical activities without the slightest hint of Motion. The illusion is complete until the aircraft encounters turbulence or hits an up-or-down draft. But the inability to detect Motion in a smoothly moving System is no illusion. As an established scientific reality, it has never been disproved by any mechanical test.

Light cannot detect Motion either

This brings Relativity to where Galileo and Newton left it. There is just one more test that cannot distinguish Motion from non-Motion. That test is based on the most unique *VELOCITY OF LIGHT*. For reasons clear from the history of physics, Light was not considered as a test. Because it was believed to be a wave propagated at a constant velocity through a Space-filling substance called "ether," its measurable Velocity had to be influenced by the Velocity of the Observer's equipment moving through that same ether-filled Space. The failed attempts by Michelson and Morley to detect such an influence were ultimately explained by recognizing *Light as another test unable to distinguish Motion from non-Motion* (due to the Principle of Relativity). Those reluctant to embrace Relativity were finally forced to accept it.

A marriage of exclusives

Now *if Motion cannot be distinguished from non-Motion, the two phenomena ordinarily regarded as contrasting and mutually exclusive must then form a totally new and unique kind of a physical state.* This, indeed, had been recognized much earlier by none other than Newton and has been known as the *INERTIAL SYSTEM*. Here we no longer have either one *(Motion)* or the other *(non-Motion)* as there is *NO DIFFERENCE* between them. The only requirement for inclusion is that *Motion, if present, has to be uniform and linear,* that is, at a constant Velocity and along a straight line. All these requirements are satisfied for both Motion and non-Motion if there is *NO ACCELERATION (or deceleration), actually the one and only requirement.* For special reasons, *ABSENCE of ROTATION*, a state that includes acceleration, is also a requirement. Einstein generalized this indistinguishability by making the Relativity Principle *all-inclusive: The laws of nature are identical in all Inertial Systems.* For the benefit of those just too dense to understand that Light must be included as a test, he added *a corollary: The speed of Light is the same in all Inertial Systems*. That's all there is to it. The science of mechanics was never before placed upon a simpler and a more solid foundation. The streamlining of mechanics did come at a cost: The idea of an absolutely constant and even flow of Time throughout the universe had to be given up for good. More about it later in this book.

B - 4

Then who is moving?

Identification of the Inertial System was our first application of the Equivalence Principle. The second application concerns the *role of the OBSERVER* to whom some things clearly appear to move and who now has to decide who *(or what)* is moving and who *(or what)* is standing still. The whole question may seem utterly preposterous as we always have plenty of visual and other external clues as to Motion. If no mechanical or electromagnetic tests can determine which system is in Motion, what can we use as a standard? It is impossible in principle *(as we'll see)* to build a universal localization system *(similar to the Earth-bound global localization system)* capable of determining exactly where we are in Space and if our location is changing, e.g. if we are moving. Michelson and Morley *(Chapter X)* tried to do that and failed. As a concession to our personal ego-centricity and historical geo-centricity, *each and every inertial (non-accelerating) Observer can legitimately claim to be standing still* and being *the true STATIONARY REFERENCE POINT,* the one and only fixed point in the universe. As far from ideal as it may seem, it is the only option operationally available to anyone *(and to us, too).*

A clash of opinions

With two observers, both can claim to be standing still and both can point to the other as moving! What we have here is a case of instant *DISAGREEMENT*, a real clash of opinion. *It is impossible for Observers in different Inertial Systems to come up with identical observations* not only pertaining to the identity of who or what is moving but also when it comes to the more specific measurements of Distances and Time intervals. All this we'll see soon enough. And that is one more reason why everything gets so awfully complicated. And complicated it really is. And there are still more surprises to come.

As all observations must be considered true and the theory kept logically consistent as well, we must remove the conflict by recognizing all observations as being OBSERVER-LINKED, same as OBSERVER-DEPENDENT, one more obstacle to overcome. *All observations, and that includes also all measurements, must, therefore, be tagged as to WHO made them.* Without such identification, all measurements are incomplete. Fortunately, a suitable terminology has been developed that automatically attaches this identification to all measurements. And we better adhere to this practice throughout our exercises. Observer Dependency is something with which we'll have to live happily *(or put up with)* ever afterward.

Observations come in "Frames"

The Observer now makes all observations and measurements from his unique point of view that includes everything not in Motion relative to him. Everything *NOT* in Motion makes up his Inertial System, his *FRAME of REFERENCE.* This Frame of Reference is the Observer's own *HOME FRAME*. A moving Observer's Home Frame is located in stationary Observer's *(and our) OTHER FRAME (of Reference)*, not in his *(and our)* Home Frame. If it sounds confusing, it certainly is. Don't worry. It'll get much worse before it gets better. Get used to this set of strange concepts and do your best to understand it. This way you'll avoid a lot of bewilderment later on.

Basic observable elements

In Special Relativity, we deal chiefly with *SPACE, TIME AND MOTION*. But can we define them? Various dictionaries and encyclopedias describe them in different ways but always with an orientation that carries an unmistakable stamp of Relativity linking Space and Time so inseparably that considering them in isolation from each other is impossible. The resulting union is known as *SPACETIME. Motion is defined NOT as a change of position in some absolute Space but only in relation to the Observer (who is) viewing that Motion.* Absolute Motion a la

Newton is, therefore, rendered meaningless. Time is a particularly difficult concept. The Encyclopedia Britannica has ten and a half pages on Time, discussing it in various categories o f meaning. Physical sciences have gradually narrowed down Time just to a single category. Operationally, a definition of this kind has come to us in form of a quip by Einstein who said that *"Time is what is measured by clocks."* What a strange definition, almost absurd *!!!* It is very much like saying that water is what is measured by buckets and wheat is what is measured by bushels. But on second thought, this definition of Time does have precision that the other, similar definitions of water and wheat lack because of the following implied disclaimers: first, clocks do not measure anything but Time and, second, *only the measurable aspect of Time is important in physical sciences, everything else being irrelevant*. In this way, we focus our attention upon the one and only physical attribute of Time we can describe by simply measuring it. It is fortunate that Time, unlike Space, does not have more than one dimension. Spacetime in four dimensions is tough enough a nut to crack and we definitely don't need to get lost in more dimensions.

How to link Space with Time?

In our study, the linkage of Space, Time and Motion is given full recognition by the way they are represented on paper, e.g. on our *SPACETIME MAPS*. A *marker* specifying *a definite location in Spacetime* is called an *EVENT*, another notion for us to accept and use.

An Event is something that just *happens* or is planned to happen. But we need to refine this crude concept further by *limiting its extension in Time and in Space.* In our precision-obsessed study we'll be using ultra-short flashes of Light or blips of Radar Signal from very small sources located accurately somewhere in Space *(but again and always)* in relation to the Observer to mark specific Events in both Space and Time. In this way *the Space and Time extensions of the Event are approximated to zero* and *Events become similar to mathematical points on a line, surface or Space of any number of dimensions.* Also, *Reflections* and *Receipt of reflected Signals* are Events. All those Signals link Events to Observers who can use them to measure *Space and Time (Coordinate) SEGMENTS (Distances and Time intervals)*. With such handy tools at our disposal, we can stage science experiments in any place at all, even while seated in a comfortable armchair.

Time can measure Distances

In order to keep *Space on equal terms with Time*, we must allow either one to be expressed in terms of the other. Conventionally, however, we measure *Space in terms of Time.* An easily understood example should help out. Imagine a filling station attendant at the outskirts of a small town sitting on a chair waiting for customers. One finally arrives and asks: "how far is the next town from here?" The attendant answers: "about an hour, my friend." After a while, another comes from the same direction asking the same question. This time the answer is: "a good five hours." His answer to the third traveler is: "twenty hours if you can make it." It is not at all unusual to give Distances in terms of the Time needed to reach them. What is unusual here are the widely differing answers. But it all makes good sense once we know that the first traveler was in a car, the second on a bicycle and the third one on foot. As you see, Distances can be given in terms of Time of travel. So *Time can be used as a measure of Distance* after all.

Lighttimes as Distance units

But are there better measuring sticks than cars and bicycles? Well, *Light has exactly the same Velocity to all inertial Observers* under any and all conditions *(as we'll see)*. It is, therefore, a true *ABSOLUTE* in the universe and should be *ideally suited to calibrate both Space and Time*. The use of Lighttimes to convey Distance measurements is not new. It is well estab-lished in astronomy. For instance, Alpha Centauri is 4.3 Lightyears away. Our Sun is about 8.4 Lightminutes from us. The moon is one Lightsecond away. Jupiter can be one Lighthour or a lot more away depending upon its orbital position around the Sun. The enormous Distances in astronomy are measured in thousands, millions and even billions of Lightyears. Lighttimes have become standard Distance units and are part of the *International Standard Nomenclature of Relativity*. Expressing Time in terms of Distance traveled by Light *(Lightdistances in contrast to Lighttimes)* is not a common practice. Doing it the other way has become the norm.

Let us now establish Lighttime standards that relate Times to Distances in terms of Velocity of Light. We know that Light travels *(almost exactly)* 300,000 km in one second. This round figure is easy to convert into smaller units. Let us use the standard meter instead of kilometer as a starting measure and construct a Distance-Time correspondence table:

300,000,000 m or meters *(Earth-to-Moon Distance)*	1 Lightsecond
300,000 m *(New York to Boston)*	1 millisecond
300 m *(length of a sports field)*	1 microsecond
0.3 m *(1 foot length)*	1 nanosecond
0.000,3 m *(0.3 millimeters, paper thickness)*	1 picosecond
0.000,000,3 m *(size of a virus)*	1 femtosecond
0.000,000,000,3 m *(a large molecule)*	1 attosecond

The *standard practice to give both Distances and Times in Time values* may seem a little odd at first but soon enough you'll get used to it. It's all a matter of habit. Additionally, unspecified Distances in equations are given in form of algebraic symbols, X or x, Time by T or t. The simplest way to express velocity is to give it as a fraction of the speed of Light. For inst-ance, if you walk at a velocity, $v = 3.6$ km per hour, you actually go at 3600 m per 60 x 60 = 3600 seconds which by calculation is 1 meter per second. This turns out to be 1 divided by 300,000,000 m/sec or 0.000,000,003,333 of c, the commonly used symbol for the speed of Light, or 0.000,00-0,003,333 *(or $3.3x10^{-9}$)* without a specific unit designation attached. The above fraction, v/c can, therefore, be written simply as v, but always understood as a fraction of c.

Get set, ready, go visit Spacetime

In the forthcoming examples and exercises making up most of the chapters, *we'll study Motion not in the 3-Dimensional Space we are familiar with but in Lineland, a Space that has only one Dimension.* Because all observations are Observer-dependent, the Observer has to be in that same one-dimensional Space as well. All Motion, therefore, takes place directly toward or away from the Observer. One-dimensionality of Space has the single purpose of keeping our mathematics as simple as possible. *Most of our problems here are conceptual and, therefore, intuition-straining* but we also have to keep math as simple as possible. We'll get by with *high school algebra* although *trigonometry* does occasionally intrude. Trigonometry is really neat but is more for the sophisticated amateur. Here we'll go real slow, on foot, step by step the longer and the more laborious route but with the advantage that all road markers are always clearly placed. The Space values in our illustrations will be represented simply by Segments of the X-Coordinate. This is possible by keeping the Y- and Z- Coordinate Segments equal to zero. Time values are given by Segments of the T-Coordinate.

There is also this problem with the complex number system. It can be faced either boldly head-on or avoided altogether by cowardly hiding it behind a simple mathematical trick *(which we'll see how it is done)*. **Complex number system is the mathematical alphabet of Special Relativity** and the peculiarities of Spacetime come wrapped in mathematical dress that appears strange, a little confusing to say the least. So it is important that we get used to this peculiar alphabet. What we gain is nothing less than a new-and-improved ***description of Motion in Space and in Time which no longer behave like absolute, Newtonian, entities.*** Imaginary numbers will ultimately help us make better sense of Special Relativity. Sooooo......., good luck and happy adventure. See you in Spacetime*!*

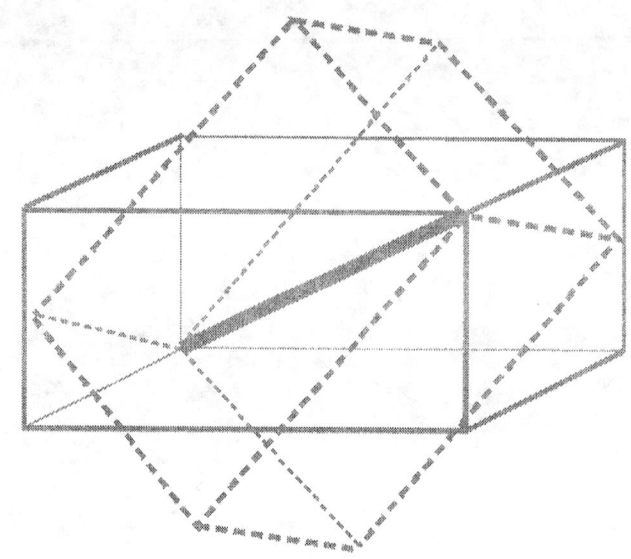

I - THE INVARIANT

A position in Spacetime

An observable Event always happens at a certain LOCATION in Space and at a certain MOMENT in Time. It can be *completely specified within four DIMENSIONS,* three of *SPACE* and one of *TIME.* This broader definition of a *SPACETIME LOCATION* makes it possible to *express all physical measurements in four dimensions* instead of the single dimension featured in our conventional Distance and Time measurements. Attempting to force all our Distance and Time concepts into a four-dimensional mold may seem draconian but we'll find out soon enough that over one-dimensional measurements, Observers in high-speed relative Motion *ALWAYS disagree. Four-dimensional measurements*, on the other hand, *are most reliable, completely dependable, exactly conserved* and, as we'll see, *absolutely the same to all Observers.* These well-deserved superlatives refer to what is known as *the INVARIANT,* a true *ABSOLUTE* in this new world of "relative" quantities, *the fundamental, 4-dimensional entity in Special Relativity.*

In this chapter, we'll develop the Invariant first in a *2-D Space (surface)* then in the *3-D Space* and, finally, in the fully *4-D Spacetime.* By this step-by-step process, *the Invariant emerges as a geometric structure that will be our key concept in all subsequent chapters.*

Invariance means 4 Dimensions all the way

For gaining insight into the geometry of four dimensions, there is nothing better than beginning with a few ordinary measurements in our familiar 3-dimensional Space *but under NEW RULES.* Working through a few examples will teach us a *universal method* applicable to all conceivable circumstances. Once we know how to work with three Space dimensions, adding *TIME as the FOURTH DIMENSION* on equal terms *(with Space dimensions)* will simply follow the same set pattern.

Let's do measuring exercises

Let us take a ruler and measure the length, *L* of some oblong objects placed in various positions such as shown below. With a ruler on top of, or next to, these objects, we can determine their Lengths in the usual, simple manner:

F1.1

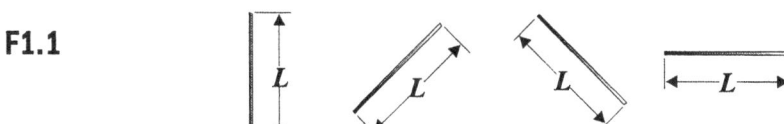

That was too easy. So let us re-measure the same objects but this time under certain *CONSTRAINTS.* Suppose we can place the ruler only into horizontal and vertical positions. Still, there would be no difficulty measuring the first and the last objects which themselves are in those positions. But we cannot measure the other two in slanted positions. These we can measure only *INDIRECTLY* by measuring their *PROJECTIONS,* or shadows, cast upon the horizontal and vertical rulers. Doing so gets us two values for each slanted object as illustrated below:

F1. 2

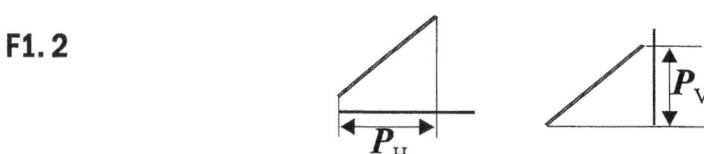

For convenience, we can join our vertical and the horizontal rulers at a right angle to form something like a carpenter's angle with suitable inch or metric markings on each arm. With this **Combination Ruler,** we can measure Projections of any slanted object in a number of ways, still observing the horizontal-vertical restriction as before:

F1.3

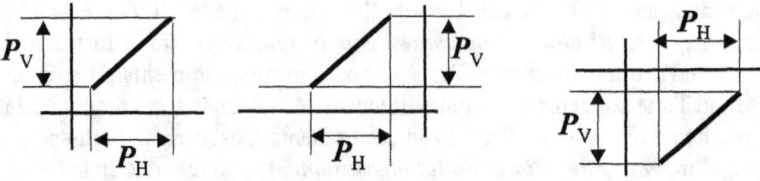

Coordinating with Coordinates

Let's now show the results in a line drawing using the horizontal arm of the ruler as the **X-COORDINATE** and the vertical arm as the **Y-COORDINATE,** all within the terminology devised by Descartes, representing the object to be measured simply as **LENGTH, L**:

F1.4

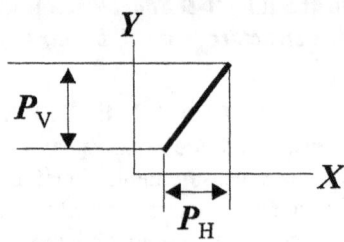

Next we convert the lines **X, Y** and **L** into a right angle triangle that was already suggested in the Figure 1.4:

F1.5

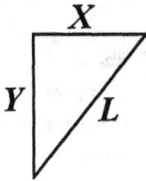

Pythagoras to rescue

By recruiting the help of Pythagoras, we can now calculate the value of **L** according to his well-known theorem:

$$L^2 = X^2 + Y^2$$
$$L = \sqrt{X^2 + Y^2}$$

Constructing Frames of Reference

Next we extend the projection lines until they cross beyond the object, to obtain a rectangle made up of parallel lines, two equal to **X** and two equal to **Y**. The measured object now forms the **DIAGONAL, D** of the rectangle which we will call our **FRAME OF REFERENCE,** in this case a **REFERENCE RECTANGLE:**

F1.6

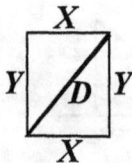

Any number of Frames are possible

The Reference Rectangle is fashioned around the object to be measured so that the object's Length, *L* becomes its ***Diagonal, D.*** This generalization remains true even if we ***rotate*** our Coordinates clockwise or counterclockwise around a pivot formed by the lower left corner of the Reference Rectangle and choosing any angle of rotation we may wish. ***With rotation, the values of the Projections change*** but whatever we do, ***the Length, L always remains the Diagonal*** of any new Reference Rectangle we may create. By using different amounts of rotation, we can create any number of Rectangles. ***The Length of the object always forms the COMMON DIAGONAL of unchanging magnitude shared in the following illustration by all three different Reference Rectangles:***

F1.7

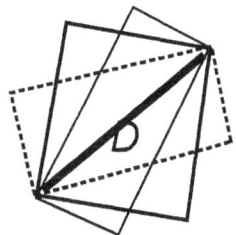

Now, let's take a look at those three different Reference Rectangles sharing the Common Diagonal. To distinguish the separate *X* and *Y* Coordinates, these are labeled X_1, X_2 & X_3 and Y_1, Y_2 & Y_3 respectively:

F1.8

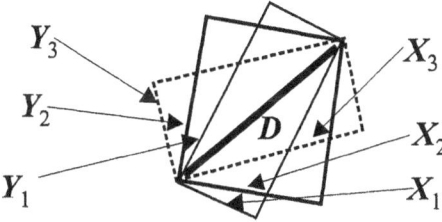

Again, according to Pythagoras, the value of the Common Diagonal remains the same in all Reference Rectangles even though the individual *X* and *Y* measurements come out unequal:

$$L^2 = X_1^2 + Y_1^2 \quad or \quad L = \sqrt{X_1^2 + Y_1^2}$$
$$L^2 = X_2^2 + Y_2^2 \quad or \quad L = \sqrt{X_2^2 + Y_2^2}$$
$$L^2 = X_3^2 + Y_3^2 \quad or \quad L = \sqrt{X_3^2 + Y_3^2}$$

so:

$$X_1^2 + Y_1^2 = X_2^2 + Y_2^2 = X_3^2 + Y_3^2$$

and *(but)*:

$$X_1 \neq X_2 \neq X_3 \quad and \quad Y_1 \neq Y_2 \neq Y_3$$

If, however, the Right Angle Ruler is ***NOT rotated*** but is ***shifted parallel*** either to the horizontal or vertical Coordinate, the measured Projections upon the *X* and *Y* Coordinates will not be changed in magnitude so that:

$$X_1 = X_2 = X_3 = X_{etc} \quad \text{and} \quad Y_1 = Y_2 = Y_3 = Y_{etc}$$

F1.9

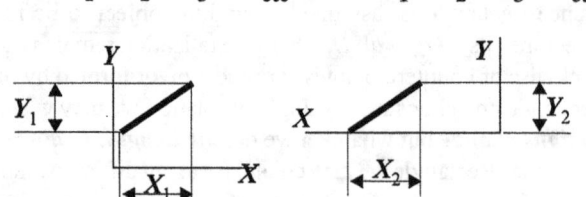

Translation or Transformation

A PARALLEL SHIFT of Coordinates does not do anything except shift the location *(on those Coordinates)* where the Diagonal is projected. It *does not change the magnitude of the Coordinate Projections.* This parallel shift of Coordinates *(resulting only in a change of location where the Diagonal is projected)* is known formally as *TRANSLATION.* It is a term you'll see from time to time in technical literature about Classical Mechanics and it is useful to know what it means. In Special Relativity, a different positional shift of Coordinates is involved. The Frame of Reference here becomes *ROTATED* around a pivot conventionally formed by the lower left corner of the Reference Frame. This process is known formally as *TRANSFORMATION.* Here *the projection values are changed*, or *transformed*, as we saw in Fig. 1.8. Making our Frame of Reference four-dimensional and assigning Time to the Fourth Coordinate creates an additional complication. More about it soon enough in this chapter.

If either the X or the Y Coordinate is aligned parallel to the Object, L, then the Projection of L upon that Coordinate will exactly equal the Length of the Object, L and the Projection upon the other Coordinate will be zero giving us $X = L$ and $Y = 0$, or $X = 0$ and $Y = L$:

F1.10

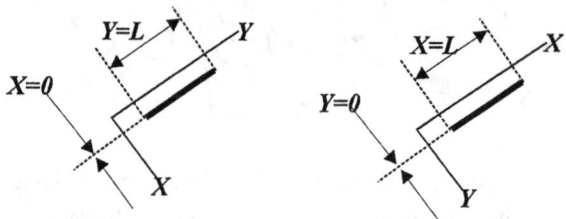

And finally: the Invariant

At this point in our study, we are finally prepared to make a *sweeping generalization: The Coordinate positions and values can vary a great deal while the main object of interest, the Diagonal of the Reference Rectangle, the D, always keeps the same value.* Therefore, the Xs and Ys *(the Coordinate Projections upon them)* are our variable or *VARIANT* quantities while the Diagonal, or D, is the *INVARIANT* quantity. These terms clearly point out as to what quantity can change and what always remains unchanged *(or constant, or conserved or invariant).* This is the important *PRINCIPLE of INVARIANCE* that in Special Relativity is involved in all measurements.

The equality of the Invariant in all Inertial Systems is an important bridge linking all Frames of Reference thereby allowing us to negotiate mathematically from Space and Time measurements in one Inertial System to those in another. How? Let's find out. But to get our conceptual framework ready, we have to construct a 3-dimensional Frame of Reference to which *Time can be added as the fourth Coordinate* as we convert a 3-dimensional Frame of Reference into a *4-dimensional Frame of Reference.*

3-D Frames converted into 4-D

As we live in 3-dimensional Space, we need to know how to work with all of its three dimensions. To start, let's place the X and the Y Coordinates upon the 2-D surface facing us. The third Coordinate, Z will then be directed horizontally towards us. Each Coordinate will be at right angles to the other two. The three Coordinates along with the additional vertices parallel to the three original ones form a fully enclosed square box with six sides, a ***HEXAHEDRON*** that will be our ***three-dimensional FRAME OF REFERENCE***, the ***Reference Hexahedron.*** These Coordinates are now part of a square box: X as the width, Y the height and Z the depth. The initial point of orientation where the original X, Y and Z are joined, is the ***POINT of ORIGIN***, or the point, O. From there, we now place a three-dimensional Diagonal, D directed to the opposite, right upper front corner:

F1.11

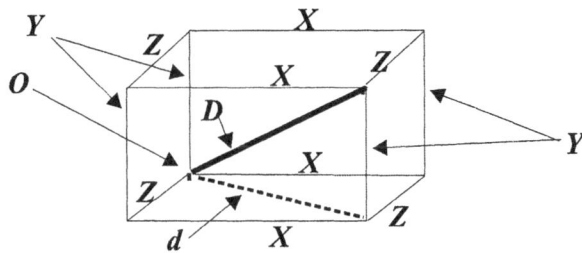

The Length of the Diagonal can be derived from the values of its Projections which in this particular case are equal to the edges *(vertices)* of the box. First, we project the 3-D Diagonal upon the rectangular bottom of the box to form a Diagonal on that 2-D surface, the d *(not capitalized to keep it distinct from D)* which divides the bottom of the box into two equal right angle triangles, both mapped as XZd. Now we can calculate the value of d in the usual manner:

$$d^2 = X^2 + Z^2$$
$$d = \sqrt{X^2 + Z^2}$$

On Fig. 1.11, D is the ***Diagonal*** of the right angle triangle, ***DdY,*** here obliquely and diagonally placed, where:

$$D^2 = Y^2 + d^2$$

Substituting the derived value of d into the last equation, we can finally solve for D:

$$D^2 = Y^2 + \left(\sqrt{X^2 + Z^2}\right)^2$$
$$= Y^2 + X^2 + Z^2$$
$$D = \sqrt{X^2 + Y^2 + Z^2}$$

Quite analogous to the process of drawing more than one Reference Rectangle around the Common Diagonal, we can also form more than one Reference Hexahedron around any Length serving as the Common Diagonal using the more remote lower left corner as pivoting point, or point O. To avoid cluttering up our next illustration with too many Hexahedrons, only two will be shown surrounding the Common Diagonal:

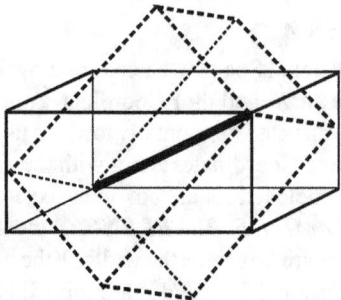

How to keep adding Coordinates

The universal three-dimensional method of measuring Lengths we just developed gives us complete freedom for placing our Coordinates any way we wish. No particular orientation is preferred and all orientations give exactly the same value for the Common Diagonal, the Invariant. So, let us pause and reflect upon what we have achieved: **(1)** in *ONE*-dimensional world: $L^2 = X^2$; **(2)** in *TWO*-dimensional world: $D^2 = X^2 + Y^2$; and **(3)** in *THREE*-dimensional world: $D^2 = X^2 + Y^2 + Z^2$. The value of D is, of course, the square root of the sum of squares of the X, Y and Z values.

It is apparent that adding another dimension merely requires adding another coordinate measurement squared under the square root sign. Thus in a hypothetical world of many dimensions, the value of the Diagonal will be:

$$D^2 = X^2 + Y^2 + Z^2 + P^2 + Q^2 + R^2 + ... + (etc.)^2$$

and

$$D = \sqrt{X^2 + Y^2 + Z^2 + P^2 + Q^2 + R^2 + ... + (etc.)^2}$$

4-D Frames, at last

Actually, we do not need that many Coordinates, only one additional for time, T will suffice:

$$D = \sqrt{X^2 + Y^2 + Z^2 + T^2}$$

It would be nice to demonstrate visually how Time Coordinate fits in with the three Space Coordinates. Unfortunately, the fourth Coordinate must be oriented at right angles to each Space Coordinate. As hard as we may try, it is extremely difficult to add a fourth Coordinate this way and it is impossible to show it in a drawing. As with most abstractions, it certainly is difficult to imagine something like a four-dimensional square box. But we need it badly to get going in four dimensions, so let's just accept the concept and adopt it as our *4-D REFERENCE HYPER-HEDRON*. At least, the name is descriptive and we don't even need to know how it looks.

"Compressing" 4 Coordinates into 2

To simplify the problem of visual representation, let's align our X-coordinate to be parallel to the Common Diagonal *(or whatever the object we want to measure)* so that $Y = 0$ and $Z = 0$. Of the Space Coordinates, we are then left only with the Projection of D on the X-Coordinate so that its value will be the same as the Diagonal: $D = X$. Now we have eliminated Y and Z from the picture and are left only with two Coordinates, X and T:

$$D = \sqrt{X^2 + 0^2 + 0^2 + T^2} \quad \text{and} \quad D = \sqrt{X^2 + T^2}$$

Time as an obligatory border of any Frame containing Motion

Time is easy to measure. After noting the calendar date we just need to look at the clock to mark the **MOMENT** when we start counting Time. This gives us a ***starting POSITION*** on the Time Coordinate. To measure ***TIME INTERVAL*** or the ***TIME SEGMENT*** *(a measured Segment of the Time Coordinate)*, we need another Time marker for the Moment we finish counting. These two *(Time)* markers give us the ***TIME SEQUENCE*** or the ***TIME ORDER of EVENTS,*** that is what happens **BEFORE** and what comes **AFTER.** If the two Moments coincide, these can be said to have happened at the same Moment, at the same Time, or ***SIMULTANEOUSLY.*** In case of Motion, the ratio of Distance traveled to the Time spent traveling gives us the ***VELOCITY*** of Motion. In Classical Physics, Time is usually plotted on horizontal Coordinate as the ***independent variable*** *(the variable that is independent of our actions or not under our control)*, and Distance is represented as the ***dependent variable*** *(dependent upon Time progression)* on vertical Coordinate.

F1.13

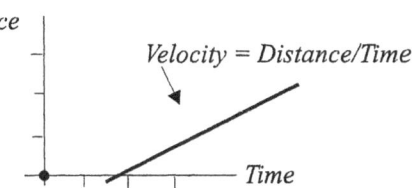

You can see *(from Fig. 1.13)* that uniform Velocity gives us a line that is straight, always with the same slant throughout. If the vehicle gets stalled for a while, the slant of the line for that period is horizontal:

F1.14

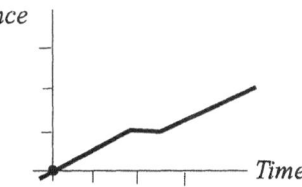

Before going any further, it is necessary to change the way we plot Time and Distance. In Relativity: Space coordinate is placed horizontally and Time vertically. Redrawing our last illustration according to this new orientation still gives us a slanted line for Velocity but slower speeds are now closer to the vertical directiobn than are faster speeds:

F1.15

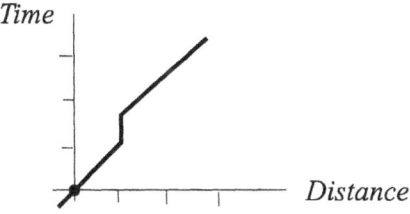

With Motion, the line still goes up and to the right *(or left if Motion proceeds in the opposite direction)*. With no Motion, the line goes straight up. With faster Motion, the line shows a greater tilt from the vertical.

Let us next draw lines to show Motion at differing speeds: a walker (*W*), a bicyclist (*B*), a racing car (*R*) and an airplane (*A*)

F 1.16

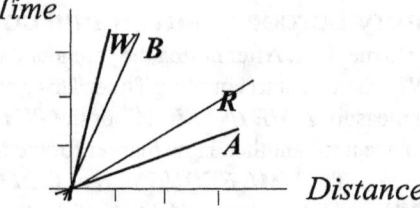

Standard Spacetime Frame of Reference

And, finally, we need to standardize our Time and Space Coordinates to make them fit into the Spacetime concept using *Light as a calibrating tool*. This standardization results in *equal Coordinate line lengths standing for both one SECOND of Time and one LIGHTSECOND of Distance*. By international convention, both Time and Distance are given in seconds. If numerical values are known, both measurements can be given just in numbers.

Light travels one Lightsecond of Distance in one *(Time)*second of Time. As shown in the following illustration, the line for the speed of Light is placed at a 45 degree angle. As nothing goes faster than Light, no movement can be represented by a slope greater than 45 degrees from the vertical:

F 1.17

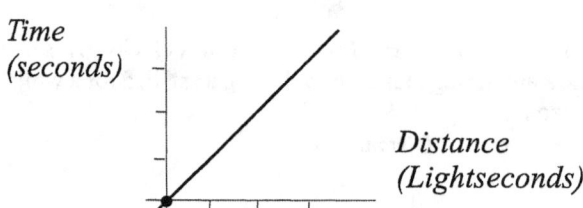

Visualizing Motion in standard Frame

The following proportional graphic illustration, a re-drawing of Fig. 1.15 of a walker, a bicyclist, a racing car and an airplane does not really distinguish between them. All slopes are so close to vertical that everything seems to be standing still, nothing at all moving.

F1.18

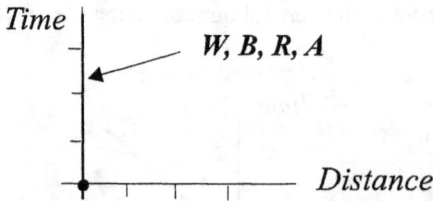

Plotting *Motion in Lightseconds per second* seems to create a disproportionate scale for Distances and for Time. A short one-second Time line is made equal to the very long 300,000-km Distance line. But this is what Mother Nature ordered and we better get used to it. Truly proportional mapping of this kind accommodates all kinds of fast Motion which are best described only by Special Relativity. To illustrate this last point let us next plot some *really fast objects like*

the Earth's Motion in orbit around the Sun, fast moving alpha particles, cosmic rays striking the upper atmosphere and Light itself:

F1.19

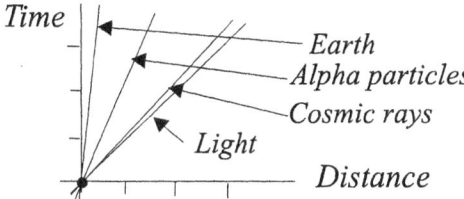

Is Time real or is it really imaginary?

In calculations, the Space Segment (*X*) values are entered into equations in Lighttime units which feature ordinary, ***REAL NUMBERS.*** But it may surprise you that Time *(T)* Segment values have to be entered as ***IMAGINARY NUMBERS*** . Why? A good question. Let's just say that our calculations can produce correct results only if ***Time is understood as an "imaginary" quantity*** and its projection values are given in imaginary numbers. As was said in the Introduction: ***Complex number system is the mathematical alphabet of Special Relativity*** and so we have to spell out our measurements accordingly.

By definition, an imaginary number is a real number multiplied by $\sqrt{-1}$. The square root of -1 is *i* $(= \sqrt{-1})$ and $i^2 = -1$. With all this made clear, we can restate our 4-D equation with the four Coordinate values:

$$D = \sqrt{X^2 + Y^2 + Z^2 + (iT)^2}$$

Note that in the above equation we have Distance *(or Length)* Coordinate Projections in Space values so, to be consistent, we have to express *T* in Space value as Distance traveled by Light during *T,* that is *c* times *T,* or *cT*:

$$D = \sqrt{X^2 + Y^2 + Z^2 + (icT)^2} = \sqrt{X^2 + Y^2 + Z^2 + i^2 c^2 T^2}$$

Now we finally have it! Look at the above equation carefully. This is the correct way to express our 4-D Diagonal but you can see that Time here is given in form of an imaginary number, something new in our math tool box. With the above equation, we have been introduced to the ***COMPLEX NUMBER SYSTEM, the mathematical alphabet of Special Relativity***. The odd member in the group of numbers, the *i* squared, becomes minus one ($i^2 = -1$), changing the above equation into the following:

$$D = \sqrt{X^2 + Y^2 + Z^2 - (cT)^2}$$

We can also give *X*, *Y* and *Z* in terms of Time needed for Light to travel those Distances instead of dressing them up in Space measurements :

$$D^2 = \left(\frac{X}{c}\right)^2 + \left(\frac{Y}{c}\right)^2 + \left(\frac{Z}{c}\right)^2 - T^2 \quad \text{because:}$$

$$T_x = \frac{X}{c} \qquad T_y = \frac{Y}{c} \qquad T_z = \frac{Z}{c}$$

We can next simplify everything by eliminating Y and Z from our basic equation. This can be done by making $X = D$ so that both Y and Z become equal to zero: $Y = 0$, $Z = 0$. In addition, we can understand that Distance is always given by the amount of Timeneeded for Light to travel that much Space so X, in Lighttimes, does not need to be divided by c:

$$D^2 = X^2 - T^2 \quad \text{and} \quad D = \sqrt{X^2 - T^2}$$

Now what's a 4-D "Distance"?

According to standard terminology, the D can be calculated and expressed by its Distance value. This gives us Spacetime "**SEPARATION**," or S. The D calculated and expressed by its Time value is the Spacetime "**INTERVAL**" or I.

The second alternative of using Time values for both Time and Distance is now the **INTERNATIONAL STANDARD.** In order not to confuse ST Interval with simple Time interval, we'll adopt the practice of reserving the word "interval" for ordinary Time segments and use the word "**INVARIANT**" for both **Spacetime Separation** and **Spacetime Interval** but always with the capital letter, I standing for both of them in all equations. The unambiguous, ordinary use of the word "interval" is thereby retained. The letter, X is used for any measured Space Segment, and T for any measured Time Interval or Segment both, however, being Projections of the Invariant on Space and Time Coordinates. The Diagonal within a Frame of Reference will then be the Spacetime Invariant and rewriting the last equations gives us:

$$I^2 = X^2 - T^2, \qquad I = \sqrt{X^2 - T^2}$$

Getting used to imaginary numbers

Because imaginary numbers are somewhat unfamiliar to many, a few additional comments are needed.

(1) By squaring a quantity preceded by i, we obtain the same quantity squared but now preceded by a minus (-) sign, the (-) sign belonging to that number or symbol standing for it. In other words, an imaginary number squared becomes the same number squared but with a minus sign in front of it. In contrast, a negative number squared is always a positive number. The only apparent change we have to get used to is seeing T^2 in the ST Equations always *(or almost always)* supplied with a minus sign. The out-of-place, eccentric character of Time squared with a minus sign in front of it is, therefore, explainable. Let's get used to it. The ***minus sign before T^2*** has been a source of a good deal of confusion to many but should now be understood as being due to an ***imaginary quantity squared.***

(2) If we take the expression, $I^2 = X^2 - T^2$ and do some actual calculations involving Motion, the absolute numerical value of T is always greater than that of X giving us I^2 as being equal to a number squared with minus sign before it. Taking a square root of this negative number gives us an imaginary number, a number preceded by i.

(3) We can avoid the minus sign before imaginary numbers squared by multiplying both sides of the equation by -1. On the right side of the = mark, we'll get $(-1)(X^2+Y^2+Z^2-T^2)$ which gives us the unwieldy $-X^2 - Y^2 - Z^2 + T^2$ that, by rearrangement, becomes $T^2 - X^2 - Y^2 - Z^2$ Many equations of this kind may not actually make up the Invariant but represent sets of measurements made in different Inertial Systems. We'll work all this out later.

What you usually see?

In most popular magazine articles and even college textbooks, practically all equations are given in the above "sanitized" form, avoiding imaginary numbers altogether: (1) T^2 as a positive number, (2) Space Coordinate values squared and negative. This is done to make the equations look reader-friendly as many folks are repelled or puzzled by the mystery of i. In the long run, however, it is best to become well acquainted with "Mr. i" to better understand the relativistic peculiarities of Spacetime geometry.

Fortunately, when calculating a coordinate Time value in one Frame of Reference from that given in another, the result comes out positive. This you'll see commonly in the following chapters and is due to *concealed multiplication of $-T^2$ with -1.* It happens whenever we move it *(the $-T^2$)* from one side of the equal sign to the other, being essentially the same result obtained by multiplying both sides of the equation with *-1*. Let's keep this concealed conversion in mind whenever we find the T squared dressed up as a distinctly positive and friendly character.

What next?

Our next task is to describe Motion of two Observers and find out how to plot it on a 2-D paper surface. This task requires considerable re-orientation of our common-sense intuition to accommodate geometric visualization of Time and Space and is taken up in the next chapter, "Axioms And Operations."

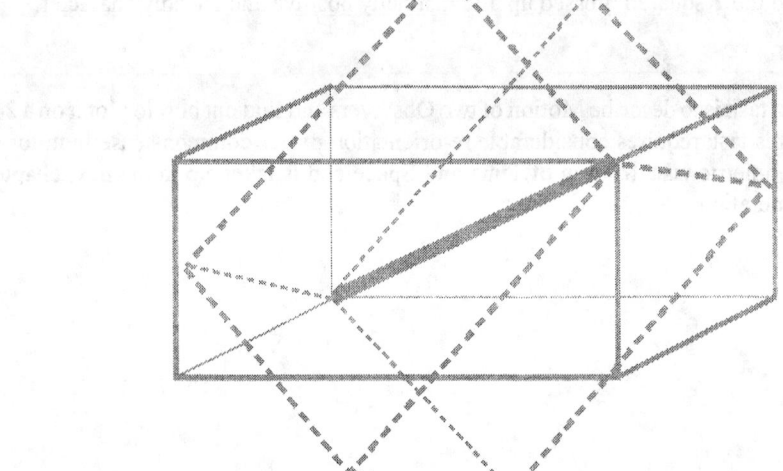

II - AXIOMS AND OPERATIONS

Key words

Let's start by defining the words in the chapter title. By ***AXIOMS***, let us mean those basic statements that are simple, explicit and reasonable. These were taken up at length in the Introduction. It is most fortunate that Special Relativity has been reduced to axiomatic ***FIRST PRINCIPLES***, also called ***BASIC POSTULATES***, from which the entire theory can be derived. That's what Einstein did and that's what we'll adopt for our method. To keep everything as clear as possible, we'll do it slowly, paying special attention to concepts that have always been stumbling blocks to intuition. By ***OPERATIONS***, we mean those rules of thumb we can use to manipulate concepts, obtain results, get things done. But before getting down to Space, Time and Motion at close range, let us first summarize what we have learned in the first chapter about Minkowski's geometry in 4-D Spacetime.

Summary of the first chapter

In 1-dimensional Space, a measured Length can be shown as a line segment mathematically expressed as:

$$L = \sqrt{X^2} = X$$

In 2-D Space, such as the flat surface of this page, Length as a Space Segment forms the Diagonal of a rectangle chosen as our ***Frame of Reference*** *(Diagonal also being the Hypotenuse of its component Triangle):*

$$L = D = \sqrt{X^2 + Y^2}$$

In 3-D Space, Length is the Diagonal of a Hexahedron, our Frame of Reference:

$$L = D = \sqrt{X^2 + Y^2 + Z^2}$$

In the 4-D Spacetime, the Diagonal is enclosed in a 4-D Hyperhedron that is difficult to visualize and impossible to draw. Time values are entered as imaginary numbers, both Space and Time measurements are expressed in Time units and the Velocity of Light calibrates all Space and Time *(Spacetime)* measurements:

$$L = D = \sqrt{X^2 + Y^2 + Z^2 + (iT)^2}$$

The last equation can be written without the *"i"* in it:

$$L = D = \sqrt{X^2 + Y^2 + Z^2 - T^2}$$

Making *Y* and *Z* equal to zero "compresses" Minkowski's Spacetime Equation into a manageable 2-D Frame of Reference with greatly simplified mathematics:

Compressing a 4-D Frame into a 2-D Frame allows us visualize everything on a 2-D surface such as a page in this book. A curved Spacetime can be plotted on a flat, 2-D Map without difficulty, a procedure we'll explore in detail in this chapter and in Chapter III.

$$L = D = \sqrt{X^2 - T^2}$$

Bringing *I*, *X* and *T* to life

After extolling the importance of the Invariant *(the Diagonal),* we need to bring it to life by giving it some palpable substance within our "compressed" *(2-D)* Reference Frame made up of one Space Coordinate *(X, while making Y and Z equal to zero)* and one Time Coordinate *(T).* Let's start by asking: ***What in our everyday world would correspond to the "DISTANCE" between two SPACETIME EVENTS?*** By dealing with Space and Time Segments possessing concrete values, we can point out parallels to our familiar, everyday situations and become better acquainted with elementary concepts applied to Motion in Spacetime. The exercises that follow will also provide us with the ***TERMINOLOGY*** needed for the rest of our study. Let's start with something in our experience in 2-D Space *(Surface)* that is most analogous to Spacetime "Distance" from the first chapter.

If in a town with a square street pattern, we first walk four blocks North then three blocks East, each block measuring 100m, then what is the Distance between the starting and the finishing points? Let's first map our walk then open our familiar math tool box to find the answer:

F2.1

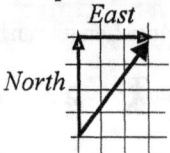

The initial equation is set up this way: **(Diagonal Distance)**^{2} = **(4 Blocks to North)**^{2} + **(3 Blocks to East)**^{2} and the results of the calculation are put to words as follows: The Diagonal Distance, D is equal to the square root of a sum made up of 4 squared and 3 squared. This gives us a 5-block diagonal from start to finish *(example chosen to allow calculation in whole numbers):*

$$D^2 = (4B_N)^2 + (3B_E)^2 , \quad D = \sqrt{(4B_N)^2 + (3B_E)^2}$$
$$= \sqrt{4^2 + 3^2}$$
$$= \sqrt{16 + 9}$$
$$= \sqrt{25}$$
$$= 5 \ = 5B = 500m$$

As you see, the Distance between starting and finishing points came to 500m. In this example, our Frame of Reference was made up of two Space Segments. Time was not needed in the equation. All very plain and simple.

If, instead, we just walk four blocks and do it in three minutes, *we combine one Space Segment with one Time Segment between our starting and finishing points. What we get is totally unusual.* Well, let's see: there is this Space Segment of 400 meters and Time Segment of 3 minutes. What does this give us? Well. Let's visualize it on the Map:

F2.2

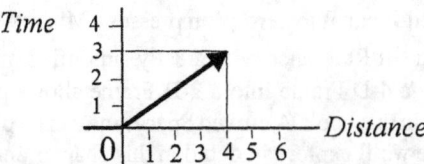

You can see that the Diagonal cannot be a Distance as we know it nor can it be called a Time duration as we understand it. What we have here is a fusion of both into this ***new-fangled,***

abstract Spacetime "Distance." Can we calculate its magnitude? Yes but in order to prepare our data for proper processing, we have to express all Coordinate measurements in standard Time values then do the math:

(1) The Space Segment of 400 meters is "translated" into a number that specifies the Time duration in seconds *(or fractions thereof)* needed by Light to travel that 400m Distance. This is done by dividing 400*(m)* by 300,000,000 *(meters per second, the velocity of Light)*. What we get is 1.333 x 10^{-6} seconds of Lighttime for the Space Segment.

(2) Converting the Time Segment of 3 minutes into seconds is easier: 3 x 60 = 180 seconds.

Finally, to calculate the Spacetime "Distance" *(the Invariant)*, we need to set up Minkowski's Spacetime equation with the Space and Time values *(projecting the Invariant on the X and T Coordinates)*. First, the "compressed" Spacetime Equation:

$$D \ (= I) \ = \ \sqrt{X^2 + (iT)^2} \ = \ \sqrt{X^2 - T^2}$$

then putting in the numbers we have calculated:

$$D \ = \ \sqrt{(1.3333...x10^{-6})^2 - 180^2}$$

$$= \ \sqrt{1.7777x10^{-12} - 32,400}$$

$$= \ \sqrt{-32,399.9999999999972}$$

which now gives us *(almost exactly)*:

$$I \ = \ \sqrt{-32,400} \ = \ i180$$

The final answer is: I equals *i180*, an imaginary number for the Spacetime "Distance" between starting and finishing points. This Distance plotted on the Map in real numbers is going straight up, not a surprising result because the number for the Space quantity was vanishingly smaller than the Time quantity:

F2.3

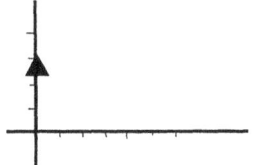

The calculated result is visually also identical to what was plotted in Fig. 1.18 *(Chapter I)* for Velocity of the Walker who asked the service station attendant for Distance information. So the "Distance" traveled here contains hardly any of the geographical component walked in 3 minutes, something we always understand as Distance on Earth or in Space, and made up almost entirely of the Time component of the equation. Moreover, the result is additionally a strange, *imaginary number*. It makes no sense at all! But haven't we found out already *(in Preface)* that Relativity does not make sense in our common-sense world? Classical mechanics still rules supreme there. The Velocities that bring out relativistic effects are completely unfamiliar to us from personal experience and are suitable only for describing the result of a very-very fast Motion! For that kind of Motion only Relativity gives us accurate *(and invariant)* results although in form of *complex numbers*, the mathematical alphabet of Special Relativity just as was noted in the Introduction.

Calculating Invariants in numbers

To continue exploring Spacetime using familiar, down-to-earth concrete numbers in a more systematic fashion, let us calculate the Invariant *("Distance")* between a number of already mapped Space-time Event pairs without the usual context of Motion from one Event to the other. In the **FIRST SERIES**, the Space Segment, X is set at 5 *(Light-)* seconds while the Time Segment, T in seconds is allowed to increase stepwise from 0 to 10. In the **SECOND SERIES**, it'll be T that is set at 5 seconds and X in Lightseconds is allowed to vary from 0 to 10. With the numerical examples mapped on the "compressed" 2-D surface $(I^2 = X^2 - T^2)$, the basic 4-D Invariant is calculated using three different but analogous computations:

(1) On the left side of the page you'll see superimposed areas corresponding to X^2 and T^2 along with their graphic subtraction.

(2) The results of $X^2 - T^2 = I^2$ factored as the sum-and-difference product, $(X+T)(X-T)$, are then given in numbers and also graphically illustrated.

(3) Finally, on the right side of the page, you'll find the standard calculation of the Invariants resulting either in *real* numbers or *imaginary* numbers.

F2.4   **FIRST, the series with X fixed at 5:**

Subtracting squares, $X^2 - T^2$:   Multiplying $(X+T)(X-T)$:   Standard calculation

$X = 5$
$T = 0$

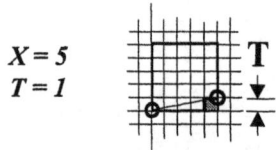

$X + T = 5$
$X - T = 5$
$5 \times 5 = 25$

$$I^2 = 5^2 - 0^2$$
$$= 25 - 0$$
$$= 25$$
$$I = \sqrt{25}$$
$$= 5$$

$X = 5$
$T = 1$

$X + T = 6$
$X - T = 4$
$6 \times 4 = 24$

$$I^2 = 5^2 - 1^2$$
$$= 25 - 1$$
$$= 24$$
$$I = \sqrt{24}$$
$$= 4.9$$

$X = 5$
$T = 2$

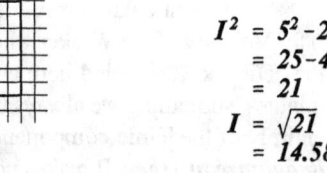

$X + T = 7$
$X - T = 3$
$7 \times 3 = 21$

$$I^2 = 5^2 - 2^2$$
$$= 25 - 4$$
$$= 21$$
$$I = \sqrt{21}$$
$$= 14.58$$

$X = 5$
$T = 3$

$X + T = 8$
$X - T = 2$
$8 \times 2 = 16$

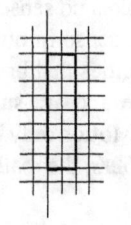

$$I^2 = 5^2 - 3^2$$
$$= 25 - 9$$
$$= 16$$
$$I = \sqrt{16}$$
$$= 4$$

$X = 5$
$T = 4$

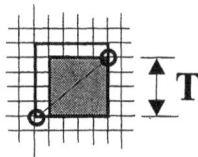

$X + T = 9$
$X - T = 1$
$9 \times 1 = 9$

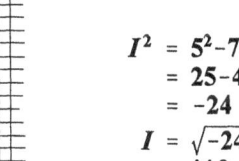

$I^2 = 5^2 - 4^2$
$\quad = 25 - 16$
$\quad = 9$
$I = \sqrt{9}$
$\quad = 3$

$X = 5$
$T = 5$

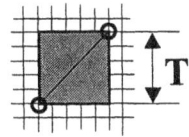

$X + T = 10$
$X - T = 0$
$10 \times 0 = 0$

$I^2 = 5^2 - 5^2$
$\quad = 25 - 25$
$\quad = 0$
$I = \sqrt{0}$
$\quad = 0$

$X = 5$
$T = 6$

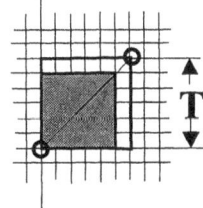

$X + T = 11$
$X - T = -1$
$11 \times -1 = -11$

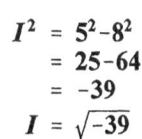

$I^2 = 5^2 - 6^2$
$\quad = 25 - 36$
$\quad = -11$
$I = \sqrt{-11}$
$\quad = i3.32$

$X = 5$
$T = 7$

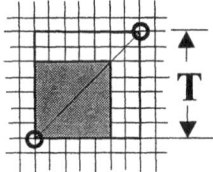

$X + T = 12$
$X - T = -2$
$12 \times -2 = -24$

$I^2 = 5^2 - 7^2$
$\quad = 25 - 49$
$\quad = -24$
$I = \sqrt{-24}$
$\quad = i4.9$

$X = 5$
$T = 8$

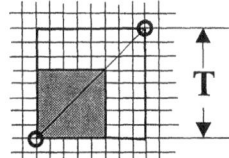

$X + T = 13$
$X - T = -3$
$13 \times -3 = -39$

$I^2 = 5^2 - 8^2$
$\quad = 25 - 64$
$\quad = -39$
$I = \sqrt{-39}$
$\quad = i6.25$

$X = 5$
$T = 9$

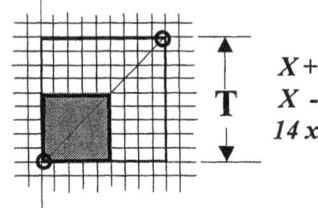
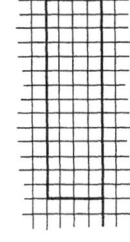

$X + T = 14$
$X - T = -4$
$14 \times -4 = -56$

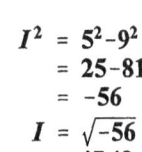

$I^2 = 5^2 - 9^2$
$\quad = 25 - 81$
$\quad = -56$
$I = \sqrt{-56}$
$\quad = i7.48$

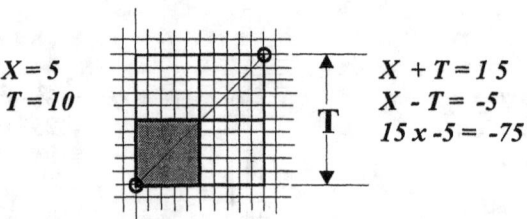

$X = 5$
$T = 10$

$X + T = 15$
$X - T = -5$
$15 x -5 = -75$

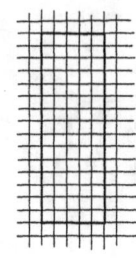

$I^2 = 5^2 - 10^2$
$= 25 - 100$
$= -75$
$I = \sqrt{-75}$
$= i8.66$

SECOND, the series with T set at 5:

$X = 0$
$T = 5$

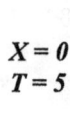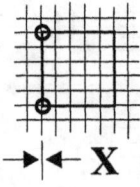

$X + T = 5$
$X - T = -5$
$5 x -5 = -25$

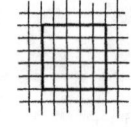

$I^2 = 0^2 - 5^2$
$= 0 - 25$
$= -25$
$I = \sqrt{-25}$
$= i5$

$X = 1$
$T = 5$

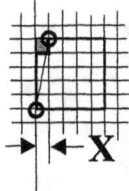

$X + T = 6$
$X - T = -4$
$6 x -4 = -24$

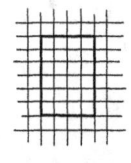

$I^2 = 1^2 - 5^2$
$= 1 - 25$
$= -24$
$I = \sqrt{-24}$
$= i4.9$

$X = 2$
$T = 5$

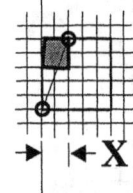

$X + T = 7$
$X - T = -3$
$7 x -3 = -21$

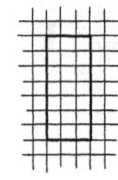

$I^2 = 2^2 - 5^2$
$= 4 - 25$
$= -21$
$I = \sqrt{-21}$
$= i4.58$

$X = 3$
$T = 5$

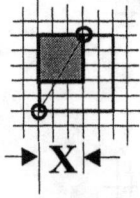

$X + T = 8$
$X - T = -2$
$8 x -2 = -16$

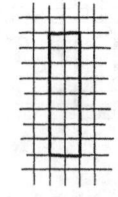

$I^2 = 3^2 - 5^2$
$= 9 - 25$
$= -16$
$I = \sqrt{-16}$
$= i4$

$X = 4$
$T = 5$

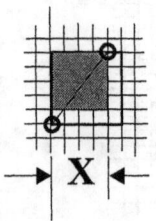

$X + T = 9$
$X - T = -1$
$9 x -1 = -9$

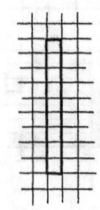

$I^2 = 4^2 - 5^2$
$= 16 - 25$
$= -9$
$I = \sqrt{-9}$
$= i3$

$X = 5$
$T = 5$

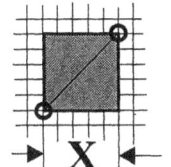

$X + T = 10$
$X - T = 0$
$10 \times 0 = 0$

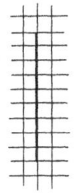

$I^2 = 5^2 - 5^2$
$\quad = 25 - 25$
$\quad = 0$
$I = \sqrt{0}$
$\quad = 0$

$X = 6$
$T = 5$

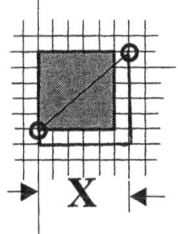

$X + T = 11$
$X - T = 1$
$11 \times 1 = 11$

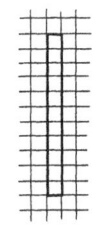

$I^2 = 6^2 - 5^2$
$\quad = 36 - 25$
$\quad = 11$
$I = \sqrt{11}$
$\quad = 3.32$

$X = 7$
$T = 5$

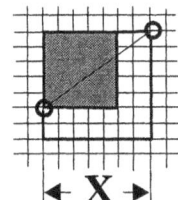

$X + T = 12$
$X - T = 2$
$12 \times 2 = 24$

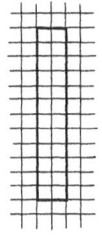

$I^2 = 7^2 - 5^2$
$\quad = 49 - 25$
$\quad = 24$
$I = \sqrt{24}$
$\quad = 4.9$

$X = 8$
$T = 5$

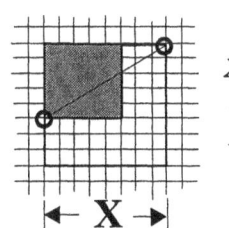

$X + T = 13$
$X - T = 3$
$13 \times 3 = 39$

$I^2 = 8^2 - 5^2$
$\quad = 64 - 25$
$\quad = 39$
$I = \sqrt{39}$
$\quad = 6.24$

$X = 9$
$T = 5$

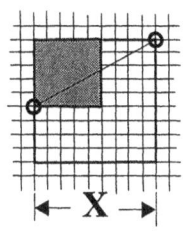

$X + T = 14$
$X - T = 4$
$14 \times 4 = 56$

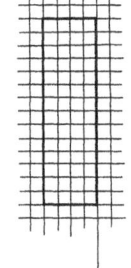

$I^2 = 9^2 - 5^2$
$\quad = 81 - 25$
$\quad = 56$
$I = \sqrt{56}$
$\quad = 7.48$

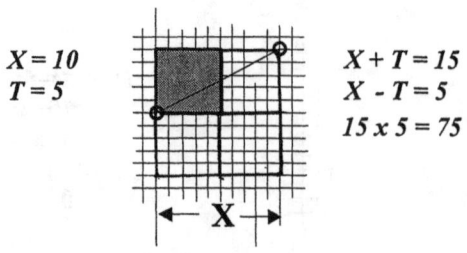

$$X = 10$$
$$T = 5$$

$$X + T = 15$$
$$X - T = 5$$
$$15 \times 5 = 75$$

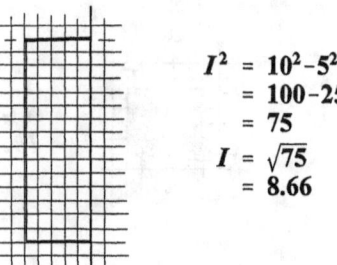

$$I^2 = 10^2 - 5^2$$
$$= 100 - 25$$
$$= 75$$
$$I = \sqrt{75}$$
$$= 8.66$$

The first crop of results

(1) *If X = T, the INVARIANT is ZERO!*

(2) *If X > T, the INVARIANT is a POSITIVE, REAL NUMBER.*

(3) *If X < T, the INVARIANT is an IMAGINARY NUMBER.*

(4) **The DIAGONAL of the 4-D HYPERHEDRON can be NUMERICALLY EQUAL to, but is usually SMALLER than, X or T, NEVER GREATER (!!!),** a most unexpected result.

The calculations point to an **underlying geometry that differs significantly from what we used to conceptualize our Common, Shared Diagonal in Chapter I** as the key concept. It is important to recall that in the 2-D *(as well as in the 3-D)* Space, the Common, Shared Diagonal was always greater, only at times equal to one of the sides of the Reference Rectangle or Hexahedron, never smaller. That's because the **geometry where the DIAGONAL is either EQUAL to, or SMALLER than, X or T, but never greater, goes only with the COMPLEX NUMBER SYSTEM**, the mathematical alphabet of Special Relativity.

The shortfall in the magnitude of the Invariant is a confounding surprise. The Invariant was defined *(in Chapter I)* as the Hypotenuse of the component right angle triangle formed by *X*, *T* and *I* *(the Invariant)*. But now, we are forced to consider an entirely different triangle as part of our Frame of Reference. The difference from conventional geometry is due to the very peculiar nature of Spacetime *(roughly described as curved but not spherical)*. This is a topic of *ST* Geometry that will be taken up in the next Chapter III, "Mapping Spacetime." In the meantime, let's just check out the ground level of Spacetime Geometry containing Motion, our next project.

Let's get moving fast - finally

With basic orientation out of the way, we are ready *to start investigating Fast Motion,* the "meat and bones" of Special Relativity. It is necessary here to repeat again that we will be studying fast Motion along a straight line at uniform Velocity. This does not mean that Special Relativity is inapplicable to scenarios in two or three-dimensional Space. Limiting our attention to uniform Motion in one-dimension of Space merely keeps the mathematics simple and the basic concepts better in focus. *All motion described in this chapter and in most of the following chapters is, therefore, taking place at a set (uniform) Velocity toward or away from the Observer (and his or her instruments).* To avoid disastrous collisions, we'll allow the one-dimensional Space to be ever-so-slightly two-dimensional so that moving objects can safely pass without crashing into the Observer. The deviation from the one-dimensional Space would then be negligible at greater Distances. Occasionally, details of relativistic behavior can be demonstrated better from a side view rather than a line-of-sight, end-on view. But side views are still meant to illustrate what's happening in 1-D Space.

The terms we need

Over a period of years, the literature of Special Relativity has adopted a geometrical representation, terminology and mathematics that constitute **STANDARD FORMALISM** of

Special Relativity. What we'll use in this book is pretty well conventional and, except for minor differences in emphasis or wording, is found in all books dealing with the theory. As made clear in the first chapter, Time is measured along the **T-COORDINATE** *(or T-AXIS for short)* and Distances along the **X-COORDINATE** *(or X-AXIS for short)*. The **Y** and **Z** values are made equal to zero merely to simplify the math. Of the four Coordinates, we are then left only with **X** and **T**. Internationally adopted relativistic terminology provides Time in **Time values** *(such as seconds)*, Distances in **Lighttimes** *(such as Lightseconds)*. **Both of them can, therefore, be simply given in Time values**. There may be some difficulty with this usage at first but we'll find out soon enough that it is easier to manipulate Time and Distance measurements using Time values for both *(that is simply numbers without unit designation)*.

In our virtual experiments, we'll be using different Observers in different Inertial Systems each moving at a set Velocity relative to each other. We could call them by their first names in alphabetical order: Alice or Anton, Betty or Benny, Carol or Carl, etc. without gender prejudice. To keep Observer designations simple, we'll identify them only by letters: **A, B, C,** etc. To indicate their Astronaut-Participant Observer status, we'll place those letters into angular brackets standing symbolically for spaceships: *<A>*, *<B>*, *<C>*, etc.. And, finally, we'll allow the use of the Observer-specific personal pronoun, "he" *(along with the words "him" and "his")* for Andy *(as <A>)* and "she" *(along with "her")* for Barbara *(as <B>)*. Additional participants will be labeled *<C>* for Carl, *<D>* for Dorothy, etc..

No Observer can really tell which one of them is moving and which one is stationary. This is in keeping with the Principle of Equivalence *(same as the Principle of Relativity)*. It is important to be reminded again and again *(and again!)* that linear Motion cannot be distinguished from non-Motion by any mechanical or electromagnetic *(Light or Radar)* tests. Therefore, **all Observers consider themselves standing still or remaining in the same place, always pointing to others as being** *(the "guilty" ones)* **in Motion.**

In our exercises, Observer, *<A>* with whom we'll usually identify ourselves will be stationary or non-moving, so his position in Space does not change but his clock still keeps ticking and it is only his position in Time that keeps constantly changing. His successive Spacetime locations are plotted on the vertical **T**-Coordinate, drawn straight up from the initial position. Such a line is known as **WORLDLINE** *(or WL for short)*. Due to being stationary, that is not moving anywhere in Space, it can also be called his **ZERO-DISTANCE LINE**. His **X**-Axis connects all Events at all Distances that take place at the same Moment thus being **SIMULTANEOUS** within his *(and our)* Frame of Reference. The line linking all simultaneous Events can be called his **ISOTEMPORAL LINE** *(or ITL for short)*, also called **ZERO-TIME LINE** *(all Events on that Line being separated by zero Time)*. The Coordinate System used by *<A>* will be his very own **FRAME OF REFERENCE,** or his **HOME FRAME**. Other participants in alphabetic order will be *<B>*, *<C>*, etc.. We, along with *<A>*, will consider them as either **(1)** non-moving or **(2)** moving and located in **OTHER FRAMES** as the case may be from *<A>*'s *(and our)* point of view.

To keep our Space and Time terminology simple and **Observer-specific**, we'll adhere to the following convention:

(1) Distance and Time measured by *<A>* are given in Capital letters, **X** and **T**.

(2) Distance and Time measured by *<B>* are given in lower-case letters, **x** and **t.**

(3) Measurements may occasionally be identified by additional subscripts.

How to visualize Relative Motion

Proceeding now with our virtual experiments, let us first show two Observers, *<A>* *(Andy)* and *<B>* *(Barbara)* in their separate space vehicles going through their prescribed paces. In the first example *(Fig. 2.5)*, we keep them at a constant Distance from each other. **The question** *as to whether they are both standing still or merely moving at the same Velocity along parallel paths is irrelevant.* There is no way to make the distinction, no method to determine their "actual" or "absolute" state of Motion anyway. For our purpose, they are both standing still and both their *WL*s are plotted vertically. In the second example, *<B>* is moving toward *<A>* and her *WL* is tilted toward *<A>*'s *WL*. In the third example, *<B>* is moving away from *<A>*. In the final example, *<B>* is moving away from *<A>*, *<C>* *(Carl)* is moving toward *<A>* and *<D>* *(Dorothy)* is moving toward *<A>* but faster than *<C>*:

F2.5

To keep our intuitions flexible, let's next assign to *<B>* the role of being the Stationary Observer and see how the other Observers in the above four examples are seen by *<B>* from her point of view. This ***ROLE REVERSAL*** helps us remember that **(1)** *<B>* as well as *<C>* and *<D>* always consider themselves stationary, that is non-moving and that **(2)** their observations are in every way true, their mutual disagreement solved by the Observer-Specificity concept.

From Fig. 2.5 we can see that the Worldlines of moving Observers are tilted to indicate that their position in Space is changing in relation to *<A>* *(and ourselves)*. We'll consider the Home Frames of *<B>*, *<C>* and *<D>* to be in *<A>*'s ***Other Frame*** *(of Reference)*. In Fig. 2.6, it is *<B>* who is in her Home Frame, those other Observers being in her Other Frame. How these terms actually help us, we'll start seeing soon:

F2.6

While positions in Space can remain the same, Time cannot be stopped. ***Time can be measured only by a clock located within one's own Home Frame,*** that is, with a clock that does not move and Time measured directly this way is called ***PROPER TIME***. The Proper Time of *<B>* can be projected on *<A>*'s Time Axis *(or Coordinate)* and the result obtained is *<A>*'s ***CO-ORDINATE TIME*** *(of <B>'s Proper Time)*. In the same way, directly measured Lengths are called ***PROPER LENGTHS***, those cast as ***PROJECTIONS*** on a Space Axis are called ***COORDINATE LENGTHS***. Distances spanning across the wide open Space can only be measured indirectly and are always ***COORDINATE DISTANCES***. Proper and Coordinate Times and Lengths will become better understood a little later in our study.

The views of each Observer need to be summarized again at this time:

(1) *A **FRAME** of Reference is a geometrical configuration: **RECTANGLE, HEXAHEDRON** or **HYPERHEDRON** (as described in Chapter I).* It is the geometrical basis of our orientation in Spacetime.

(2) *A **HOME FRAME** of Reference is always stationary. Every Observer is located in his (or her) own **HOME FRAME** and **ALWAYS** considers him(her)self **STATIONARY** !!!* Their *T* and *X*-Axes are oriented at 90 degrees to each other and all Home Frames are rectangular, or square.

(3) *The **OTHER FRAME** is how the stationary Observer (<A>) is viewing the **HOME FRAME** of a **MOVING OBSERVER** (<B>) who, incidentally, also considers herself **STATIONARY** in her own square **HOME FRAME**.*

Also, keep in mind *(repeated here again to pound it hard into our minds)* that:

(4) *All observations and measurements are OBSERVER-DEPENDENT, also called OBSER-VER-SPECIFIC, and are identified with a specific Observer who is making this or that observation or measurement.*

(5) *The results of an observation or measurement cannot be duplicated by other Observers if there is RELATIVE MOTION between them, more so when the RELATIVE VELOCITY of Motion is very high.*

Measuring inter-Observer Distances

Our next task is to measure Distances between Observers. In interstellar Space, these can never be done directly *(due to great Distances or fast motion)*. The use of ultra-short flashes of Light would work but are technically not as easy to use over long Distances as are Radar Signals. So the most reliable method of measurement is by ***RADAR PROBES*** used by Observers in their *(stationary)* Home Frames. So properly equipped with the "latest" technology, let <A> send a blip of a Radar Signal toward <B> at Moment, *S* or the *S-EVENT (S for "Send")* and shortly thereafter he would receive back at Moment, *R* the *R-EVENT, (R for "Receive")* the reflected Signal. If all these Events are automatically recorded then the data in the computer printout can be studied any time later. The initial results *(<A>'s WL with the Events S and R)* can now be shown:

F2.7

The observations mapped in Figure 2.7 are very basic and the drawing itself looks deceptively simple. On <A>'s ***WL***, there are just two Time markers, *S* and *R*, containing the ***SR TIME SEGMENT*** between them. But there is more there than meets the eye. From this meager data, we can extract more information as shown next:

Let us say that a Signal was sent at 00:00:00 and its Reflection was received 2 seconds later. Therefore, the Time segment, ***SR*** = 2 sec. We also know that the speed of the Radar Signal is equal to that of Light, or *c*, and that ***the amount of Time that the Radar Signal spent traveling toward <B> must be equal to the Time spent on its return*** *(to <A>)*. This is because <A> *(by definition in his own Home Frame)* is not moving in Space. The Signal bouncing off <B> at Moment, ***Q*** must have happened at half-way point between *S* and *R* on <A>'s ***WL*** at a Moment *(Event)* designated as ***q***. The *q*-Event *(Moment)*, therefore, comes one second after *S*.

F 2.8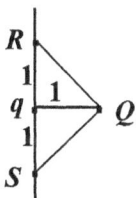

Now we can conclude that:

$$Sq = qR = qQ = \tfrac{1}{2}SR = 1 \text{ second } (or\ 1\ Lightsecond\,).$$

Note that ***DISTANCE*** is represented by a corresponding ***SPACE SEGMENT*** on <A>'s Space Coordinate, ***X*** and can be entered into equations as a Space Segment value. Its numerical result can be given simply in seconds without indicating the Time units *(such as seconds)*.

The Radar Signal arrival at *<B>*'s *WL* is the *Q-EVENT*. Because information *is* obtained by *<A>*'s Radar Probe, it provides *<B>'s* Distance from *<A>* at his *q*-Moment. But it does not say anything about *<B>*'s state of Motion. For all we know, *<B>* could be stationary as to *<A>*, moving towards, or away from, *<A>*. These three possibilities are shown next:

F2.9

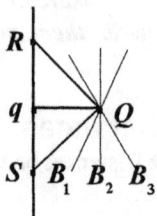

Simultaneous = Isotemporal

The Time Segment, *Sq* in the above example was 1 second and equal to *qR*, thus also equal to half of the *SR* Time Segment. The *qQ*-Distance projected on *<A>*'s *X*-Axis would also be 1 second *(Lightsecond!)*. The *q* and *Q*-Events occurred at the same Moment in *<A>*'s Home Frame meaning that the *q* and *Q* were not just simultaneous as such but were *SIMULTANEOUS (only and specifically) to <A>,* all due to Observer Specificity that governs all observations*!* The *Q*-Event in *<B>*'s Home Frame *(<A>'s Other Frame)* occurred at Moment, *q (according to <A>)* and was located on *<A>'s X-AXIS (in his Home Frame).* Events simultaneous to *Q* in *<B>*'s Home Frame would then have to be located on *HER (<B>'s)* separate and different *x*-Axis *(or Isotemporal Line)* which, as we already know, does not coincide with *<A>*'s *X*-Axis and would, therefore, be isotemporal *(only)* to *<B>*. To top off the confusion, we can say that *q* and *Q* are *NOT* simultaneous *(isotemporal)* to *<B>!!!*

If you are overwhelmed by such hair-splitting detail pertaining to Observer-Specificity, digest the above paragraph very carefully keeping in mind as to *exactly WHOSE* observations or measurements we were talking about. Remember: *ALL OBSERVATIONS (and MEASURE-MENTS) are OBSERVER-DEPENDENT or OBSERVER-SPECIFIC !!!* Note that *<A>* cannot directly measure *<B>*'s Time, *OQ* and that *<B>*'s Distance from *<A>* is *NOT* the same as *<A>*'s Distance from *<B>* (that is x).This measurement process we have not yet surveyed. These Distance measurements by *<A>* and *<B>* always differ if there is relative Motion between them.

The Radar Probe data shown in Fig. 2.8 and 2.9 consists of the timing of *S* and *R* on *<A>*'s *WL* initiated in Fig. 2.7. The *qQ* Distance as obtained by *<A>* is designated as his Space Segment, *X (qQ = X)*. To determine if *<B>* is moving, *<A>* would have to use another Radar Probe to obtain a second *qQ* and compare it with the first *(Radar Signal paths not shown this time):*

F2.10

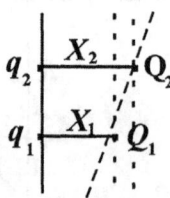

The Space Segment difference, $X_2 - X_1$ if other than zero, would indicate that *<B>* was moving relative to *<A>* and that the Space Segment, X_2-X_1 would be the Distance traveled by *<B>*. Equal Distances, $X_2 = X_1$ would mean *NO MOTION*:

$$X_{\text{path traveled}} = q_2Q_2 - q_1Q_1 = X_2 - X_1 \text{ and}$$

$$T \text{ between the two Distance measurements } = Oq_2 - Oq_1$$

The **VELOCITY**, v of $<B>$'s Motion then equals the Distance from X_1 to X_2 *(or X_2 - X_1)* divided by Time lapse from q_1 to q_2 *(or $q_2 - q_1$) and* would be:

$$v = \frac{X_2 - X_1}{q_2 - q_1}$$

Standard **ST** Motion experiments

A good way to correlate $<A>$'s and $<B>$'s Distance and Time information is to create an Event so that both their Space and Time locations are momentarily *(almost)* identical to, and shared by, both $<A>$ and $<B>$. This is achieved by allowing $<B>$ to pass $<A>$ at a very minimal, near-collision Distance. *At the MOMENT of PASSING (at the Passing Moment or Passing Event)*, **both occupy momentarily (almost) the same location in Space** and are momentarily stationary to each other, that is momentarily without relative Motion between them *(as further examined in greater detail in Ch. XI)*. So both can **synchronize their clocks** to the same setting, say 00:00:00 at the exact Passing Moment, the anchoring Moment, **O**, or the **O-EVENT**. Now, $<A>$ has established **three Time markers: O, S** and **R** of which only **O** is shared by $<B>$. From this new set of data, we can extract even more information:

F2.11

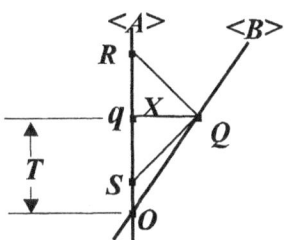

First, during the Time span, **Oq** (= **T** *according to* $<A>$), $<B>$ traveled the Distance, **qQ** = X. Notice again *(and pay attention to this!)* that **(1)** X (in Lighttime) **equals qQ which is equal to half of SR,** that **(2)** **SR** in turn **is the difference between OR and OS,** that **(3)** **OS is ½ of OR minus OS, that is ½ (OR-OS), and** that **(4)** **T is half of the sum of OR and OS: ½ (OR+OS).** Also make sure that you know the identities of X and T. Thus:

$$X = qQ = Sq = qR = \frac{1}{2}(OR - OS),$$
$$\text{and } T = Oq = \frac{1}{2}(OR + OS)$$

Now, the Velocity, v *(of $<B>$ relative to $<A>$)* would then be:

$$v = \frac{\frac{1}{2}(OR - OS)}{\frac{1}{2}(OR + OS)} \qquad \text{or} \qquad v = \frac{OR - OS}{OR + OS}$$

A lot of new stuff was poured out on these last pages *!!!* Are we going too fast? Stop, think and look again at everything. Stare at the illustrations and equations as long as it takes until these begin to talk to you and make sure you understand all this before proceeding. What we just saw were important concepts we'll be using throughout the chapters. Understanding everything *NOW* will be immensely helpful.

It is again instructive to find out *how stationary <A> sees <B> using her (<B>'s) own Radar Probe from her different (but also stationary, like <A>'s!) point of view (<A>'s and <B>'s roles reversed!).* An initial example of Role Reversal was presented on Fig. 2.5 and 2.6. In Figure 2.12(A) below, <A> in his Home Frame considers <B>'s use of *HER* Radar Probe to measure *HIS (<A>'s)* Distance from *HER (not <B>'s Distance from <A>!).* Here <B> is situated in *HIS (<A>'s)* Other Frame but because *HE* is viewing the roles-reversed setting from *HIS* point of view, the Lightlines *(Radar Probe Signal paths)* are mapped in *HIS (<A>'s)* Home Frame*!!!* Next in Fig. 2.12(B), <B> uses her Radar Probe to measure <A>'s Distance from her while positioned *in HER (<B>'s) own Home Frame.* As she is considering everything from *HER* own point of view, the Radar Signal paths are mapped in *HER* Home Frame *!!! The rule is: Radar Signal paths are ALWAYS mapped at 45 degrees from the WL and ITL of the stationary Observer (in his or her Home Frame):*

F2.12

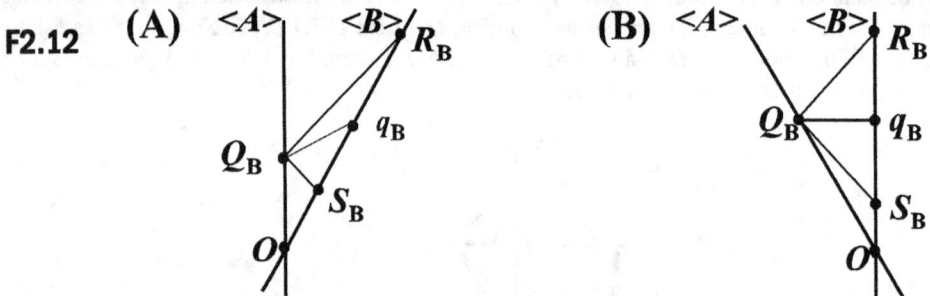

Because <A> has determined <B>'s direction of Motion according to *HIS* Home Frame, it is clear that *AFTER* the Passing Event, <B> was *CLOSER (to <A>)* at S_B than she was at R_B *(<B>'s Radar signal had more Distance to travel AFTER reflecting off <A>).* And <A> also knows that <B> considered herself stationary in her own Home Frame and, therefore, marked off her own q_B at half-way point between S_B and R_B. as shown both in Fig. 2.12(A)&(B).

In the next Fig. 2.13(A), <A> superimposes his and her Radar Probe results to make his *(<A>'s) q* and <B>'s *Q* coincide. In Fig. 2.13(B), on the other hand, it is <B>'s *q* and <A>'s *Q* which are made simultaneous according to <B>:

F2.13

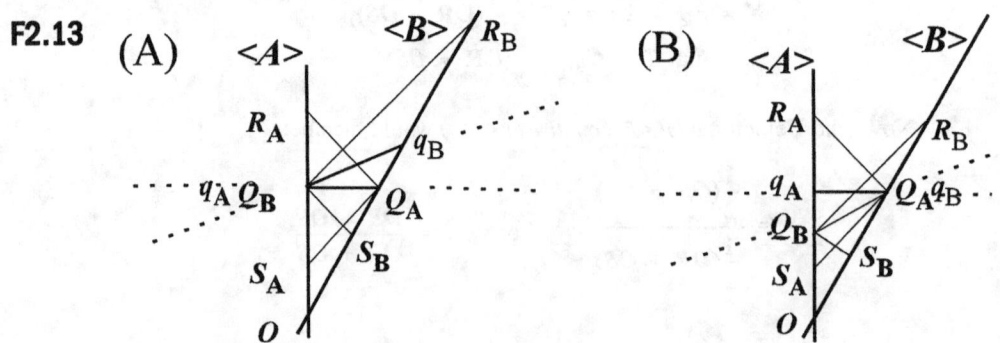

Relativity of Simultaneity

What was shown in Fig. 2.13 is extremely important: *<A>*'s *qQ* is represented by a line that is **HORIZONTAL** while *<B>*'s *qQ* is **SLANTED**. Dotted extensions were added to both *ITL*s to make their different directions more obvious. The *qQ*-lines connect Events, *q* and *Q* for both *<A>* and *<B>* so that their separate *q*- and *Q*-Events are **SIMULTANEOUS** or **ISO-TEMPORAL** either to *<A>* or to *<B>* **but not to both (*<A>* and *<B>***) *!!!* Therefore, *what <A> considers to be simultaneous does not agree with what <B> considers to be simultaneous*. This disagreement clearly demonstrates that *two Observers in Motion relative to each other NEVER agree on what is simultaneous, and what is isotemporal.* This lack of agreement is *(again)* due to Observer-Specificity *(same as Observer-Dependency)* and is *always making what is simultaneous to mean as being "RELATIVE" only to a specific Observer,* not to all Observers. This very important qualification is called

RELATIVITY OF SIMULTANEITY !!!

For additional demonstration, let us draw the separate *ITL*s for both *<A>* and *<B>* but this time making these lines go through *<A>*'s and *<B>*'s common *O*-Event. You should see that all lines are now going through the common Reference Event, *<A>*'s and *<B>*'s *ITL*s clearly *NOT* coinciding:

F2.14

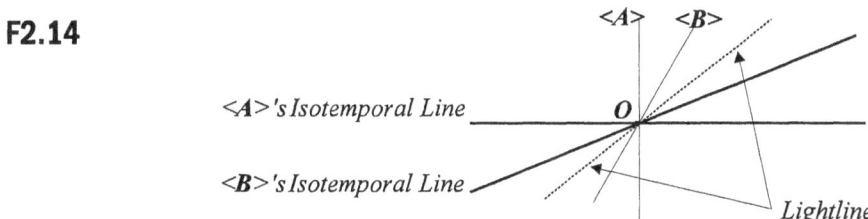

It is also appropriate at this time to illustrate Relativity of Simultaneity by *Einstein's favorite virtual experiment* using a railroad train, his favorite fast moving vehicle, to illustrate the nature of this Relativity of Simultaneity. No space vehicles existed or could be realistically imagined in those days. In the example below, we are viewing the experiment from the side rather than from line-of-sight direction of the 1-D Motion for better display of detail:

F3.15

In the fast-moving railroad car, *<B>* is standing next to a Light Source located in the exact middle of the car. At an agreed-upon Moment when *<B>* in the specially equipped car is passing *<A>* who is standing on the ground near the tracks *(and when the Light Source happens to be at the exact position of passing)*, a flash of Light is emitted from the Light Source. The Light travels toward the Leading and Trailing Walls of the car. The railroad car has *(either)* an open or transparent side facing *<A>* and both *<A>* and *<B>* have accurate instruments for detecting and recording the Moments as to exactly *WHEN* the Light is emitted and exactly *WHEN* it reaches the Targets mounted on the Leading and Trailing Walls at equal Distances from the Light Source.

The Observer, *<B>* in the moving car considers herself stationary and her instruments show that the Light Signal reaches both wall Targets *SIMULTANEOUSLY*. To the ground-based Observer, *<A>*, however, the car is moving and its Leading Wall is moving away from the Light Signal speeding toward it while the Trailing Wall is moving toward the approaching Signal. As a result *(but only according to <A>)*, *the Light hits the Trailing Wall BEFORE reaching the Leading Wall.* To *<A>*, these hits do not happen at the same Moment *(or simultaneously).*

A similar disagreement over Simultaneity occurs in case the Light Source is kept on the train but the two Targets are placed on the ground near the tracks. A Light flash is again emitted when the Source reaches the exact middle between the Targets *(now on the ground)*. The Observer on the ground registers simultaneous hits while the Observer on the moving train finds the ground-based Targets moving toward and away from the oncoming Signal. The roles of the Observers are now reversed and so are the results. It makes no difference whether the Light Source is on the ground or on the moving train because *the velocity of Light (relative to both Observers) is not influenced by the Motion of the Light Source or of the Observer!!!*

Again, to complicate matters, *BOTH Observers consider themselves stationary* and *THE OTHER as moving.* Both hits will be *simultaneous only to the Observer in whose Home Frame these Targets are located.* It is not that one of the Observers must be wrong. Actually, *BOTH are right !!!* Remember, the results are *Observer-dependent (or Observer-specific),* a circumstance that produces this *Relativity of Simultaneity* which also means that *all results are "relative" to the participating Observers !!!* So much for that. Now that this is clear *(or is it ???)*, let's move on to the really important consequence of the *absolute equality of Invariants to all Observers.*

Invariants are the same, that is equal, to all Observers

Returning to *<A>*'s Radar Probe shown in the Figure 2.11 reproduced below for your convenience, we'll see that *<B>*'s Distance from *<A> (as determined by <A>)* was half of the 2-second *SR* duration. The *qQ* was, therefore, 1 *(second)*. We do not know *<B>*'s own $S_B R_B$ measurement *(Fig. 2.12)* and are not yet in position to calculate what it would be. But, fortunately, *<A> can derive <B>'s x and t values from the data he (<A>) can acquire firsthand by his own (<A>'s) Radar Probe.* How this is done we'll see next.

Let's look at the Radar Probe used by *<A>* to measure *<B>*'s Distance *(from him)*. Here *<A>* observed *<B>* move from the *O*-Event to *Q*-Event which are the two *ST* Events that define this particular Spacetime Invariant. For both *<A>* and *<B>*, the Time, *OQ = t* is directly measurable only to *<B>* and so for her, it is not a Projection. To refresh our visual memory, look again at the Figure 2.11 *(see below)* before proceeding:

F2.11 *(here reproduced)*

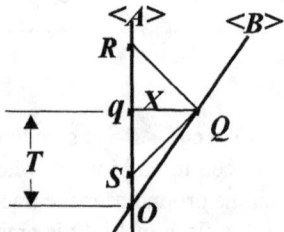

That the Common Diagonal (the Invariant) is of equal magnitude in all Frames of Reference comes now handy. The Invariant is defined by Events, *O* and *Q*, each on *<B>*'s *WL* with only *O*-Event also located on *<A>*'s *WL*.

Let's restate the equality of the Invariants in terms of the two different Observers, *<A>* and *<B>*:

$$I^2 \; = \; X^2 - T^2 \; = \; x^2 - t^2$$

so we can state that: $X^2 - T^2 \; = \; x^2 - t^2$.

By Radar Probe, <A> measured <B>'s Distance, qQ indirectly as a **Projection** on his Space Coordinate. The measurement *(see Fig. 2.11 showing that T= Oq and X= ½ SR)* gave him the X value in the equation. The Time, T, measured directly by <A> as Oq, was also a **Projection** of <B>'s OQ on his *(<A>'s)* Time Coordinate. While <A>'s position in Space did not change from one Moment to the next *(according to <A>)*, it was <B>'s position relative to <A> that did change and, as said, was measured *(indirectly as a Projection)* by <A> as X. But the Invariant, OQ was <B>'s Time, t because both O and Q happened to be on **her WL** and <B> did not change its position in Space. So, both O and Q *(according to <B>)* were at the same Space location. Therefore *(and pay attention to this !!!)*, <B>'s **x remained equal to zero** *(by definition)* !!! *(x = 0)*. This much <A> should know even without measuring it. And <B>'s Time *(or Oq = t)* was not known to <A>!!! How much more confusing can it get ?!!!

Incidentally, by using letters X, T, x and t, we have specified as to who actually did or could have done the measurements. Observer Dependence, again, at work *!!!* Don't forget it*!*

Now, let us identify again all the symbols in the equation, $X^2 - T^2 \; = \; x^2 - t^2$:

(1) X was <B>'s *(Coordinate)* Distance, qQ, from <A>.

(2) T was <A>'s Proper Time, Oq but in this case it was also <A>'s Coordinate Time of <B>'s Proper Time, OQ. Make sure you can distinguish Oq from OQ !!!

(3) x was <B>'s Space Segment from O to Q but because <B> considered herself stationary, x, by definition, was equal to zero, thus making it a known quantity for <A>.

(4) t was <B>'s Proper Time from O to Q, or OQ, measured directly by <B> *(and available as an indirect Coordinate Projection for <A>)*, therefore, the one-and-only unknown for <A>.

Now we can solve for the only unknown *(t)* in our equation because x by being equal to zero, stopped being an unknown, so:

$$X^2 - T^2 \; = \; 0^2 - t^2$$

$$X^2 - T^2 \; = \; - t^2$$

and

$$t^2 \; = \; T^2 - X^2$$

$$t \; = \; \sqrt{T^2 - X^2}$$

Crunching the now simplified equalities

There is a treasure hidden in the above Minkowski's **ST** Equation. If we first multiply, then divide, the right side of the last equation by the square root of T squared ($\sqrt{T^2}$), then the magnitude of the right side of the equation will not be changed but it *(showing what it is equal to)* is changed to:

$$t = \sqrt{T^2 - X^2} = \frac{\sqrt{T^2}}{\sqrt{T^2}}\sqrt{T^2 - X^2} = \sqrt{T^2}\frac{\sqrt{T^2 - X^2}}{\sqrt{T^2}} = T\sqrt{\frac{T^2}{T^2} - \frac{X^2}{T^2}} = T\sqrt{1 - \frac{X^2}{T^2}}$$

As $X/T = v$, $t = T\sqrt{1 - \left(\frac{X}{T}\right)^2} = T\sqrt{1 - v^2}$, v given as a fraction of c.

Restated: $t = T\sqrt{1-v^2}$ and $T = \dfrac{t}{\sqrt{1-v^2}}$

So, from now on, let us also remember that: $\sqrt{T^2 - X^2} = T\sqrt{1-v^2}$

We have just produced an expression, $\sqrt{1-v^2}$ which was the *"FUDGE FACTOR"* *(that Lorentz derived from information about the differing electron Velocities)* that helped to correct and reconcile experimental discrepancies for the momentum $(p = mv)$ of electrons in cathode ray experiments *(see Introduction)*. This expression is a very useful tool for converting one Observers's measurements into those of the other Observer. In general literature about Relativity, it is used in form of its *inverse value (which we will not be using)* and as such, it is called **GAMMA** (γ) or the **LORENTZ COEFFICIENT**, also called the **LORENTZ FACTOR.** In scientific literature, it is used almost exclusively in this inverse form. Those with insufficient math expertise often have trouble with inverse values. For that very practical reason, what we'll be using in this book is the *non-inverse expression*, the one shown above **which we'll call the LORENTZ MULTIPLIER** and name it by the Greek letter, **BETA** *(β)*. Also, remember that our equations are further simplified by agreeing that **(1)** *v ALWAYS stands for the FRACTION OF c* and can be expressed as 0.5 *(or half of c)* and 0.999,9 *(almost that of c)*, etc. and that **(2)** *Velocities less than that of Light can simply be written as v instead of v/c.*

The use for Lorentz Multiplier

It should be clear that $<B>$'s Velocity relative to $<A>$, known from $<A>$'s measurements, must be the same Velocity obtained by $<B>$ from her own numerically different data *(differing from that obtained by $<A>$)*. The Velocity between them is shared and must, therefore, be of the same magnitude to both with the only difference that their Velocities are pointed in opposite directions.

Our newly-found knowledge about Lorentz Multiplier gives us a shortcut to the value *(unknown to $<A>$)* of $<B>$'s qQ measurement of $<A>$'s Distance, or x which is not the same as *(or equal to)* X. Let us see how we can calculate it:

As $v = \dfrac{x}{t} = \dfrac{X}{T}$ so $\dfrac{x}{t} = \dfrac{X}{T}$ and $x = \dfrac{Xt}{T}$

But we have already calculated that $t = T\sqrt{1-v^2}$, therefore:

$x = \dfrac{Xt}{T} = \dfrac{X}{T}T\sqrt{1-v^2}$ *(v expressed as a fraction of c, that is v/c)*

And canceling out T in the Numerator and Denominator gives us:

$x = X\sqrt{1-v^2}$ and, also $X = \dfrac{x}{\sqrt{1-v^2}}$

Refusing to be confused by reversal of roles

Before proceeding to the next step, we have to clarify the confusing identity of x *($<A>$'s Distance from $<B>$ as measured by $<B>$)*. This is not equal to zero because it measures how much $<A>$ has moved during the Time Segment, t *(!!!)* in $<B>$'s Home Frame. In Fig. 2.12(**A** and **B**) page II-14, $<A>$'s Distance from $<B>$ at the Q-Moment is the Space Segment, $q_B Q_B$ measured by $<B>$ on her *ITL* located between the two *($<A>$'s and $<B>$'s) WLs*, q_B both on $<B>$'s *WL* and *ITL*. In Figure 2.13(**B**), there you see another $q_B Q_B$ which, however, is not identical with the $q_B Q_B$ in Figure 2.13(**A**). Make sure you see and understand the difference! So $<A>$'s Distance from $<B>$

(according to <B>) seems awkwardly counter-intuitive. Two pairs of illustrations in Figure 2.16 should help clear up the mental fog. Please, scrutinize the illustration, F2.16 relying totally on the visual argument and labeling:

<B> in <A>'s Other Frame   <B> in her own Home Frame

F2.16

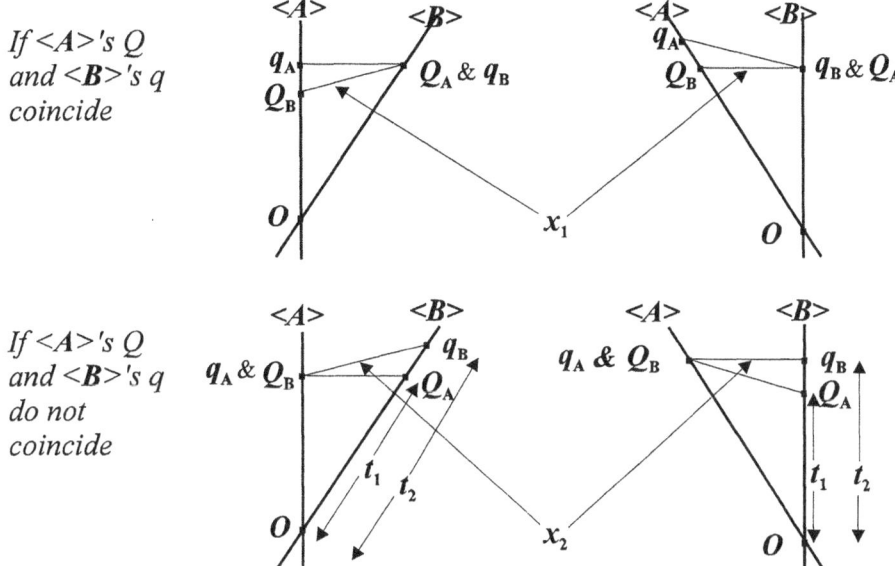

If <A>'s Q and <B>'s q coincide

If <A>'s Q and <B>'s q do not coincide

Now that *(in Fig. 2.16)* we have two *x*'s: x_1 and x_2, it should be clear that the two Maps placed horizontally side by side are identical except that they are viewed from <A>'s and <B>'s separate points of view. The diagonally placed Maps are analogous but not identical. The illustrations and labels in Fig. 2.16 show that if Q_A and q_B coincide, the moving Observer's x_1 and t_1 are always smaller than stationary Observers's X and T. If, on the other hand, <A>'s Q and <B>'s q do not coincide, the x_2 and t_2 are greater than X and T.

By scrutinizing Fig. 2.16, it should become clear as to which x is which and that not only $x_1 < x_2$ but also $x_1 < X < x_2$ *!!!*. Study the above visual examples well until their identities begin to make sense. These inequalities are more completely taken up in Chapters IV and VII.

The quantitative relationships shown above remain the same if we merely switch the stationary Observer's role from <A> to <B>. Then <B>'s two readings of <A>'s Distance from her *(x_1 and x_2)* are mapped as horizontal lines and <A>'s reading of <B>'s Distance as slanted lines. On the next illustration, the two different *x*s *(of <B>)* and the one X *(of <A>)* touch <A>'s and <B>'s *WL*s remain in the same corresponding positions even when their stationary and moving roles are reversed:

F2.17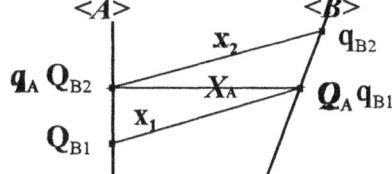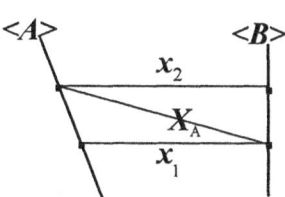

Putting equations to work

The equality of \<A\>'s Velocity from \<B\> and \<B\>'s Velocity from \<A\> can be stated as follows:

$$v = \frac{X}{T} = \frac{X\sqrt{1-v^2}}{T\sqrt{1-v^2}} = \frac{x}{t}$$ which, of course, is another way to show that $v_A = v_B$.

So, you should see that \<A\> and \<B\> are in perfect agreement about the relative Velocity between them. But there is a *quantitative difference between T and t as well as between* x_1 *and* x_2. Note also that \<A\>'s and \<B\>'s Space and Time Segments are not equal but that v is equal for both which is another way of saying that the ratios of X to T and x to t are equal.

Let's rewrite the equalities derived earlier in this chapter:

$$x = X\sqrt{1-v^2} \text{ and } t = T\sqrt{1-v^2}$$

$$X = \frac{x}{\sqrt{1-v^2}} \text{ and } T = \frac{t}{\sqrt{1-v^2}}$$

Our Lorentz Multiplier is always smaller than 1.0 ($\sqrt{1-v^2} < 1.0$) so any quantity is reduced if *multiplied* and increased if *divided* by the Lorentz Multiplier.

We need to keep in mind that Isotemporal Lines on which the X and x are plotted never coincide, but tend to cross if there is relative Motion between \<A\> and \<B\>. Time values are always plotted on each Observer's *WL*s which are not parallel *(do not coincide)* if there is Motion. The Invariant *(which has the same value for both \<A\> and \<B\>)* is defined by its *End-Events (the Events marking its Ends)*. *It always takes two Spacetime Events to define a given Distance*, X *(or* x*), or Time*, T *(or* t*).* And the *ST* Segment between any two *ST* Events is, therefore, always an Invariant *!!!* So X *(or* x*)* can be the Invariant but only if T *(or* t*)* in the "compressed" Reference Frame is made equal to zero. Similarly, T *(or* t*)* in the "compressed" Reference Frame can be the Invariant but only if X *(or* x*)* is equal to zero. This is to say that *any individual parameter can be the Invariant if in the same Home Frame all the other Coordinate measurements are zero.* Coordinate measurements are Projections and cannot be Invariants in the particular setting and are *not* equal to zero in any "compressed" *ST* Frame of Reference *!!!* The difficult sentences above *(confusing only because of Observer-Specificity)* are best restated in form of equations.

When $T = 0$, $I = \sqrt{X^2 - T^2} = \sqrt{X^2 - 0^2} = \sqrt{X^2} = X$, the Invariant

When $X = 0$, $I = \sqrt{X^2 - T^2} = \sqrt{0^2 - T^2} = \sqrt{-T^2} = iT$, the Invariant

When $t = 0$, $I = \sqrt{x^2 - t^2} = \sqrt{x^2 - 0^2} = \sqrt{x^2} = x$, the Invariant

When $x = 0$, $I = \sqrt{x^2 - t^2} = \sqrt{0^2 - t^2} = \sqrt{-t^2} = it$, the Invariant

There is one more thing of interest here. If we calculate t from T & X, or T from t & x, the result will be a "normal" or real number *(without the "i")* but this is so only because of a "concealed" multiplication of $-T^2$ and/or $-t^2$ by *-1* which happens whenever we move them from one side of the "=" mark to the other side.

Motion "causes" slowing ("Dilation") of Time

We should next calculate $<B>$'s Time, t for a selection of actual Velocities first with v = 0, then v=0.25, v=0.5, v=0.9, v= 0.999 and, finally, v=1.0.

First, $v = 0$:

$$t = T\sqrt{1-v^2} = T\sqrt{1-0^2} = T\sqrt{1} = 1.0\,T$$

And v=0.25

$$t = T\sqrt{1-v^2} = T\sqrt{1-0.25^2} = T\sqrt{1-0.0625} = T\sqrt{0.9375} = 0.9682T$$

Then with $v = 0.5$:

$$t = T\sqrt{1-v^2} = T\sqrt{1-0.5^2} = T\sqrt{1-0.25} = T\sqrt{0.75} = 0.866T$$

Next, $v = 0.9$:

$$t = T\sqrt{1-v^2} = T\sqrt{1-0.9^2} = T\sqrt{1-0.81} = T\sqrt{0.19} = 0.4359T$$

Now, $v = 0.9999$:

$$t = T\sqrt{1-0.9999^2} = T\sqrt{1-0.9998} = T\sqrt{0.0002} = 0.014\,T$$

And, finally, $v = 1.0$:

$$t = T\sqrt{1-1^2} = T\sqrt{0} = 0T = 0$$

What the above calculations show is this: At half the Velocity of Light, moving clocks are only slightly slower. It takes $0.9c$ Velocity to make Time progress at nearly half of its normal rate. For Time to really slow down, Velocities have to be rather close to that of Light. *At the Velocity of Light, Time progression stops completely!*

The twin paradox

The slowing of Time is called "***TIME DILATION***" *(Dilation = Dilatation = Stretch)*. This effect is rather dramatic at very high Velocities and makes a great impression on any lay audience if something called the "*twin paradox*" is described. For instance, if $<B>$ manages to travel very close to the speed of Light then Time for her would come *(almost)* to a standstill, allowing $<B>$ to travel enormous Distances to a far-away solar system, visit the inhabitants there and return back to Earth within her own lifetime. In her travel, she would be spending Time mostly for acceleration and deceleration, for conducting interstellar diplomacy and she would hardly waste any Time for what would otherwise be a long and tedious journey.

To illustrate the twin paradox in concrete numbers, let $<B>$ do the impossible by traveling at the speed of Light and visit a planet orbiting a star some 8000 Lightyears away. Her itinerary would then read something like this *(Time measured by on-board clock programmed to show "Earth Time")*:

Accelerating after takeoff from Earth	1 month
Travel Time	0 Time
Decelerating before landing	1 month
Greeting natives and conversing in sign language	1 week

Accelerating after takeoff	1 month
Travel back to Earth	0 Time
Decelerating before landing	1 month

Adding up <B>'s Time spent on her mission gives us 4 months and 1 week. On Earth the Time lapse would be a little more than 16,000 years *!!!*

If <B> was <A>'s twin then on her return, her brother would no longer be alive and the civilization on Earth, if still existing, would be totally changed. And <B>, on the other hand, would hardly be older. We'll see later *(in Chapter VIII)* that actual Velocities cannot really reach that of Light so that a Space traveler always ends up spending a prohibitive amount of Time getting from one solar system to another. Even Velocities equal to $0.9999c$ still require considerable amount of Time so that only *Startrek "warp" Velocities* would make it possible for <B> to return home within her brother's lifetime. Any place outside our own little solar system looks rather inaccessible for now.

For Light, Time does not exist *!!!*

Light is peculiar in its obligatory Motion at the fastest possible speed while spending none of its own Time doing that. It can go from one edge of the universe to the other *(if such edge actually exists)* and do it in absolutely zero Time. Perhaps Light somehow exists *outside our Space and Time Coordinates* and only *trans-sects* our Space dimensions in order to interact with our universe and be observable. *The path of Light from our new, relativistic perspective is absolutely the shortest Spacetime route (actually zero Spacetime "Distance") between any two of our Space locations. Not moving in Space (as it normally happens in all Home Frames) is paradoxically always the longest possible Spacetime route.* In the next illustration, you'll see a selection of paths from #1 to #6 from one Space location to the same place in Space a little later. The non-moving #1 Path is the "longest," the curved ones becoming progressively "shorter" until the shortest available path is traversed in no Time at all *(for the Light)* by reflecting off a mirror at *Q* then making its return to exactly the same location in Space but to another Moment in Time, Path #6. Strange indeed *(!!!)*:

F2.18

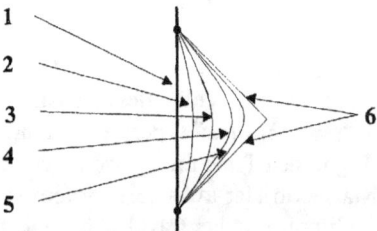

Not moving at all is actually equivalent *(or identical)* to moving at uniform speed. It is impossible to observe any difference between these two states. But in everyday life, we are never in doubt because we always have ample external clues to go by. Ancient civilizations considered the Earth to be stationary and even now we have no direct visual evidence that it moves. Despite Copernican revolution, our common-sense intuition still serves us well enough for everyday purposes. So in some sense, we are all members of the Flat Earth Society.

Does Relativity rule over Angular Motion?

All our virtual experiments were staged in 1-dimensional Space so that everything always happened along a straight, thin line of sight. An increasingly nagging question that needs answered is this: ***would conclusions drawn from Motion observed in 1-D Lineland be applicable to 2-D Flatland and 3-D Spaceland,*** that is, if we observe something moving not in our line of sight but "sideways," in ***Transverse Motion?***

The "Light Clock"

As this last question is important, let us relate the question to a concrete situation: ***If an object in uniform Motion does not hit the Observer but passes at a safe Distance*** *(at which Moment it is transiently in pure Angular Motion where the line-of-sight Velocity happens to hit zero),* ***would the effects of Special Relativity be different at the very Moment of Passing?*** This kind of transverse or "sideways" Motion is exactly what is involved in the performance of the so-called ***"Light Clock" experiment.*** This amounts to a standard demonstration of Relativity "at work" at the Passing Moment and will give us a firm answer to the last question.

Let's imagine a Light Source on a platform inside a suitable vehicle capable of traveling at half the Velocity of Light. Directly above the Light Source *(at a Lightnanosecond Distance)* there is a Light-sensitive Receptor. These two *(Source and Receptor)* can be connected to a device that, for both <A> and <B>, will register the Time it takes for the Light to travel from the Source to the Receptor. With atomic clocks currently available, this should be easy in concept. Without Motion, what ***both*** *(the passenger in the vehicle and the bystander)* would see is a ***vertical path*** of the ultra-short flash taking 1 nanosecond to reach from the Source to the Receptor. With suitable mirrors, we could make the Light bounce back and forth any number of times maintaining 1 nanosecond interval between the "ticks" for as long as we wish. But to keep everything simple, let's just plot one such "tick" on our Map:

F2.19

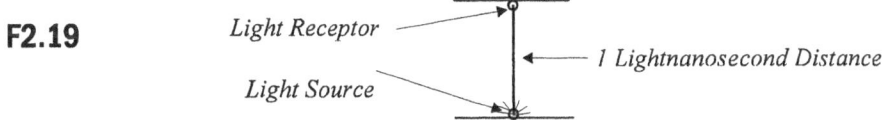

Light Receptor

Light Source

1 Lightnanosecond Distance

Now let's take the vehicle to a suitable Distance from the bystander (<A>) and let it go ***whoosh*** at *v*=0.5 without hitting him while performing just one such "tick" at the exact Moment of Passing. Now, what does the bystander, <A> record compared to what is recorded by <B>? To the passenger *(<B>)* in the vehicle, nothing has changed. The path of Light is still vertical and the Time registered on her on-board atomic clock is still 1 nanosecond. OK so far? To the bystander, however, the Light path appears ***slanted*** in the direction of motion. The following illustration shows it clearly and the accompanying right-angle triangle has its sides labeled so that the helpful math is clearly suggested:

F2.20

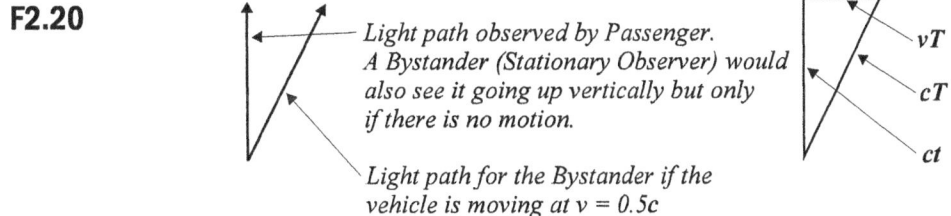

Light path observed by Passenger. A Bystander (Stationary Observer) would also see it going up vertically but only if there is no motion.

Light path for the Bystander if the vehicle is moving at v = 0.5c

vT

cT

ct

Note that for the ***Moving Observer*** *(the passenger),* the experimental setup did not move. So the Velocity, *v* would be zero *(v = 0)* and the Length of the vertical Light path would remain the same 1 Lightnanosecond. The ***Bystander,*** however, ***observes a slanted Light path that is longer in Space than the vertical path,*** here forming the Hypotenuse of a right angle triangle.

Now Light Velocity is the same for both Observers, so the travel path as well as the Time the Light spends going from the Source to the Receptor must be different for both, taking longer to travel the slanted path. Let <A>'s Time be *T* and <B>'s Time *t*. The path traveled by the vehicle *(as observed by <A>)* corresponds to the horizontal side, *vT*, of the triangle. For <B> who considers herself stationary, *vt = 0* and for her, there is *NO* triangle. Now Time, *T* as well as *t*, are then proportional to the observed Distances traveled by Light which is *cT* for <A> and *ct* for <B>. From the right angle triangle in Fig. 2.20, we can calculate the Time, *t* in terms of *T* and vice versa:

$$ct = \sqrt{(cT)^2 - (vT)^2} \;=\; \sqrt{c^2T^2 - v^2T^2} \;=\; \sqrt{(c^2 - v^2)T^2} \;=\; T\sqrt{c^2 - v^2} \quad \text{or}$$

$$ct = T\sqrt{c^2 - v^2}$$

Dividing both sides of the equation by *c*, we get:

$$\frac{ct}{c} = T\frac{\sqrt{c^2 - v^2}}{\sqrt{c^2}}$$

And

$$t = T\sqrt{1 - \frac{v^2}{c^2}} \;=\; T\sqrt{1 - \left(\frac{v}{c}\right)^2}$$

The expression, $(v/c)^2$ is given as a fraction of *c* squared, but it can also be written as v^2, the same way we can write *v/c* as *v*. Therefore, the above result simplifies into:

$$t = T\sqrt{1 - v^2} \quad \text{and} \quad T = t\frac{1}{\sqrt{1 - v^2}}$$

You can see that the Light Clock confirms our known Lorentz relationship between *T* and *t* in 3-D Spaceland exactly the way we established it in 1-D Lineland. For the moving Observer, <B>, in whose Home Frame the Light Clock is located, the Time measured from the Light-flash to the Moment the Light-flash reaches the Receptor is shorter for <B> than what is clocked by <A>, the stationary Observer in whose Other Frame the Light Clock experiment is being performed. The difference is given by the Lorentz Multiplier. So, *Special Relativity applies to all cases of uniform, linear motion in 1-, 2-, or 3-D Space without regard as to direction of Motion to any Observer be that line-of-sight, transverse or oblique.* The only advantage of the Lineland setting is the simpler mathematics compared to what we must use in the more complex settings in Flatland and Spaceland.

Reaching out to the limits of Spacetime

Because the Velocity of Light is the same in all Inertial Systems and it behavior is identical to all Systems, its Path separates 4-D Spacetime into **(1)** domains of the Past, Present and Future; **(2)** regions where cause-and-effect relations are or are not operative, and **(3)** areas that are accessible and not accessible to our observational reach. All this and more will be taken up in the next chapter, Mapping Spacetime.

III - MAPPING SPACETIME

Lost in Spacetime

Without a street map, you can still find your way in a strange town. But once you enter Spacetime, home to Relativity, you are hopelessly lost without a handy **SPACETIME MAP.** In "Axioms And Operations," we got our feet firmly on the ground by blazing some local Spacetime travel paths. What we need now is to first examine the surface features on the Map and then examine the outer reaches of the Realm.

Let's remember what we've learned

What we've learned so far in this book can be committed permanently to memory by restating everything once more *(and for the last time)* in outline form:

(1) The Special Theory of Relativity is the conceptual and mathematical consequence of the **Principle of Relativity:** A state of uniform Motion cannot be distinguished from a state of rest as both are **equivalent** *(if not entirely identical).*

(2) The state of uniform motion and the state of rest are examples of a single system, the **Inertial System,** united by the **absence of acceleration.**

(3) In uniformly moving systems *(moving along straight lines at constant velocity without acceleration)* all known laws of mechanics and electromagnetism are completely valid as they are known to operate in non-moving systems and the velocity of Light is the same in all Inertial Systems.

(4) Special Relativity describes Motion that is linear and proceeds at uniform Velocity.

(5) All motion is relative without relevance as to what or who is moving and consists simply of a change in Distance between the Observer and whatever is being observed, be that an object or another Observer.

(6) All Observers in Inertial Systems *(Inertial Observers)* consider themselves to be stationary. Everything non-moving is located in their **Home Frames** *(of Reference).* Everything in Motion relative to the Observer is considered to be in their **Other Frames** *(of Reference).*

(7) Observers in different Inertial Systems in Motion relative to each other do not agree about what or who is moving or about the results of their Space or Time Segment measurements. For that reason, all observations and measurements are **Observer-related or Observer-dependent** and must always be identified with specific Observers *(or Inertial Systems)* to be meaningful. It is meaningless to consider Motion as Absolute*!!!*

(8) Time and Space are intrinsically linked and cannot be considered separately from each other. This closely linked union is called **Spacetime.**

(9) Space Segments are measured by the amount of Time required by Light to travel that Distance and are expressed in Time units.

(10) We'll pick <*A*> to be the stationary Observer in all our virtual experiments. All others are called *(alphabetically)* <*B*>, <*C*>, <*D*>, etc.. To avoid gender prejudice, the duties of <*A*> and <*C*> are assigned to men and those of <*B*> and <*D*> to women.

(11) An **Event** is something happening at a Location in Space and at a Moment in Time. A very brief Event serves as a **marker of a Spacetime Location** *(or Event).*

(12) The cornerstone of Special Relativity, the all-important **Invariant,** can be viewed as a 4-D geometric structure analogous to the Diagonal of a 3-dimensional box-like Frame of Reference to which Time has been added as the Fourth Dimension.

(13) The Invariant is of the same magnitude to all Inertial Observers regardless of their relative Motion. In general literature, you'll see it referred to as a **Spacetime Interval or Separation.** In order to allow the use of the word "interval" in the ordinary, common-sense meaning, we'll let the term, **"Invariant"** to stand for both "Interval" and "Separation."

(14) M inkowski's Spacetime equation for the Invariant, with *X, Y & Z* as Space Coordinates *(Axes)* and *T* as Time Coordinate *(Axis)*, is as follows:

$$I = \sqrt{X^2 + Y^2 + Z^2 + (iT)^2} = \sqrt{X^2 + Y^2 + Z^2 - T^2}$$

(15) Making *Y* and *Z* equal to zero **compresses 4-D Spacetime into 2-D** that can be easily mapped on a flat 2-D surface. All our virtual experiments are in this "compressed" setting and Minkowski's Spacetime equation becomes compressed to $I = \sqrt{X^2 - T^2}$.

(16) The Invariant can be equal to, but is never greater than, any of its Projections on the Observers's *(four Space and Time)* Coordinates.

(17) The Space and Time measured by *<A>* in his Home Frame are given in upper-case *(capital)* letters, *X* and *T*. The Space and Time measured by *<B>* in her own Home Frame, that is in *<A>*'s Other Frame, are given in lower-case *(small)* letters, *x* and *t*. This terminology tags all measurements to specific Observers. Only the Invariant, *I* does not need such a tag as it is equal *(in magnitude)* to all Observers.

(18) On all "compressed" Space and Time Coordinates, 1 second of Time and 1 Light-second of Space are represented by equal line lengths.

(19) The inequalities in the Time and Distance measurements are *(usually)* *T* > *t* so that $t = T\sqrt{1-v^2}$, and also *X* > *x* so that $x = X\sqrt{1-v^2}$, *T* = *t* and *X* = *x* only if there is no Motion.

(20) All Observers view their own Home Frames *(of Reference)* as stationary. The Home Frames of moving Systems are Other Frames for stationary Observers.

(21) In all virtual *ST-* experiments, we identify ourselves with the Observer, *<A>* whom we always *(with rare exception)* consider stationary.

(22) The successive positions of an Inertial Observer or Object in Spacetime traces out a path plotted on a straight line on the Time coordinate and is known as Worldline, *WL* for short.

(23) A stationary Observer's location in Space does not change and his *WL* goes up vertically on the *ST* Map. His *WL* is also his Zero-distance Line, the reference line from where all Distances are measured.

(24) A stationary Observer's Space Coordinate *(X-Axis for short)*, connects all Events at all Distances on the *ST* Map that are **simultaneous** *(or isotemporal)* and that line is, therefore, also his **Isotemporal Line** (or *ITL)*, drawn as a horizontal line at 90 degrees to his *WL*.

(25) The Home Frame of the stationary Observer, made up of *X-* and *T-*Axes, is a Coordinate System that is perfectly square on the *ST* Map, each Coordinate positioned at 90 degrees from *(all)* the other Coordinate*(s like Y and Z)*.

(26) The *WL* of anything that moves is tilted toward, or away from, *<A>*'s *WL*.

(27) The moving Observer's *x*-Axis *(same as ITL)* and *t*-Axis *(same as WL)* are both tilted symmetrically toward the Lightline between them.

(28) Lightlines are positioned at a 45 degree angle to the stationary Observer's *WL* and *ITL* within his Home Frame.

(29) Distance measurements *(in all Frames)* can be done only indirectly by the Radar Probe. What we obtain this way are Projections on the Space Coordinate, also called **Coordinate Distances.** **Proper Distances** can be measured directly by a measuring tape or ruler and only in the Observer's Home Frame.

(30) All Time measurements read directly off the clock within the Observer's Home Frame are Proper Times. Time cannot be measured directly within Other Frames.

(31) A Proper Time in one Inertial System can also double as Coordinate Time Projection of another Observer's Proper Time. The same cannot be done with Proper Lengths which cannot double as Coordinate Lengths.

(32) Synchronization of clocks in different Inertial Systems can be accomplished by near-collision ***Passing Events*** which are momentarily simultaneous to both Systems *(as both can transiently occupy nearly the same Spacetime location).*

Motion warps all Spacetime surfaces

An attempt to map Spacetime according to Euclidian geometry reveals *a problem that is not unique to Relativity.* It has been a characteristic of Earth Maps all along. For instance, equal land Distances on Earth's equator and on the arctic circle are always mapped by unequal line lengths on conventional flat-surface maps. In the same way, equal line lengths on the map of the Earth mark off unequal Distances on the equator and on the arctic circle. This is so because *flat maps of the Earth are NOT true to scale !!!* A true-to-scale map of the Earth is impossible unless drawn on a globe whose surface is analogous to that of the Earth. All flat maps of the Earth are said to be "Mercatorial" *(named after a famous mapmaker).* They are useful nevertheless as long as their limitations are kept in mind. Our Spacetime Maps are similarly affected.

The disproportion in Spacetime Map lines is even worse than in Earth Maps because ***Spacetime is not only CURVED but also NON-SPHERICAL.*** This warping makes reading of Distances and Times difficult. We have seen it already but have always studiously ignored it. Now we'll have to take a closer look.

Let's recall that inside the Reference Rectangle *(on 2-D surface)* and in the Hexahedron *(in 3-D Space)*, the ***Diagonal*** is the ***Hypotenuse*** of the component right-angle triangle where it *is always the longest side.* But the *calculated value of t (represented by that same Hypotenuse or Diagonal)* is *never greater than X or T,* the other two sides of the triangle. We must, therefore, conclude that *the geometric method introduced in the first chapter is not exactly accurate because it does not allow true-to-scale mapping of Spacetime.* Worse still, *it cannot be drawn true to scale on a globe surface* either.

The trouble with trigonometry

In the literature about Special Relativity, there are two different ways we can visualize the component triangle *(in the Frame of Reference)* drawn on the ***ST*** Map. For proper orientation, let's again draw the basic Spacetime Map of the astronaut, *<B>* traveling, let's say, 1 sec at velocity, ½*c* and as viewed by *<A>*:

F 3.1

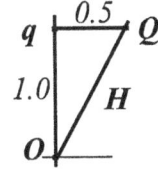

For better understanding, let's draw the triangle again and label its sides, *X, T* and *t.* Note that *t* forms the Hypotenuse of the triangle:

F 3.2

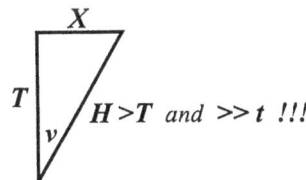

Trying to apply conventional trigonometry to this triangle gets us quickly into trouble. In Special Relativity, *v* is the velocity between *<A>* and *<B>*. It is usually represented by the angle, alpha (α) between T and t and is given by the trigonometric function, sinα. But a good look at the above triangle tells us that in no way can we describe X divided by T as sinα. Here, sinα should be X divided by the hypotenuse, t and the X/T in the Minkowskian triangle should be called a tangent *(or tan)* instead. Now that two different trig functions should be applied to X/T, we should also assign a different Greek letter such as ϕ to the angle in Minkowskian-shape triangle representing the geometric angle for v in these two different cases. While Minkowskian mapping is the standard in Special Relativity, the triangles formed by XTt cannot be true-to-scale. That's because ***the measured or calculated values of the moving parameters*** *(such as t and x)* ***are not proportional to the line lengths standing for them.*** To handle both stationary and moving parameters, ingenuity has given us ***two methods for drawing Frame-of-Reference triangles*** made up of ***two stationary Coordinate*** values X, T and one *(t)* that belongs to the ***moving System***. These two different triangles *(that* can be made up of X, T and t) are, therefore, **(1)** *not-true-to-scale* and **(2)** *true-to-scale.*

(1) THE NOT TRUE-TO-SCALE TRIANGLE

In chapters I and II, we simply pretended that ***MINKOWSKI's*** geometry was the right geometry. Sure, the T and X Coordinates of stationary <A> were oriented at the required 90 degree angle to each other *(and also to the Y and Z Coordinates we avoided using)* and ***their line lengths were always TRUE TO SCALE***. But the line length of the slanted t, the Time counted by the moving clock and ***represented by the Hypotenuse is NOT TRUE TO SCALE*** *(not proportional to the Time actually measured)*. ***Still the not-true-to-scale method of mapping the two intersecting Inertial Systems has remained the most insightful in practice.*** The angle that represents Velocity, v should best be labeled with the Greek letter phi (ϕ) although in most books and journal articles the letter alpha (α) is the letter used. The trigonometry in that case is non-Minkowskian. Strictly speaking, on a Minkowskian Map, the proper trigonometric function applicable to the velocity is ***tanϕ*** and it also equals, as expected, X divided by T:

$$v = tan\phi = X/T$$

When $X = T$, the velocity obtained is the Velocity of Light, and $v = tan\phi = X/T = c = 1.0$, a result that is also correct!

The nice thing about $v = tan\,\phi$ is our ability to keep all Cartesian Coordinates oriented at the theoretically correct 90 degree angle to each other and map all Velocities without changing this very basic arrangement while the Mapline lengths of X and T also remain true to scale without limit into the hypothetical faster-than-Light range.

F3.3

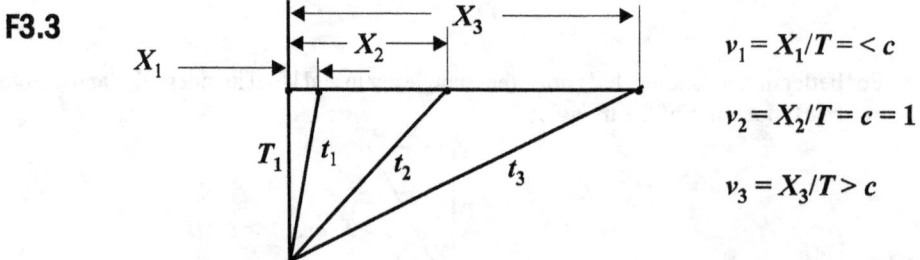

$$v_1 = X_1/T = < c$$

$$v_2 = X_2/T = c = 1$$

$$v_3 = X_3/T > c$$

At the velocity of Light, the Time, T in the triangle becomes equal to X and the magnitude of t, despite of its actual line length in the triangle, becomes equal to ***ZERO***. As you see,

Minkowskian geometry is only partly true-to-scale. Anything that moves introduces *LONGIT-UDINAL distortion* in the Map lines. Because all moving parameters are not located in <*A*>'s square Home Frame, these, therefore, cannot be drawn true to scale.

(2) THE TRUE-TO-SCALE TRIANGLE

The triangle most commonly used to express velocities is one where *ALL* sides of the right-angle triangle are equal to the actual quantitative values, measured or calculated. *The Cartesian arrangement of Minkowski's geometry is thereby abandoned but without showing it on the Map.* The *X* and *T* Coordinates, if actually plotted true to scale, would no longer remain at 90 degrees to each other. The deviation of the true-to-scale mapping from Cartesian mapping is initially not easy to see but becomes increasingly noticeable at higher Velocities. At half of *c*, the actual location of the 90 degree angle *(in the true-to-scale triangle)* becomes more obvious and is seen located between *X* and *t* instead and *NOT between X and T (in <A>'s Home Frame)*:

F3.4

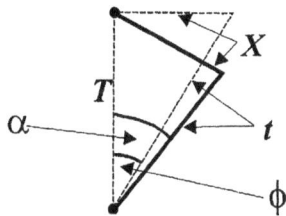

At Velocities even closer to *c*, the difference between the two representations becomes unmistakable:

F3.5

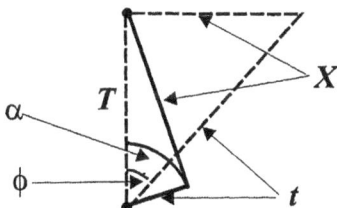

At the Velocity of Light, when *X* = *T* and *t* = 0, the *X* and *T* *(which are now equal)* become superimposed on each other while the third side, *t* disappears altogether from view and makes the triangle collapse into a single line:

F3.6

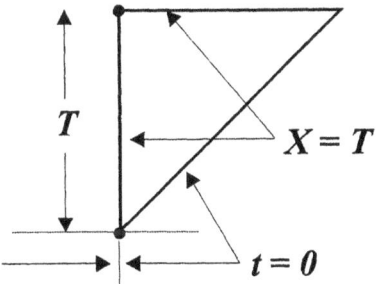

In the true-to-scale triangle, the angle facing *X* *(in Fig. 3.4, 3.5 and 3.6)* is the one called alpha (α) and the trigonometry is appropriately given by the *sine* function: $v = \sin\alpha$.

In the above examples, *T* was kept in vertical position and *X* was allowed to became slanted but the series of *sin α* functions can also be shown with *X* kept oriented horizontally and *T* allowed to become slanted *(Fig. 3.7)*, varying with Velocity.

Incidentally, the true-to-scale method makes it impossible to visualize Velocities above that of Light. This reflects the theoretical impossibility of Light ever exceeding *c*. Also note that the 90 degree angle is always between *X* and *t*:

F3.7

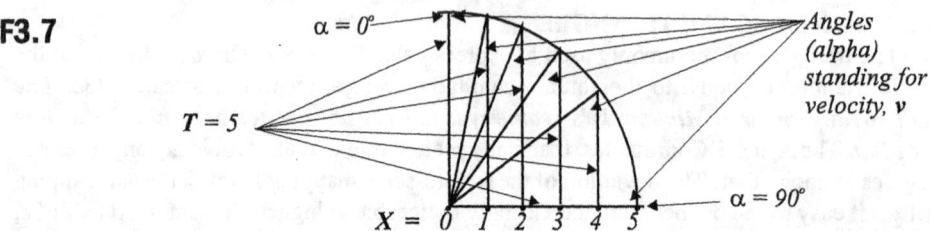

By the way, you may have already figured out that: $tan\phi = sin\alpha = X/T = v$ and that $\phi \neq \alpha$!!!

We can add here another useful detail: with true-to-scale triangles, $sin^2 + cos^2 = 1$ and, consequently: $cos\,\alpha = \sqrt{1-sin^2\alpha} = \sqrt{1-(v/c)^2}$, or simply $\sqrt{1-v^2}$ *(the Lorentz Multiplier!)*, clearly an advantage in the more complex calculations. As an alternative, we could **use hyperbolic trigonometry** but switching over to this less familiar trig makes our math much too difficult to most readers.

Downsizing troublesome trigonometry

In our study, we'll be using Minkowskian geometry at all times and depend almost exclusively on algebra as our math tool. Geometric visualization helps us understand Special Relativity but now we must also keep in mind that Velocities can be shown in two entirely different ways. This is especially important if calculations are done in trigonometry. In the scientific literature what we most often see is that **the Map** *(combining both Home-Frame and Other-Frame lines)* **is typically Minkowskian while the trigonometry carried out** *(involving both Frames)* **is non-Minkowskian** *(and true-to-scale)!!!* Strictly speaking, **tan** should be applied to Minkowskian triangles and **sine & cosine** to true-to-scale triangles. Without knowing that this kind of inconsistency is the accepted practice, it is easy to become mixed up about how *(in most cases)* trigonometry is *(mis)*applied. Omitting trigonometry *(that probably is insufficiently familiar to many readers anyway)* almost completely in this book circumvents this particular source of confusion.

Let's do mapping

With all the preliminaries now cleared away, we can finally start mapping Spacetime. With line drawings in "Axioms And Operations," we have done some of it already. Our initial *("compressed")* data consists of Time *(T on <A>'s WL)* and Space *(X on <A>'s ITL)* measurements. We'll map them the standard Minkowskian way on <A>'s **WL** and **ITL** which are part of <A>'s Home Frame. A non-stationary Observer's Time and Distance (*t* and *x*), on the other hand, are located in the Stationary Observer's (<A>'s) Other Frame. This particular **MIXED MAPPING method** allows intuitive visualization of all the possible relationships between <A> and <B> so that mathematical descriptions *(equations)* will just about jump out at you.

Let's restate the main point of the last paragraph: Map lines for nonmoving parameters are always Euclidian, square, and true to scale. **Only the map lines standing for Space and Time values of moving systems are not true to scale.** Motion, therefore, always introduces distortion of scale but in a predictable way and this fortunate circumstance makes our basic Minkowskian Maps eminently intuitive.

Constructing Spacetime Grids out of "longitude and latitude lines"

Let us now map \<A\>'s Home Frame by plotting his **WL** (**T**) and his **ITL** (**X**). We can see that \<A\>'s **WL** is always vertical and his **ITL** is always horizontal, both at right angles to each other. The **O**-Event on the Map is where the **WL** and **ITL** cross. The vertical **WL** connects all Events with zero Distance between them *(no motion in Space from one moment to the next!)*. The horizontal line connects all Events that are simultaneous *(or isotemporal)* with the **O**-Event and have zero Time duration between them *(from one Space location to the next)*:

F3.8

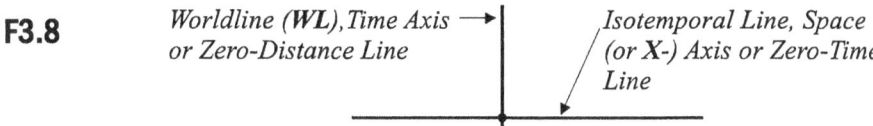

On the **X**-Axis (**ITL**), let's mark off equal increments of Distance by equal line segments:

F3.9

Next, on the **T**-Axis *(which is the WL)*, let's mark off equal increments of Time by equal line segments:

F3.10

By connecting all Distance markers *(Events)* on the **X**-Axis (**ITL**) with Event locations in the Past and Future that are isodistant *(at the same Distance)* from the **T**-Axis (**WL**), we obtain vertical ***Isodistance Lines*** parallel to the **WL**. In the same manner, connecting all Time markers on the Observer's **WL** to all other Event locations simultaneous *(or isotemporal)* to those markers, we obtain a series of parallel, horizontal ***Isotemporal Lines***. All vertical Isodistance Lines are at right angles to all horizontal Isotemporal Lines. What is obtained this way is a stationary Observer's ***SPACETIME GRID*** that maps out motionless Spacetime by abstract Isodistance and Isotemporal *("longitude and latitude")* Lines.

F3.11

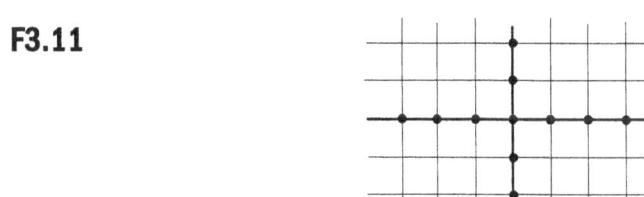

This Spacetime Grid can be extended indefinitely in all directions so that all Events in the Past, Present and *(potential)* Future are included. This Grid is square, rectangular and laid out on the flat paper surface just the way it appeared to Euclid, Ptolemy, Galileo and Newton. It is the way it looked to our ancestors who thought that the Earth was flat and it is the way it *(at close range)* even appears to us! The classical, Euclidian character of Spacetime is obtained for things that do not move or when Motion is very slow compared to the Velocity of Light. *Very fast Motion,* on the other hand, *produces different Spacetime Grids.*

To repeat: all Home Frames are Euclidian and square as everything in Home Frame is stationary, non-moving! Now that this is clear, let's find out how one Observer's square Grid appears to another Observer if there is Motion between them and ***BOTH*** consider their own Grids as stationary and genuinely square *!!!*

Combining Home and Other Frames of Reference

For combining two Inertial Systems, we start with just one privileged point of view: <A> observing <B>'s square Grid but within his (<A>'s) own square Grid. So <B>'s Grid would be located in <A>'s Other Frame. We can map this by first plotting the basic *WL*s and *ITL*s of both <A> and <B> but without showing the Radar Probe Signal paths. Lightlines, however, need to be shown and these remain in the same position for both but are always mapped in the stationary Home Frame:

F3.12

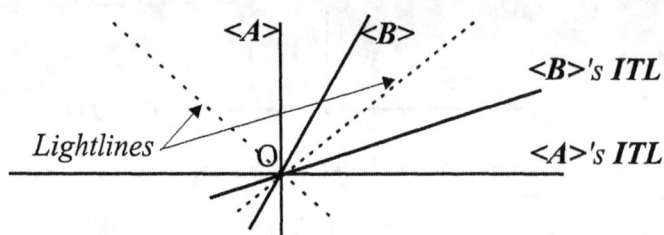

Next, on <B>'s *WL*, let's mark off equal Time Segments by equal line lengths *(Lengths and Distances)* and draw *ITL*s *(parallel to each other)* through these Time markers:

F3.13

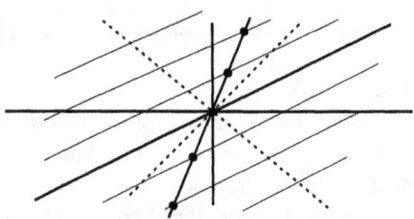

On <B>'s *ITL*, let us mark off equal Distance increments and draw Isodistance Lines parallel to <B>'s *WL* through them:

F3.14

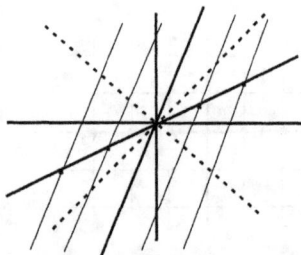

Combining the last two Maps, we obtain <B>'s *ST* Grid but now drawn in the way seen by <A> in his (<A>'s) Other Frame. In other words, what is in <B>'s Home Frame is in <A>'s Other Frame. What is different here is this: <A>'s Home Frame is a Grid that is square, rectangular; <**B**>'s Grid in the Other Frame, however, has a ***RHOMBOID SHIFT*** *(or orientation)* of parallel lines, all ***tilted symmetrically toward the Lightline*** between them:

F3. 15

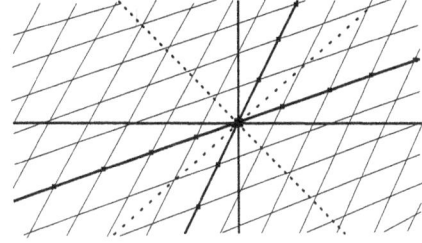

Re-mapping <**B**>'s Grid for a velocity closer to that of Light brings out Rhomboid Shift even more prominently:

F3.16

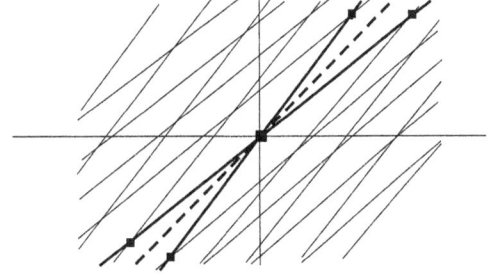

It needs re-emphasized that <**B**>'s Grid, square in her own Home Frame, becomes deformed if viewed and mapped by <**A**> in his Other Frame. But, again, let's reverse the roles of the Observers and find out how <**B**> sees <**A**>'s Home Frame according to her *(<B>'s)* Other Frame. From her *(<B>'s)* point of view, it is now <**A**> whose Grid is slanted and rhomboidal but turned the other way because in <**B**>'s view, <**A**> is moving in the other, opposite direction:

F3.17

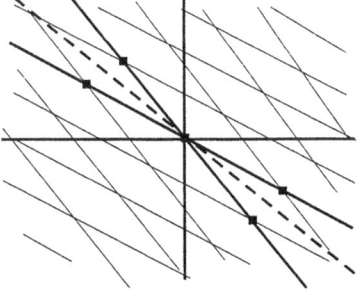

Dividing Time into Past, Present & Future

Let's next proceed with the really ambitious task of mapping the entire Spacetime. By Light *(or Radar)* Signals we can mark off basic Spacetime domain borders once and for all. But to have some real data to work with, we have to go back to an earlier Moment in the Past in order to have some Events already on record. Now let's use, say, two seconds' worth of records of what actually happened. With <**A**> not in Motion *(we are always taking his side)*, his **WL** is placed vertically on the Map. Let us assume that he had positioned two Light *(or Radar)* Signal Sources on each side at a Distance of one Lightsecond from him *(within the 1-D Lineland)* and that both units were triggered simultaneously:

F3.18

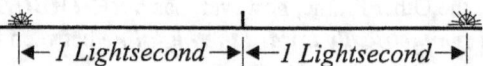

The record for the first second shows that both Signals reached <*A*> *(one from each side)* at the same Moment *(that is simultaneously)*:

F3.19

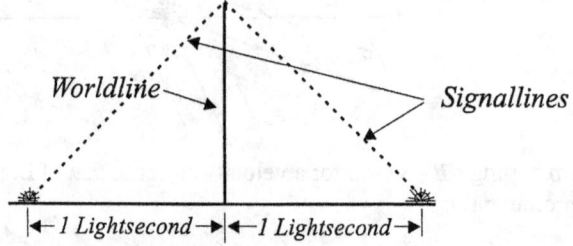

After the Detectors at <*A*> registered simultaneous hits, each Signal then continued in the same direction but now moving away from <*A*>:

F3.20

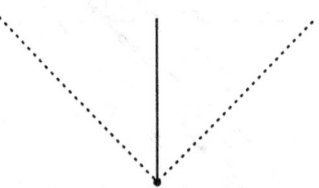

Note, that on this 2-D paper surface, the 1-D Lineland Space (*X*) Coordinate is on the horizontal Axis and the Time (*T*) coordinate is on the vertical Axis. But if we omit the Time Coordinate altogether, the two-part 1-D scenario could be illustrated as follows:

F3.21

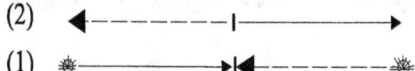

After combining the two halves of the above 2-second record on a Spacetime Map, we see <*A*>'s *WL* criss-crossed in its middle by the ***two LIGHTLINES intersecting at the anchor-ing point, the O-Event:***

F3.22

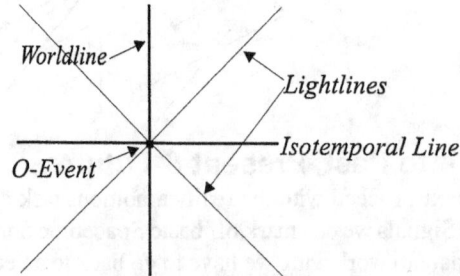

Although we started out with two Lightlines crossing at *O,* there is another way to look at our Map. We can say that instead of two Lightlines crossing, we have here **(1)** *a set of two converging and* **(2)** *a set of two diverging Lightlines.* The converging Lightlines are the paths of Signals capable of carrying information to us *(our eyes or instruments)* located at the *O*-Event and can, therefore, be called our ***SIGHTLINES***. The diverging Lightlines can carry information from us in form of Signals to other Observers in the Future after the *O*-Event. These lines can, therefore, be called our *SIGNALLINES*. With the *O*-Event as a *fixed* Spacetime location, *these Lightlines in effect divide the entire 2-second Spacetime into three basic domains,* those of the *PAST*, of the *PRESENT* and of the *FUTURE*. These divisions, however, are always specifically oriented to the *reference point, the O-Event*, the Spacetime location as the record shows, we *(or <A>)* occupied just once at that singular, brief Moment.

From the *O*-Event, Lightlines can be extended not just for a second or two but indefinitely in both directions: backward as far into the Past as our universe existed and forward into the Future as long as it continues to exist. This classification of basic Spacetime domains seems clear and easy to accept. But dividing Spacetime by a single set of Lightlines into these three domains brings out some novel features we need to explore. It pertains not only to *WHAT can causally impact on us at the O-Event and beyond* but also *WHAT we can actually observe WHEN we are observing something.*

We at the *O*-Event are bumped into by all kinds of gross, microscopic and subatomic objects caused by Events in the Past. Because material bodies always travel slower than Light, these influences could not have been located in Space at Distances greater than possible for Light from there to reach us at the *O*-Event. The Distances from where these material bodies can act upon us are thereby limited to *the domain between the converging Lightlines, the domain of the Past.* All Events located between these converging Lightlines are, therefore, *TIME-CONNECTED* or *TIMELIKE* needing only a certain amount of Time to reach our *O*-Event location. Happenings at the *O*-Event can, in turn, act upon Events between the diverging Lightlines, *the domain of the Future.* Those Events are also Time-connected, or Timelike, to the *O*-Event.

No object, not even Light in the broad domain of the *Present (located outside these converging and diverging Lightlines)* **can reach us at the O-Event** and are effectively isolated from us by Space. Let's repeat: Light Signals from this domain *(outside the converging-diverging Lightlines)* cannot connect with the *O*-Event. So all Events outside these converging/diverging Lightlines are *SPACE-CONNECTED* or *SPACELIKE* to our *O*-Event location. *INFORMATION (in form of Light or Radar Signals)* and *radiant ENERGY can travel only along Lightlines* at Velocities never slower or faster than Light and are, therefore, *LIGHT-CONNECTED* or *LIGHTLIKE* *(but again as always)* **in REFERENCE to the O-Event.** In addition to being Light-connected, they are also on our Sightlines and *SIGHT-CONNECTED*. Events that are materializing at the border of the Present and the Future are on our Signallines and are, therefore, *SIGNAL-CONNECTED from the O-Event toward the Future.*

The area on our Map completely outside the converging/diverging Lightlines, is clearly not part of the Past or Future. To call all of it the Present seems at first a little problematic to most persons unfamiliar with Special Relativity. But the area, if not part of the Past or Future, must constitute nothing less than the entire domain of the *Present* as there are no other domains in Spacetime. So *Lightlines SEPARATE the PAST from the PRESENT and the PRESENT from the FUTURE (!!!) of that O-Moment.*

Now we can confidently map and label all the three domains:

F3.23

As you may have guessed it by now, *the Spacetime location, O can be understood in two entirely different sense* depending on what line of argument we want to pursue:

(A) STATIC SPACETIME RECORD

Let's first consider the *ST* Location as a fixed point on our Spacetime Map where the Time Axis and Space Axis of a selected previous Moment, the *O-EVENT* cross. This gives us a convenient anchoring point that helps us analyze any set of established *(and recorded)* Spacetime Events. Proceeding from this central point, we can measure *(or calculate)* the various Space and Time Segment values on our Space and Time Axes. This in turn can provide us with complete data about any significant set of Events in the Past, Present and Future *(of that O-Event again!!!)* thereby *linking all interrelated Events of the Past with that fixed O-Event and in turn that fixed O-Event with all the Events in the Future (of that O-Event!!!).* In this way the *O*-Event is something already of record *(from the Past)*. Alternatively, we can pick a Moment in our Present and make it "frozen," so that everything occurring thereafter is understood as the *Future of that designated O-Moment.* The data mapped in the domain of the Future is then clearly *RETROSPECTIVE* but now *we can relate any Event of interest to what happened before, what happened at the same Time and what happened afterward.* And we can do all this without getting out of our favorite reclining chair.

(B) STUCK FOREVER AT *"NOW"*

The second and the only other way to view the *O*-Event is radically different but more familiar to us because it is always the Moment we happen to occupy *RIGHT NOW !!!* It is the *NOW-MOMENT.* To think of it, *that Moment is never fixed in Time* and is on a continuous course sliding forward in Spacetime along the Time Axis *(or Worldline)*, more correctly thought of as a process that is continually extending our Worldline upward on the *ST* Map and, consequently, letting Events in the Present slip into the Past. We at the *NOW*-Moment are always edging towards the Future that also keeps sliding unstoppably forward right in front of us. Future viewed this way is the domain we never reach. In fact, *we are eternally stuck at the curiously inconstant as well as permanent (!) NOW-Moment !!!*

Is there really a Future "waiting" for us?

The never-stopping *NOW*-Moment makes everything complicated. If Time has never progressed beyond the *NOW*-Moment and if we can map only Events with a definite past history of existence, we can never map anything in anticipation. This restriction, if strictly enforced, would then exclude the entire Future. Future, therefore, would never be a fact *(of existence or)* of record. *So in the real world, Future never gets here, always gets pushed ahead of the forward-moving center of our Present, the NOW-Moment.* Whatever happens *NOW* would materialize only in the Present and gets instantly relegated into the Past below the *(forward-moving) NOW*-Moment. We can only keep rolling forward as if riding on a wave crest of the Present toward the nebulous, indefinite, uncertain, formless Future we never catch and like a mirage always stays ahead of us. We may mistakenly believe that *Events are somehow prepared to "happen,"* that is, coming *ready-made,* pre-formed and simply waiting for us to ride the Arrow of Time through them, thereby regarding them as *pre-existent Future Events.* But, as we have pointed out, *NOTHING can actually exist inside the Future,* waiting for us to "catch up" with them. Events materialize only in the Present and *we observe them only at the dividing line between the Present and the Past, a line on the Map that is also never "frozen."*

To think of it, whatever we know is only of Events that have actually happened and were, or could have been, directly observed. This requirement limits the recorded *(or recordable)* Spacetime exclusively to the domain of the Past that has been, or could have been, passed over once by our Sightline. So *the Past is absolutely the only domain about which we can have any kind of information, data or knowledge:*

F3.24

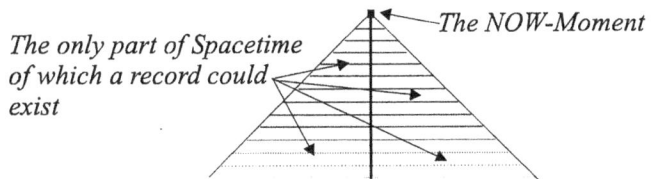

The only part of Spacetime of which a record could exist

The NOW-Moment

Can we ever "see" the Present?

That we do not see anything simultaneous to our *NOW*-Moment is not so obvious. But stop and think: What we see at our arm's length is how it *WAS* about two nanoseconds ago *(Light from there takes about two nanoseconds to reach our eyes)*. The moon you see is how it *WAS* one second ago. The currently observable sunspots are *how they WERE* 8½ minutes ago. A star a few million Lightyears away may not be there any more, having perhaps collapsed into a neutron star or a black hole during the past few million years.

What is there *NOW* is something we simply do not know. But these yet unseen Events do not remain unobservable forever. Observing the Events momentarily located on the forward-sweeping *ITL* is possible only in a given order, starting with those nearest to us. Observability then progresses outward to the more and more distant Events at the speed of Light as our *Sightline swiftly sweeps outward from the (upward-moving) NOW-Moment along the ITL (if kept fixed!)*. An Event seen at the Distance of one Lightnanosecond *(1 Lighnanosecond = 30 cm = 1 ft)* may not be exactly there anymore when we see it. The appearance and perhaps even the very existence of things are more and more questionable the more distant they are. As Time is speeding forward along the Time Coordinate, the reach of our vision to simultaneous Events keeps extending further outward into Space and, as noted, at the speed of Light. Thus all the Events once on the *momentary ITL* at the *NOW*-Moment become observable only little by little, never all at once but sequentially as our Sightline sweeps over them at the fleeting Moment when those Events in the unobservable Present slide into the Past. So, all things as they were *THEN* become momentarily sightconnected *NOW*.

In the next illustration, Events momentarily on the *ITL* become visible in the numerical sequence, # 1, 2, 3, etc. starting at *(and very near)* the *O*-Moment from where the visibility extends swiftly outward into Space as the persistently moving, continuous *O*-Moment becomes Sight-connected with more and more Events once on our *ITL* but unseen then:

F3.25

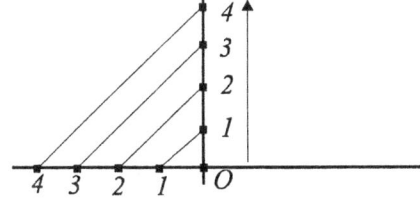

Events in the domain of the Present cannot bump mechanically *(or electromagnetically as Light or other forms of energy)* into us at the momentary *O*-Event. Only the Past and the Future are Time-like or Time-connected and cause-connected. Events and even Light in the Present are, therefore, *ISOLATED* from the *O*-Event rather *DEFINITELY, COMPLETELY* and *ABSO-LUTELY*. But then, again, as we have seen, *NOT FOREVER*. These Events can and will eventually influence us at some Future Time and the possibility for this influence starts the Moment our Sightline sweeps over them:

F3.26

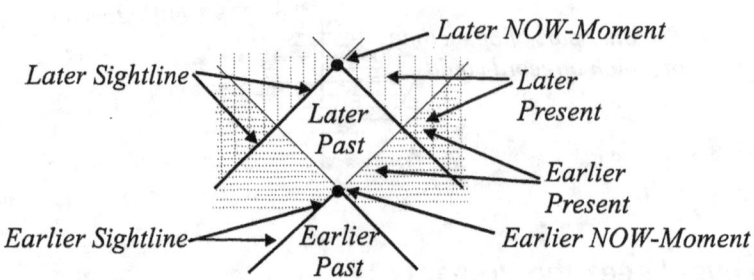

Can we see the Past? An interesting question

So what if we cannot observe the Future or the Present. What about the Past? An odd question! But think! Not seeing the Future is not a problem, we are used to it. We may have reservations about not seeing the Present but a moment's reflection should also tell us that whatever happened in the Past, has already happened and its momentary images have become extinguished, replaced by newer images. *The Events that gave rise to the visual messages about them were already relegated to the Past by the Time the Light Signals from them reached us.* What we believe we see is just the visual information reaching us *AFTER the fact* at the *NOW*-Moment. And we see only Events *as they WERE* only momentarily at the exact Moment they became relegated from the Present to the Past.

Miracle of the Present and more

The inevitable conclusion of our visibility survey, as surprising as it may be, is this: *We cannot see the Future, we cannot see the Present and we cannot (any more) see the Past!!! What we see is only what was once transiently located on the SIGHTLINE*, the instantaneous, narrow *TRANSITION ZONE* that separates the Past from the Present and is part of neither, nothing else*!!!* Seeing the Present, therefore, is an *ILLUSION,* a perceptual *MIRAGE !!!*

But what about the *NOW*-Moment? We surely *SEE* something *RIGHT HERE and NOW (!!!)* and certainly *KNOW* what is going on around us. And that's not all: What we do know we actually know in considerable detail. Our certainty about Future Events is to a great extent also unerring. We can predict what will certainly happen in the Future. We know for instance not only that there will always be another day, another week, another year, etc. but the body of knowledge called science also predicts with astounding accuracy and detail what will happen under certain specified conditions in the Future. This knowledge is based upon observations made in the Past, the Past we know and understand. So if we know much of the Past and know *(much about)* the Future, we must certainly know something about the Present that is in between.

The *remarkable awareness of the Past, Present and Future* is begging to be explained. And the explanation is as simple as it is obvious. It is due not only to the ever-present, sensible and *causal coupling of Events* but also due to the *persistence of our external world without drastic changes* from Moment to Moment and to *the cyclical recurrence of Natural Events* most of them strung together by causal chains that make things predictable and sensible *(making sense)*. It is *the pervasive relatedness, persistence and predictability that blurs the conceptually sharp divisions between the Past, Present and Future thereby making their separate identities very much indistinct.* So, remembering the Past, observing the Present and knowing the Future seems persuasively self-evident and not a mere illusion after all.

The smearing of *ST* domain borders has always been alive and well in the folklore as some folks more than figuratively seem to be "living in the Past." A common attitude about the blurred identity of the Future is also mirrored in a well-known wisecrack: "the Future ain't the way it used to be."

More trouble with the Present

A question that keeps coming up is this: If we are eternally stuck at the *NOW*-Moment and cannot even see anything in our Present domain, *can anything actually exist above our ITL?* The question is about something that is not even observable to us *(at the of NOW-Moment)*, something beyond our reach of vision. Perhaps at least the Signallines *(Lightlines)* can exist above our Now-Moment and above the Isotemporal Line diverging as they do outward/upward at the speed of Light a little faster than we ourselves can ever move, maybe going even ahead of us in Time? To explore this question, let's take a *NOW*-Moment and produce a flash of Light or a blip of a Radar Signal that gives us direction in Spacetime and also serves as a beacon pointing towards the Future.

So with a flash of a Signal at the *O*-Moment, or anywhere else on the *ITL*, let's allow it to go outward/upward in all directions, say, for 1 second:

F3.27

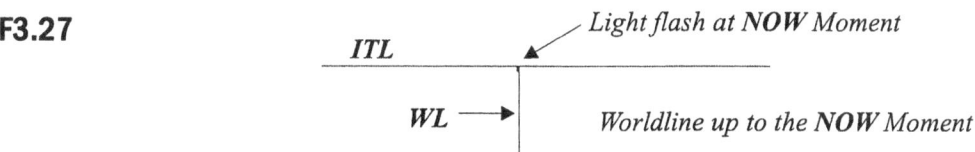

Where will we and the outgoing Signal be at the end of 1 second? Let's see. As the Signal moved outward for a 1 Lightsecond of Distance in Space, we also moved forward in Time on our *WL* by one second at which Moment *BOTH the Signal AND ourselves* became relocated on a *new ITL :*

F3.28

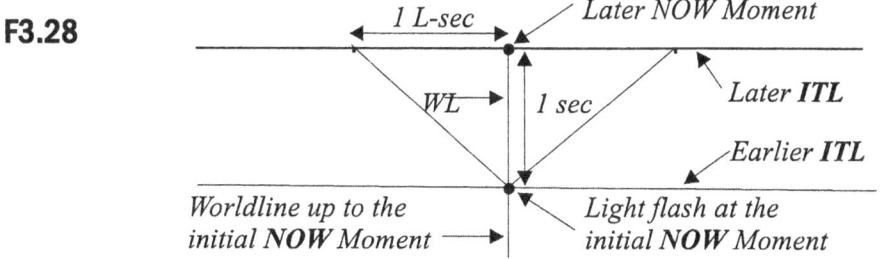

So, nothing, not even the fastest possible *(Light/Radar)* Signal can ever go from the *O*-Event beyond our constantly up-moving *ITL ahead of our WL (Time Coordinate)*. As the Present sweeps upward on the *ST* Map and forming, and remaining attached to, new *ITLs,* it allows nothing at all to reach Spacetime locations beyond it, beyond the latest *ITL* of our continuous *NOW*-Moment*!!!* And that's not all*!*

We tend to confuse the incoming Sightlines for the knife-edge-like zone of Simultaneity. In everyday life and for all nearby Events, these two lines *(Sightlines and Isotemporal Lines)* are perceived to be identical, indistinguishable and, therefore, the same. Our common-sense intuition incorporates this very lack of distinction. We may believe that a distant Event is simultaneous with any nearby Event if we just happen to see them at the same Moment. We do not ordinarily make an allowance for the Time needed for the Signal to reach us from the originating Event. Also, we do not always realize that even in our immediate surroundings, *we see every Event only a fraction of a second AFTER it happened,* one nanosecond per foot *(ca. 30cm)* of Distance. *The Moment when it is observed is, therefore, not simultaneous with its occurrence.* The discrepancy becomes glaring only if we consider Events at astronomical Distances such as a supernova bursting forth thousands or millions of Lightyears away. Even the coronal flares of our own nearby Sun are already 8½ minutes into the Past when sighted on Earth.

Still the area in the middle, the indefinite, fuzzy domain that contains our Present *NOW-*Moment is "Spacelike" to our *NOW-*Moment *(or the O-Event).* That the entire Spacelike domain is the Present is not exactly what we are accustomed to accept. We take the *NOW-*Moment as the mid-point of a knife-edge-like moving narrow, momentary barrier across the Time between the Past and the Future, an abstract line cut through Spacetime by the transient Moment simultaneous with what we take to be the ever-present *NOW* with us here. This kind of thinking *reduces the Present to a thin Isotemporal Line, a Moment of Simultaneity that never stays put, is never frozen in Time and hardly ever occupies Space or Time.* Our Spacetime, therefore, seems strangely divided just into Future and Past:

F3.29

If we hold on to this naive division of Spacetime into just Future and Past, Special Relativity still forces us to give the domain of the Present a considerable amount of spread. This is so because all the potential *ITLs of all Observers* moving at all the possible Velocities *(between 0 and c)* and all passing us at the *O-*Moment would be located in some part of this domain outside the converging/diverging Lightlines. In the next illustration, you'll see *<A>*'s, *<B>*'s and other Spacetime Travelers' Past and Future separated by this knife-edge-like line representing the Present in a small sample of markedly different Velocities and directions:

F3.30

So the wide area outside the converging-diverging Lightlines does get filled with *all the possible ITLs of Observers moving at all the possible Velocities* passing through the *O- Event/-NOW-Moment.* That's why *we have to accommodate all of them into the domain of the Present* after all:

F3.31

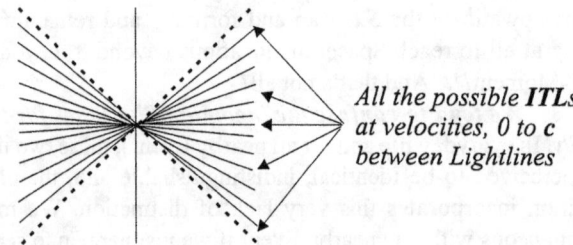

All the possible ITLs at velocities, 0 to c between Lightlines

Fans, Cones and Spheres

On conventional *ST* Maps, Lightlines delineate the Past, Present and the Future. In the "compressed" two-dimensional *(Space and Time)* domains, such as mapped on the flat surface of this page, these form *LIGHT FANS*. If Space is shown in two dimensions then Light flashes produce *LIGHT CONES*. In three dimension of Space, Light traveling omni-directionally delineates not cones but shells in successive, expanding layers starting from the center and spreading outward like *LIGHT SPHERES*. Visualizing 4-D Spacetime this way is of considerable help to intuition. But with the Time Arrow directed omni-directionally, it is more difficult to map it on a 2-dimensional surface. A three-dimensional Space representation is still possible in form of a cut "onion" with Time represented by multiple arrows:

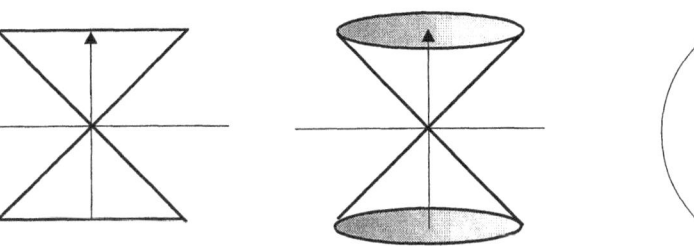

Despite difficulties with definition, the more sophisticated division of Time into Past, Present and Future domains should finally be easier to grasp. There is no trouble if we constantly monitor the unfolding Future, keep our *O-Moment* frozen in Time at a place on the Map we have arbitrarily selected, keep a record of observations and relegate everything *ahead of THAT O-Event* as *FUTURE of THAT MOMENT*. *Our actual existence, however, contains NO frozen NOW-Moments* and there are *no WLs or Lightlines extending forward beyond the NOW-Moment.*

How to read Temporal Order

It is important to repeat again that merely seeing two Events at the same Moment does not establish their Simultaneity. By the same reasoning, two Events that are simultaneous cannot be seen at the same Moment. So *Simultaneity is not something that is directly observable.* It is a totally abstract concept. It is defined operationally by the position of Isotemporal Lines. So *the system of ITLs and NOT the order of Signals received is what decides Simultaneity and also the Temporal Order of Events.*

All Events occurring on the *ITL* are simultaneous, but we must also realize that Events at different Distances on a single *ITL* can never be observed *(visually)* at the same *NOW*-Moment. For the same reason, Events occurring at different locations in Space but observed at the same Moment did not originate at the same *(isotemporal)* Moment in Time. The Lightline that brings us visual information is our *SIGHTLINE* and *NOT* our Isotemporal Line. Remember: *Simultaneous Events (at different Distances) CANNOT be observed simultaneously, at the very same Moment !!!*

These extremely important points explained above are unequivocally illustrated below in Figure 3.33. You should see e *(Fig. 3.33A)*, that Signals from simultaneous Events are not received at the same Moment and simultaneous reception of combined Signals *(R #1,2,3, etc. in Fig. 3.33B)* can come from several separate, non-simultaneous Events *(Q #1,2,3, etc.)* isotemporally marked by the corresponding sequential *q*-Events *(q#1,2,3, etc. on <A>'s WL):*

F3.33

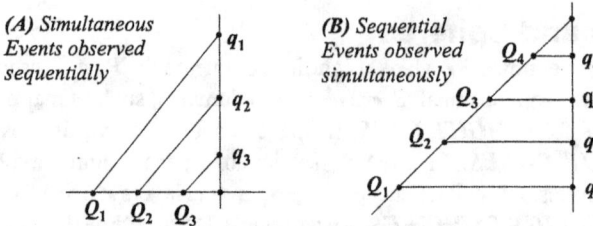

(A) Simultaneous Events observed sequentially

(B) Sequential Events observed simultaneously

The point *(repeated again and again)* is this: *ITLs* established by the Radar Probe are the only reliable means by which Temporal Order is determined.

What about the *Temporal Order of Events?* What happens *BEFORE* and what happens *AFTER* ? Can a Spacetime Map indicate Temporal Order? We have already worked out the answer earlier in this book. But to reinforce the lesson learned, we must keep hammering it constantly into our heads so that this difficult concept will finally become completely natural to us. Now that we have been so persuaded, it should not be surprising any more *(as we have already found out)* that **in the domain of the Present, the TEMPORAL ORDER is NOT FIXED !!! Therefore, Temporal Order of any pair of SPACE-CONNECTED EVENTS is not the same to all Observers. This absence of fixed order is not a perceptual illusion either.** There is an actual, *empirical variability of Time Order* of any two Space-connected Events. Remember Relativity of Simultaneity *!!!*

It is Space-connectedness itself that is the cause of Time Order mixup! Two Space-connected Events can be ordered one way, in reverse order or simultaneously depending on the Velocity of Motion of the Observers doing the observing *(and passing each other at the O-Event).* *It does not mean that Time Arrow direction itself goes into Reverse (more about it in the Chapter, "Faster Than Light"). Time-connected Events on a WL anywhere at all, even if located in the unobserved Present, always retain their Time Order to all Observers who themselves follow Time-connected Worldlines.* That *<A>*'s and *<B>*'s *WL*s do not coincide we already know. And we also know that their *ITLs* do not coincide. It is these differences between the *ITLs* of Moving Systems that produce Time order mixup and can be blamed on *Relativity of Simultaneity (as first learned in "Axioms And Operations").*

This extremely important point about the *absence of fixed Time order of Space-related Events* can be illustrated by placing two Events, *P* and *Q outside <A>*'s *WL (although O and Q-Events would do just as well).* Note that *if two separate Events at separate Space locations are mutually Time-connected, then they also remain Time-connected and in the same Time Order to all Time-connected Observers.* But the Time Order of two Space-connected Events are not necessarily in the same Time Order to two Time-connected Observers *!!!* So the difference in Time Order with *Space-connected Events such as P* and *Q* is shown on the next *ST* Map *(Fig. 3.35).* There we have three Observers passing though Event, *O.* You'll see that **(1)** *P* and *Q* are **simultaneous** to *<A>*, but **(2)** *P* comes **before** *Q* for *<C>* and **(3)** *P* comes **after** *Q* for *<B>*:

F3.34

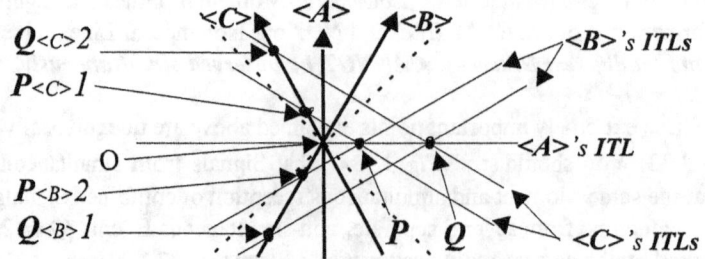

Let us conclude our discussion of the Time Order of Space-connected Events by re-re-re-stating: ***Time-connected Events are always Time-connected*** to all Time-connected Observers. ***Space-connected Events, however, can be simultaneous, in one Time Order or in Reverse Order depending on the state of relative Motion of the respective Time-connected Observers.***

Two kinds of measured Time

For more than once, we have referred to ***two kinds of measured Time***. The distinction is important enough to deserve extra attention. On the next Map of *<A>*'s Radar Probe *(Fig. 3.35)*, there is this Time, *Oq* = *T* on *<A>*'s *WL* which *<A>* can measure directly by means of a clock in his Home Frame. The Time, *OQ* = *t*, on the other hand, is on *<B>*'s *WL* and can be measured directly only by *<B>*, but not *(directly)* by *<A>*, exactly as *Oq* cannot be measured directly by *<B>*:

F3.35

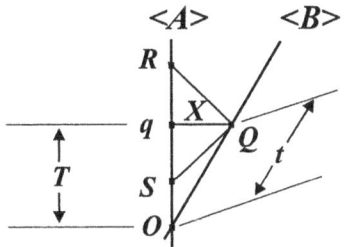

So Time measured directly by any Observer is always ***PROPER TIME*** within that Observer's Home Frame. Thus *Oq* is Proper Time in *<A>*'s Home Frame and *OQ* is Proper Time in *<B>*'s Home Frame. But that's not all. *OQ* can be ***PROJECTED*** upon *<A>*'s *T*-Axis *(WL)* where it comes out exactly, of all things, *Oq*, the same as *<A>*'s Proper Time. So *Oq is not only <A>'s own Proper Time but it is also his COORDINATE TIME (information or measurement) of <B>'s Proper Time, t.* This *(Coordinate)* information allows *<A>* to calculate *<B>*'s Proper Time, *OQ* (= *t*) by means of our Lorentz Multiplier, $\sqrt{1-v^2}$ which is his *(and our universal)* conversion tool. Is this clear*???* If not, keep reading this paragraph over and over until it is. To muddy the waters, unfortunately, *OQ* is *NOT* *<B>*'s projection of *<A>*'s *Oq* on her *t*-Axis. How come? You have seen the answer before and you'll see it again very soon. To further define the difference between Proper Time and Coordinate Time, let us say that:

(1) *<A>*'s Proper Time, *T* is *Oq* which is also his Coordinate Time of *<B>*'s Proper Time, *OQ*. In addition, it can be the Invariant for *<A>* as his *(<A>'s)* Space coordinates (*X, Y, Z*) are zero. In this case the Spacetime "Distance" is between Events, *O* and *q*. The same can be said of *<B>*'s Proper Time, *t* if her Space coordinates (*x, y, z*) are zero. So *a Time measurement, T or t, can be a Proper Time which is also an Invariant and additionally a Coordinate measurement of a Proper Time in an Other Frame.*

(2) A directly measured *X (with Y, Z and T being equal to zero)* can be a Proper Distance or Proper Length *(as we'll see in Chapter IV)* not a Coordinate Distance or Length. Coordinate Space or Time Segment, however, is never the Invariant, so *<A>*'s *X*, if obtained indirectly by Radar Probe, can never be the Invariant. It is all as simple as that.

The distinction between Proper Time and Coordinate Time is important. Radar Probes can measure only Coordinate Time and Coordinate Distance. Direct measurement of Time can never be done in remote regions across great Distances *(of any magnitude)* regardless of the state of Motion of what we want to observe and we always find a discrepancy between Time or Space values in one system and those values in the other in the presence of relative Motion between them. The inequality *(T > t)* between *<A>*'s Coordinate Time and *<B>*'s Proper Time is reversed if *<B>* obtains her Coordinate Time of *<A>*'s Proper Time. In this case, *<B>*'s Coordinate Time of *<A>*'s Proper Time *(Oq)* is not *OQ !!!* In a later chapter *(Lorentz Transformation)* we'll examine situations where neither *<A>* nor *<B>* can directly measure Time between *O* and *Q (Q not on <B>'s or <A>'s WL)* and both have only their Coordinate Times *(and Coordinate measurements of Space Segments)* to compare.

Equal Times for all Observers

Now we come to one more interesting facet of Time. In the presence of Motion, Coordinate Time, *Oq (that is T)* of *OQ*, does not equal *<B>*'s Proper Time, *OQ (that is t)*. The qualitative difference in our experiments has always been the same: $T > t$. If *<A>* and *<B>* have $v = 0$ *(no Motion)* between them, then $T = t$ and *the line lengths on the ST Map are equal*. If we now map a series of *<B>*'s Proper Times, *OQ*s on her *(<B>'s)* WLs at successively increasing Velocities but this time keep them *equal to <A>'s Proper Time, Oq*, the Minkowskian *(not-true-to-scale)* Mapline Lengths *(on the ST Map)* needed to plot them become progressively longer and longer while their numerical values remain equal to *T, t =T*.

As Time progression stops completely at Light velocity, the line length for mapping *t* at that Velocity, $v = c$ becomes infinitely long. In the illustration below, we have connected with a curved line all *OQ*s *(<B>'s Proper Times)* kept equal to *Oq (<A>'s Proper Time)* at several *(<B>'s)* increasingly greater Velocities. The line obtained this way can be called *EQUITEMPORAL LINE (equal-Times Line)*. The shape of the curve obtained is hyperbolic, its outgoing side arms on the Map gradually coming closer and closer to Lightlines set at 45 degrees from *<A>*'s *WL*.

F3.36

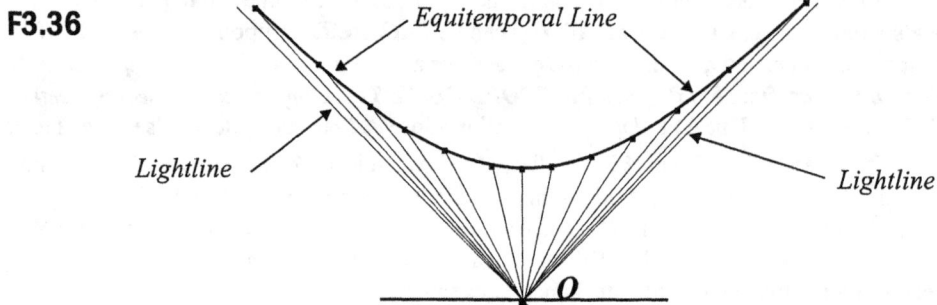

Equitemporal Line

Lightline   *Lightline*

O

Interestingly, we can apply Kepler's second law of planetary Motion to this hyperbolic Equitemporal Line as shown near the end of Chapter XII.

Drawing Spacetime "Circles"

The same hyperbolic curve results from making *<B>*'s *x*s equal to *X (that is x=X)* and plotting a series of equal *(<B>'s) x*-measurements *(x=X)* from the *O*-Event at progressively greater Velocities then connect all the *Q*-Events thereby obtaining the *EQUIDISTANCE LINE* whose hyperbolic side arms also approach Lightlines at near-Light velocities:

F3.37

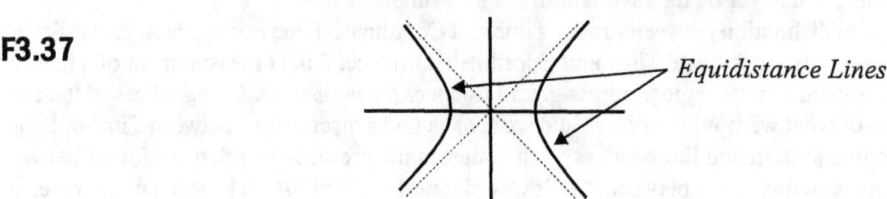

Equidistance Lines

Combining Equitemporal with the analogous Equidistance Lines in the domain of the Present, we obtain a ***hyperbolic ST "circle"*** that maps all *<B>*'s equal Times and Distances from the *O*-Event. The "circle" is warped, however, due to Spacetime curvature. As you can see, it shows, as explained before, that the Spacetime "terrain" is curved but not spherical. It is hyperbolic in all quadrants rather than round *(spherical)*. It is **this "circle"** that **best characterizes the unusual warp of Spacetime geometry** making true-to-scale Spacetime mapping on a 2-D as well as on a spherical surface extremely difficult.

F3.38

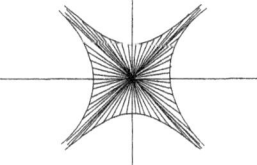

The same "circle" drawn true to scale on a 2-D surface produces a ***4-leaf cloverleaf:***

F3.39

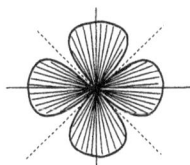

Mapping everything after the Big Bang

The Equitemporal curve leads to an interesting cosmological concept. If we extrapolate the all-inclusive Relativity Principle *("the laws of Nature are the same in all Inertial Systems")* to be valid in all parts of the universe, we are also saying that **all gross structures therein must have lasted the same length of Time after the Big Bang Event and having consequently aged evenly everywhere in our universe** *(shades of Newton's Absolute Time !!!).* As everything on that line would be located in the domain of the Present and, as we have seen, always outside our reach of observation, so Absolute Time, even if true, is not something that could be verified by observation. Our sciences, including Special Relativity, ultimately depend on observation for verification and we really cannot use the Equitemporal Line for anything except hypothetical statements.

The inability of ever knowing if the concept outlined above is empirically, realistically and actually true should not bother us because, as we have seen, our Present is entirely outside our immediate observational range anyway. But cosmologists have never been cowardly about conjecturing beyond the reach of observation and they should be the ones to know the limits of legitimacy. If we accept a Moment, *O* as representing the Big Bang origin of our universe some 14-15 billion years ago, something no-one has observed, we could also say that all gross structures of our universe must now be 14-15 billion years old *(on Earth's Time scale)* and be located on the 14-15-billion-year Equitemporal Line *(taking or leaving a billion or two-three).*

This Equitemporal Line should then be the Absolute though unobserved Border of our universe creating more Time by moving forward with our NOW-Moment (in its center) on our Time Axis and adding more Space by "spreading outward in a direction along our Space Axis."

Time progression could then be imagined and mapped as a continuous upward sweep of the Equitemporal Line containing our ***NOW***-Moment in its center and ***Space expansion*** as an additional outward sweep of the upward-moving Equitemporal Line. Without the Future ever existing beyond the *NOW*-Moment, the diverging Lightlines *(Signallines)* could not exist in the Future either and could not possibly be placed on the *ST* Map of our *existing* universe. As we should know by now, the Future can never be considered as existing *NOW* and cannot, therefore, be mapped anyway *(except in anticipation)*. *The Equitemporal Line would then be the absolute border of our finite universe where Space has 3 dimensions.*

The universe as it exists at the abstract *NOW*-Moment is *(as repeatedly pointed out)* *NOT OBSERVABLE*, not yet and whatever description may apply to it *NOW (isotemporal or equitemporal)* is a matter of pure speculation. Similarly, *at no Moment now or in the Future can we actually see the entire universe as it existed at a single Moment in Time either isotemporally or equitemporally (in the Past). Our Sightline never falls upon the entire Equitemporal Line, past or present, but it does fall on just single parts of all past Equitemporal Lines.* We'll have to wait many-many more years for a glimpse at a distant part of a recent *(Equitemporal)* Line currently situated in the domain of the Present and waiting to be scanned over by our Sightline. But by then, the universe is again much older and most of it still hidden from our eyes. Truly a no-win situation:

F3.40

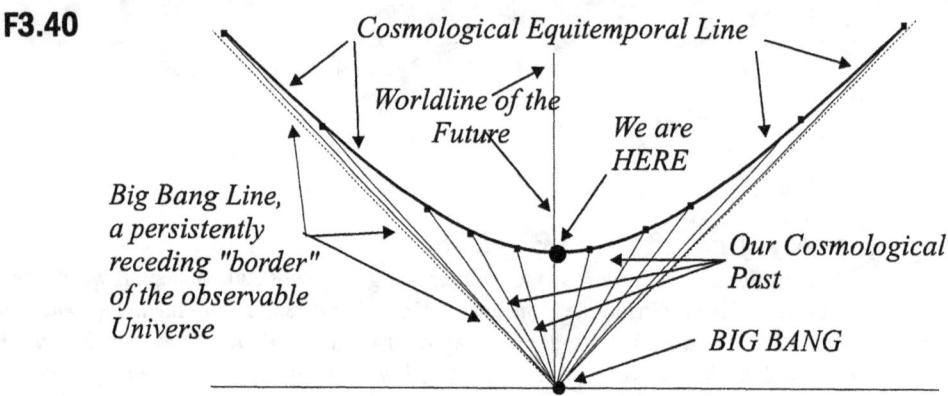

Cosmological Equitemporal Line

Worldline of the Future

We are HERE

Big Bang Line, a persistently receding "border" of the observable Universe

Our Cosmological Past

BIG BANG

Still, it is interesting that at the *NOW*-Moment, the *ISO*- and *EQUI*-temporal Lines start out in identical direction but subsequently diverge. Isotemporal Line is the most "unnatural," odd, straight line presently touching the "edge" of the universe. In contrast, the curved Equitemporal Line is more harmoniously at home in curved Spacetime, almost infinite in length on the Map but still confined to the domain of the Present and none of it observable at our *NOW*-Moment. The *Sightline*, however, is the most useful of our abstract lines because it is *the only pipeline through which all the information in our possession has ever become available:*

F3.41

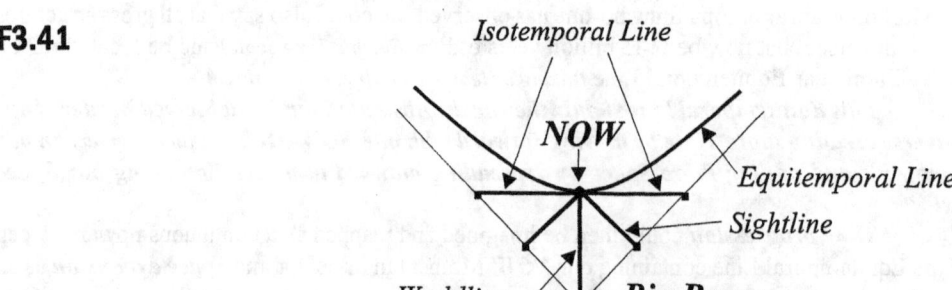

Isotemporal Line

NOW

Equitemporal Line

Sightline

Worldline

Big Bang

It is important to be repeatedly reminded of the fact that *our Sightline is the moment-arily visible transition zone where the Present is changing into the Past* at all Distances. But let's remember that *what we see (when we see it) is actually no longer existing there (in that form and shape) NOW*. We must likewise understand that what we see is *neither equitemporal (that is of the same cosmic age) nor simultaneous* with our *NOW*-Moment. Everything observable to us here and now is actually *NOT* simultaneous to us at the Moment we see it. We have been pointing out repeatedly that what is simultaneous to us at all Distances is *NOT* part of our actual knowledge and *NOT* within our reach by direct *(or indirect)* observation. We have been pretending to discuss something we believe exists there *NOW*, something called *Reality*. But everything pertaining to this foggy topic belongs to metaphysics, something not within our purview and, therefore, not further developed in this book.

"Sight reach" by telescope

The most distant reach of our Sightline touches parts of the universe still hovering close to the Big Bang that (eventually) gave birth to matter. Beyond that, there is this *invisible, transparent, still unimaginably hot, sizzling, compact, near-singular Big Bang* rushing away from us at *(nearly)* the speed of Light, steadily expanding the immensely large volume of Space and dispatching a veritable avalanche of "first-reflected" Light in our direction. What was once close to the core of the primal cosmic egg still *seems to surround the much older universe to which it gave birth*. And this Light comes to us from all the directions in form of "tired," or "chilled" Doppler-shifted photons observable as the 3 degree *(Kelvin)* cosmic background micro-wave radio frequency.

So, it is very odd that *the Early Universe close to the compact Big Bang Event that no more can exist THERE NOW (in our isotemporal or equitemporal Present), can give rise to the illusion that it still surrounds us at immense Distances and from all directions.* Time there *(at the outer reaches of our omnidirectional 3-D)* has hardly progressed at all so that what we see *NOW (with some liberties taken)* is only what happened there *THEN* some 14 or 15 billion years earlier and equally far away.

How strange*!!!* Just think about it while giving full consideration to the differences between the *Sightline, Isotemporal Line* and *Equitemporal Line* concepts. All our common-sense conclusions described above can really be stranger than fiction if we do not apply all the lessons we have learned from the information at our disposal and if we allow errors and habitual, common-sense misconceptions cloud our thinking.

Wild and wilder

Didn't you for a moment wonder if all this cosmological talk about Events long ago in the Past and phenomena far into the Future can really be within the genuine realm of science. Perhaps all this is just *wild speculation*. Are we not ever bothered by the impossibility of direct observational verification within less than billions of years? Verification may not be impossible in principle but none of us will ever live long enough to see it happen. But lack of observability, as repeatedly noted before, has never been a problem to card-carrying cosmologists. Not only have they divined the *BEGINNING* of the universe but have also prophesied its *END*. So, let's play their game and chart a simplified version of their speculations.

First, the *Big Bang:* It went simply *BANG !!!* followed by a continuous, unstoppable expansion that still keeps creating more and more Space and more and more Time. Can our best telescopes "see" the actual edge of the universe or does our vision extend only to when and where the birth of visible matter took place*?* Let's bet on neither.

Mapping Rapid Inflation

The beginning of the universe included more than just the **Bang**. Cosmologists have accepted that the origin of the universe included a phase of *"Rapid Inflation."* Briefly stated: right after *(or as part of)* the Big Bang, *the universe expanded at a rate that was incredibly fast, faster than the speed of Light !!! ???*

The question of interest for us here and now is whether or not this exceptionally rapid, superluminal expansion did violate and invalidate Special Relativity. Our answer is contained in our Spacetime Map but we'll have to limit our explanatory concepts to those we have developed so far in our study. So, Rapid *(superluminal!?)* Inflation *(Fig. 3.42)* can be visualized within our limited understanding of Relativity by zooming in on the bottom tip of the Big Bang *(also Fig. 3.40)*. There the initial *"cosmic egg"* is believed to have been a *large energy fluctuation* permitted by the quantum properties of the so-called *"false vacuum"* and, furthermore, according to the paradoxically *precise rules of the Heisenberg's Uncertainty Principle.* This energy must have initially consisted of entities that did not yet constitute matter. But photons, or whatever form of energy preceded energy-to-matter phase change, cannot exist without Motion and Motion requires availability of Space. So *Space was created by energy (in Motion)* which then continued to expand at the obligatory speed of Light. But would creation of Space take up Time? Well, not really. *Light and whatever else preceded actual matter is uniquely timeless*, its *clock stalled and Time not counting*, a "time out" before the grand game of the universe could get going.

Remember: *Light can travel any Distance without spending any of its own Time!* So the original nucleus of energy could expand, increasing its size and getting everywhere "at the same moment," *creating Space across the ever-widening front of the Equitemporal Line while its clock still stuck at Zero Time; everything in existence becoming more and more spatially removed in all directions (on the Map)* from its initial point of origin and its Worldline moving steadily in the upward direction along the Time Coordinate. Regardless of its actual line length on the *ST* Map, this Time Coordinate would *not yet indicate actual or Real Time progression*. Some cosmologists have called this Time-like *(though Time-less)* period "Imaginary Time" but we can give it also other names like "False Time," "Pseudo Time," etc. without assigning to it any "Real Time" value. Note that *as long as matter with mass was absent, the clock of the universe was not ticking.* The Big Bang could have lasted a lot of Imaginary Time but none of our Real Time so that Rapid Inflation could not have violated Special Relativity. From our perspective, we could say *that Big Bang occurred "BEFORE" Time began and "BEFORE" conditions became right for Relativity (both Special and General) to take control.*

What this kind of jumbled explanation is doing is *splitting the Big Bang into conceptually separate Space and Time actuating Events.* According to this description, Big Bang was something that may have started out within an infinitesimally small amount of Space known as Singularity but was completed in an indefinitely large volume of Space practically *no Time "later."* So what's zero Time among friends? Still, it would be *meaningless to say that Space was created "before" Time because, with zero Time between them, both happened "at the same Moment"* or, as we say, simultaneously. Summing it up: *Creation of Energy created Space and Creation of Matter created Time, both occurring simultaneously.* Get it ???

We can top off this confusion by asking: if no-one was there to hear or see it happen, was it really a *BIG BANG* or was it a *BIG FLASH ("Let there be Light")* instead? Perhaps it was more like a really *BIG "FLASHBANG" ??? !!!*

Now let us map the Big Bang and label its component parts according to the above discussion:

F3.42

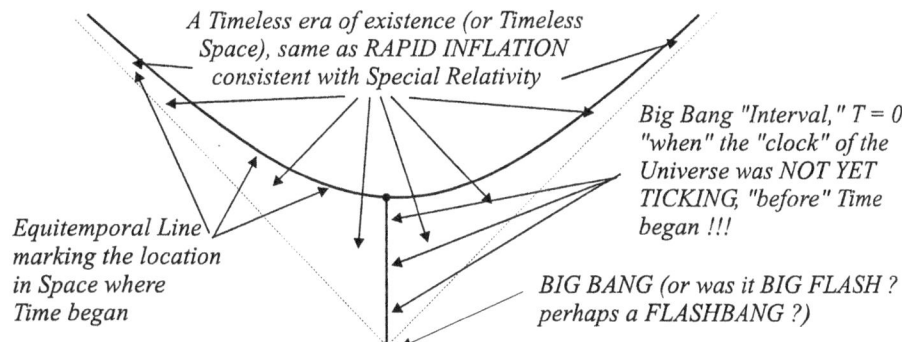

A Timeless era of existence (or Timeless Space), same as RAPID INFLATION consistent with Special Relativity

Big Bang "Interval," T = 0, "when" the "clock" of the Universe was NOT YET TICKING, "before" Time began !!!

Equitemporal Line marking the location in Space where Time began

BIG BANG (or was it BIG FLASH ? perhaps a FLASHBANG ?)

The end of Time mapped

Before the Runaway Universe burst upon the cosmology scene, cosmologists have offered us a menu of just two main choices for *the End of the Universe.* At this stage in our study, we do not know enough about Special Relativity and the Hubble's Law in astronomy to calculate Relativity-corrected Velocities and Distances the way it was done with the **COBE** satellite data investigating the cosmic microwave background radiation in the 1990ies. The missing relativity tools will be taken up in Chapters VI, VII and VIII but we'll have to keep our hands off anything concerning astronomy. So the intriguing fate of the Universe will not be touched further and here we'll deal only with the two earlier choices. The first is *continuous expansion.* This process is expected to last until the hypothesized instability of protons makes them decay spontaneously, one by one, little by little. Matter, being packets of "condensed energy," will then slowly break open and convert back to pure energy which will fade into the vacuum from whence it came. The outer border of Space keeps expanding but getting more and more indistinct until the whole universe vanishes ingloriously, fades out into the void, becomes non-existent as one poet said: "***not with a bang but with a whimper.***" Time will gradually slow down to a stop. The Sightline will never connect us with the edge of the universe. Everything will simply evaporate, disappear, eventually terminating at some indistinct ***NOW***-Moment if not sooner *(if expansion is accelerating as some evidence suggests).* The Map of Spacetime from beginning to the "fadeout," the **BIG CHILL,** would then look something like this:

F3.43

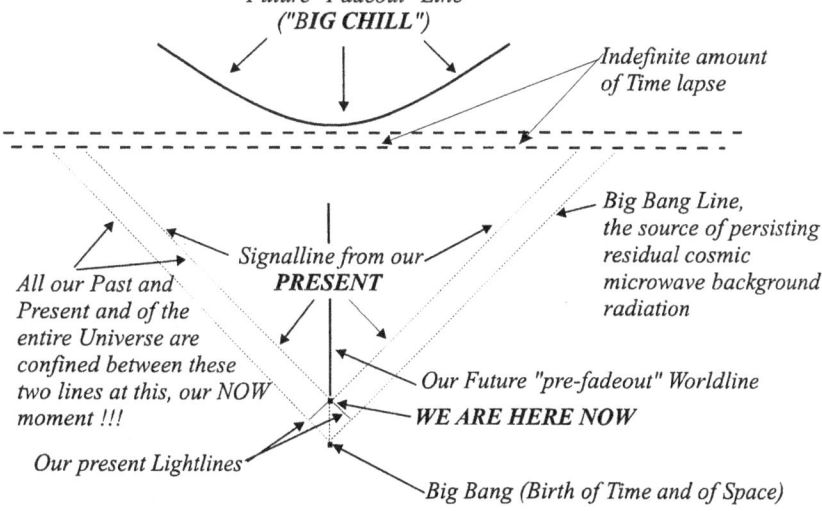

Future "Fadeout" Line ("**BIG CHILL**")

Indefinite amount of Time lapse

Big Bang Line, the source of persisting residual cosmic microwave background radiation

Signalline from our **PRESENT**

All our Past and Present and of the entire Universe are confined between these two lines at this, our NOW moment !!!

Our Future "pre-fadeout" Worldline

WE ARE HERE NOW

Our present Lightlines

Big Bang (Birth of Time and of Space)

If, however, the expansion slows down and gradually stops, then the process will enter into a ***prolonged contraction phase***. The Sightline eventually reaches the visible edge of the contracting universe and, instead of a red-shift, we'll begin to detect a blue-shift. Fortunately, the **BIG CRUNCH** would still be billions *(upon billions?)* of years away. If the Earth is still in any shape to sustain life contrary to astronomers' dire predictions, anyone still around, perhaps in a freshly-minted, brand-new planetary system, would not be able to pick up ominous signs of Impending Doom until far along the way toward the terminal, hot catastrophic crunch, a black hole that may eat up all other black holes, perhaps even itself.

The end of the universe by contraction can be visualized in the following way: **(1)** the Lightlines would begin to curve back upon the universe that no longer expands and **(2)** begins to contract instead eventually **(3)** restoring it to a new "cosmic egg," a mere energy fluctuation in the "false vacuum" from which it came:

F3.44

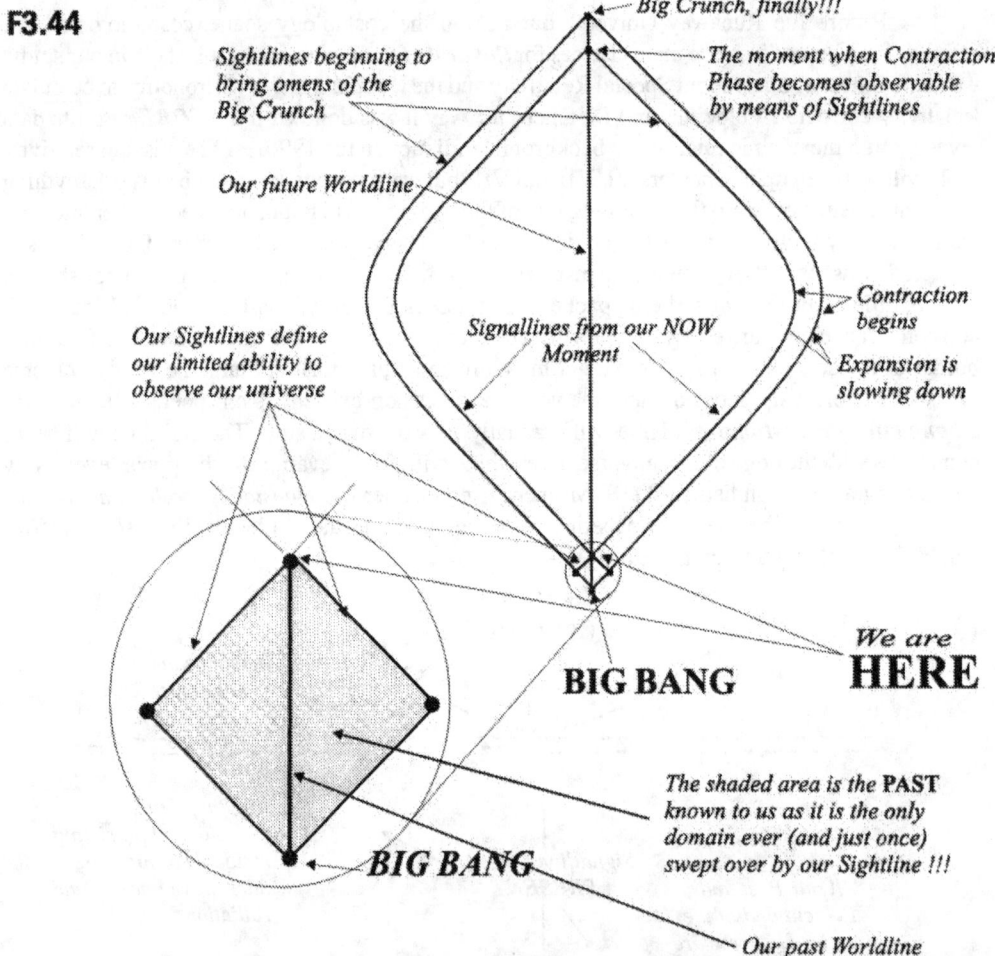

Big Crunch, finally!!!

Sightlines beginning to bring omens of the Big Crunch

The moment when Contraction Phase becomes observable by means of Sightlines

Our future Worldline

Contraction begins

Our Sightlines define our limited ability to observe our universe

Signallines from our NOW Moment

Expansion is slowing down

We are HERE

BIG BANG

*The shaded area is the **PAST** known to us as it is the only domain ever (and just once) swept over by our Sightline !!!*

BIG BANG

Our past Worldline

If the Future of the universe threatens your peaceful sleep, don't worry! It can't come soon. *We have plenty of Time left to study Special Relativity! And much else besides.*

IV - MEASURING LENGTHS

Once more a brief reminder

A golden nugget of wisdom from the important *ST* Geometry lesson of the preceding chapter needs restated: It is the Invariant that keeps the same value to all Observers. Only the magnitude of its Projections upon the four Coordinates: *X, Y, Z, T*, turn out different for different Observers in Motion relative to each other. The Invariant, specified by Minkowski's 4-D Space-time equation, is the cornerstone of our geometric approach.

In the first chapter, we learned to simplify our equations by eliminating the *Y* and *Z* Coordinate values leaving only one Space *(X)* and one Time *(T)* Dimension to work with. In our task in this chapter, we'll *eliminate also the ever-present Time Segment* and *do battle with the meddlesome Relativity of Simultaneity when measuring Space Segments (Lengths).*

Our task ahead

In Axioms And Operations, we learned to measure Distances *(X and x)* from the Observer's *(<A>'s)* location. In this chapter, we'll see how to measure *Lengths (or the Distances between the Ends)* of long objects this time *located entirely outside the Observer's own Inertial System.* Our measuring task is accomplished as usual by Radar Probes, the all-purpose tool that should be completely familiar to us by now.

Common sense ignores Time

At first glance, measuring the Length of an object should be nothing more than measuring the Distance between its most widely separated Ends. If the object is within our immediate reach, then by direct use of a ruler, measuring rod or tape, we assign our measuring device the function of the *X*-Coordinate, thereby getting its value directly while automatically assigning zero values to both *Y* and *Z*. If, on the other hand, the object is located out of our reach but pointed in our direction, our measuring task consists of determining the Distance of both of its Ends then subtracting the shorter Distance from the longer. In our exercises, we'll continue confining ourselves to just one Space dimension. Otherwise, the *Y* and *Z* Coordinate Projections become greater than zero and our math more complicated.

The correct way to ignore Time

In the first chapter, we learned about the importance of considering everything in 4 Dimensions but we did not ask why and how we should use Time in Space measurements. The Principle of Invariance may demand that we always measure in four dimensions, one of them being Time. But how come that our usual Length measurements and all those in scientific liter-ature never show Time in the final result? So how can we include Time without actually showing it in our equations? Well, there is a very good way to do exactly that and it should not surprise us. Let's follow the method we used in Chapter I to get rid of *Y* and *Z* in our *ST* equations. And here it is: *Let T be equal to ZERO !!!*

Making Time disappear is not a new trick for us. In "Axioms and Operations," we saw that *T was ZERO between two Events if these occurred SIMULTANEOUSLY.* Applying the same solution to our predicament, all we need is to *determine the POSITION of BOTH ENDS of the measured Length SIMULTANEOUSLY, that is, at the same Time.* This makes the Time Segment between the positions of *(each of)* these two Ends *(Events)* equal to zero thereby making the inclusion of Time in the measurement altogether superfluous. How simple! How clever! We have accomplished what at first seemed impossible: first including, then immediately excluding *T* by a flash of insight and a stroke of the pen *(or a tap on the computer keyboard)!* Our final result is now theoretically correct and logically consistent while remaining true to common sense and to a long tradition of classical physics.

So let us restate the measured Length in terms of the 4-D **ST** Invariant which then is quickly reduced to the simplest possible expression without the **T** *(and also without Y and Z)* in it thereby giving us:

$$L = \sqrt{X^2 + Y^2 + Z^2 - T^2}$$
$$= \sqrt{X^2 + 0^2 + 0^2 - 0^2}$$
$$= \sqrt{X^2}$$

Eureka: $L = X$ *!!!*

which is exactly the familiar result *we are* **used** *to* in the first place*!*

Whew*!* What a complicated way to come up with such an idiotically simple result. But there is a clear benefit in having done it this roundabout way: ***We have thereby established a general procedure for obtaining the Length of any object regardless of its state of Motion.*** It includes steps we can ignore in a simple setting but when measuring things that move fast, all are absolutely essential. It does not matter how we perform our measurements as long as we make **T** equal to zero. Thus properly cautioned, let us proceed with a sample measuring task.

A measuring task in Lineland

We need to start by determining the location of both Ends of the object we want to measure. In this case, and in the rest of this chapter, the object most appropriate for our study is ***a long Stick.*** It is as close to a 1-dimensional object as we can come *(in our practical world)* and essentially free of all dimensional features except Length.

We can pick our own *(or <A>'s)* position somewhere along the Stick and use that location as our point of reference. From there, we can measure the Distance of each End of the Stick then add them together. And there we have the Length of the Stick, the Distance between its Ends:

F4.1
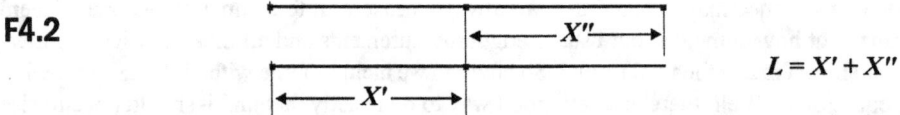

$L = X' + X''$

If the values of X' and X'' are determined sequentially, non-simultaneously *(in two steps)*, our record should also reflect this additional fact:

F4.2
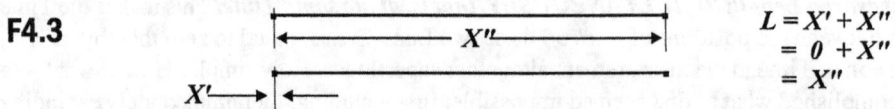

$L = X' + X''$

If the Stick does not move *(as in the above examples),* our results are not influenced by the non-simultaneous procedure. Also, when we *(or <A>, our proxy)* are located at one End of the non-moving Stick, the sequential process is not invalidated if we do not carry out the two partial measurements separately:

F4.3

X'' X' ⟶ ⟵

$L = X' + X''$
$= 0 + X''$
$= X''$

We can obtain the same result even if we are physically removed from the non-moving Stick:

F4.4

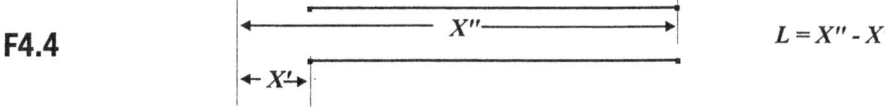

$$L = X'' - X'$$

In all these examples, the Time Coordinate was formed by our own *(or <A>'s)* Worldline and the measurements of X' and X'' were non-simultaneous. Still, the final value of the non-moving Length was not influenced by the non-simultaneous partial measurements.

We could argue that most of the time, using a tape measure at least as long as the Stick, we do not need to conduct separate partial measurements and can complete all in one simple step. That certainly is true. But what we would be doing then is adding up all component parts of the Length at the very Moment when we take our reading off the measuring tape, thereby reducing T to zero! So, any way we look at it, we cannot *(conceptually)* avoid involving the Time Coordinate in measuring Lengths. *The attention paid to Time Coordinate is superfluous only when measuring things that do not move or at least do not move fast.*

When the Stick is moving, the necessity of simultaneous determinations of both the X' and X'' becomes obvious:

F4.5

Here, $L = X_1' + X_1'' = X_2' + X_2'' = X_3'' - X_3' = X_4'' - X_4' =$ etc.
But $L \neq X_1' + X_2'' \neq X_2' + X_3'' \neq$ etc.

So, it is also obvious that $X_1' + X_2''$ or any combination of non-simultaneous partial measurements cannot equal L. In measuring moving Lengths, T must be made equal to zero so $L = X$ if, and only if, both Ends of the Stick are placed on the Isotemporal Line and we determine their positions *(of both Ends of the Stick)* and their Distance from each other simultaneously.

Preparing the Stick for Radar Scan

Because most *(if not all)* of our measuring tasks in the wide-open Space cannot be done directly, that is with a measuring tape or ruler, we need to equip the Ends of the Stick with *flanges* positioned perpendicular to each other to serve as *Reflecting Targets to our Radar Probe in the one-dimensional Space of all our virtual experiments* .

The necessity for those Targets needs demonstrated visually. Because in our virtual exercises *(in a 1-D world)* the moving Stick is essentially a 1-D object and is always seen in a line-of-sight direction, we would never be able to see its distant End or reflect Radar Signals from it because the distant End would always *remain in the shadow* of the nearer End.

F4.6

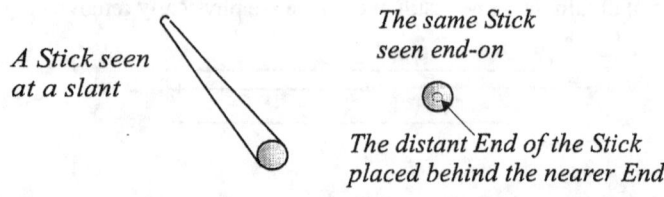

*A Stick seen
at a slant*

*The same Stick
seen end-on*

*The distant End of the Stick
placed behind the nearer End*

But once the Ends of the Stick are equipped with Reflecting Targets *(positioned at 90 degrees from each other)*, both become accessible to Light and Radar Signals in approaching as well as in receding Motion:

F4.7

*A Stick seen
at a slant and with
Reflecting Targets
(flanges) attached*

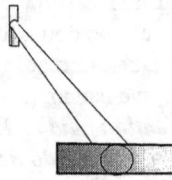

*The same Stick
seen end-on with
Reflecting Targets
(flanges) attached*

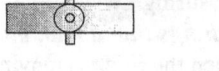

All subsequent examples of the Stick are those in uniform Motion and, unless otherwise stated, all Events are located in the 1-D Lineland and viewed from the line-of-sight perspective of the Observer, *<A>*.

Sticks mapped like ladders, 3-D objects mapped like Blue Streaks

We begin by visualizing the moving Stick at a number of separate momentary positions *(see Fig. 4.8)* as it is moving past *<A>* so that the ***various positions simultaneous to <A>*** are mapped over a period of Time as steps of a ***tilted ladder*** .

F4.8

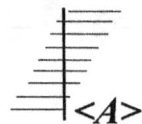

<A>

The horizontal lines in the above illustration represent successive positions of the 1-D Stick in a ladder-like stack tilted toward the direction of its one-dimensional Motion *(the Stick is coming aimed directly at us, always barely missing us)*. In this way, the essential properties of Motion are fully shown. A 3-D object, on the other hand, can be drawn as a series of overlapping pictures with each successive picture positioned up and a little forward to indicate the direction of Motion. Overlapping pictures, however, create visual clutter. The entire experiment can be shown better by drawing the picture of the object just once then drawing a series of parallel lines that connect all identical parts of the object during its "flight" and leaving the object itself un-sketched except for just once. A flying teapot, for instance, would then stand at the uppermost position of its bundle of Worldlines like a figurehead at the bow of what could be called a ***"Blue Streak,"*** the proverbial fast moving thing:

F4.9

More than one Worldline makes a Bundle

The Map lines connecting identical parts of the teapot at different Moments are the Worldlines of those parts. These lines form a Bundle of Worldlines or the **WORLDBUNDLE** which criss-cross those of other objects they pass *(avoiding collisions)*. The Worldbundles of stationary objects go straight up. The Worldbundles of moving objects are tilted. All moving objects are in the Home Frame of the Observer such as <*B*> moving with them and, as always, positioned in stationary <*A*>'s Other Frame. While <*A*> does not move in Space, he cannot avoid moving forward in Time. So all the Worldlines of <*A*>, and anything within his Home Frame, keep going straight up on the *ST* Map while the Worldlines of objects in <*B*>'s Home Frame are viewed by <*A*> as tilted due to being in his *(<A>'s)* Other Frame and in Motion.

Length defined

For the purpose of measuring Lengths, *the two most important parts of any object with Length are its Endpoints. The Distance between them defines its Length.* Therefore, only the Worldlines of these Endpoints are all we really need to show and these can stand in for any given Length on our *ST* Maps. And that is not all: *An elongated, simple object in Lineland mapped over a period of Time is observable as a Worldbundle made up of its two Endpoint Worldlines. The Worldbundle itself is an INVARIANT, this time a non-descript entity that is identical to all actual and potential Observers in any state of (relative) Motion.* The moving object at any Moment is mapped on the Isotemporal Line of any particular Observer between the points where his or her *ITL* trans-sects the object's Endpoint Worldlines *simultaneously* according to that particular Observer *(Observer-Dependency again!)*. And *the Stick's LENGTH is defined by the Length of the ITL between the Stick's ENDPOINT WORLDLINES.* Let's repeat: *The Stick is defined by its invariant Endpoint Worldlines forming the Worldbundle and the Length of the Stick is specified by the Segment of the Observer's Isotemporal Line located within the Stick's Worldbundle.* As you see, Relativity of Simultaneity has, again, stepped in and made things complicated.

Let's now look again at different versions of the ladder: **(1)** as observed by <*A*>, **(2)** as observed by <*B*> in whose Home Frame the Stick is located but shown in <*A*>'s Other Frame, and **(3)** as observed by <*B*> in her own Home Frame. Please, match the positions of the Stick with the corresponding Isotemporal Lines of <*A*> and <*B*>. Also: **(1)** The upper case letter X stands for the Length viewed by <*A*> *(always considered non-moving!)*, **(2)** lower case x being <*B*>'s moving Length in <*A*>'s Other Frame but **(3)** non-moving in <*B*>'s own Home Frame. Occasionally, the subscripts used help to further identify whose observations *(or measurements)* they represent:

F4.10

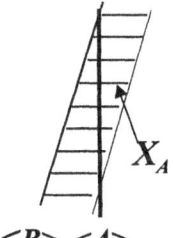 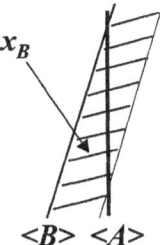 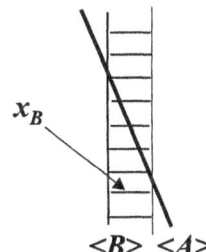

The *differing slants* of different Isotemporal Lines help explain why different "cuts" across the Worldbundle can produce different Length values for different Observers. Because in the above example, the moving Stick was in <*A*>'s Other Frame where its *ITL* was tilted, we had to modify the "ladder" to show the Stick in sequential positions in its own *(that is <B>'s)* Home Frame by tilting the lines in *<A>*'s Other Frame so that these become parallel to the (*<B>*'s) *ITL* at different Moments as viewed by *<A>* and by *<B>*.

Let's also show how the Worldbundle of the Stick can provide three different Lengths of the *(same)* Stick according to three Observers: *<A>*, *<B>* and *<C>*, first shown according to *<A>'s* perspective then according to *<B>*'s perspective:

F4.11

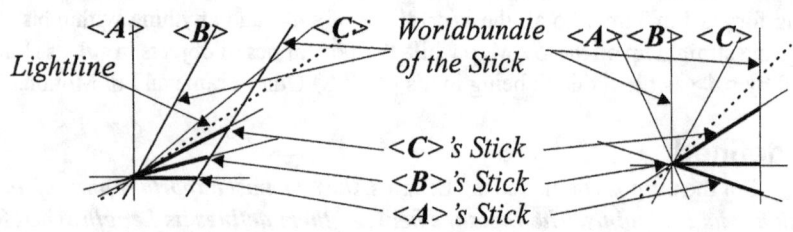

Simultaneity problem (again) intruding

Understanding Simultaneity is basic to understanding why measuring the same Length can give different results to different Observers. To review this all-important concept again, let's convert Einstein's railroad car demonstration of the Relativity of Simultaneity *(Fig. 2.16)* into a moving Stick experiment.

Instead of a railroad car, let's place our long Stick into *<B>*'s Home Frame *(<A>'s Other Frame)*, equip its exact middle with a Light Flash unit and provide each End with Light Signal detectors:

F4.12

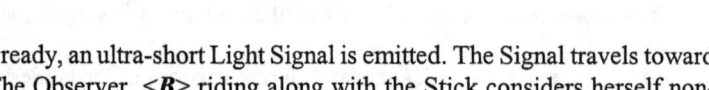

With everything ready, an ultra-short Light Signal is emitted. The Signal travels toward both Ends of the Stick. The Observer, *<B>* riding along with the Stick considers herself non-moving and finds that the Light Flash reaches Targets at both Ends at the same Moment *(or simultaneously)*, as shown in the following illustration:

F4.13

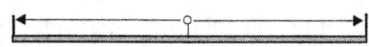

For *<A>*, the Events proceed differently. As the Signal travels toward both Ends, the Target attached to the Trailing End travels toward the oncoming Signal and both meet somewhere in between. While the Signal also travels toward the Leading End, the Leading Target keeps moving further forward and away from the oncoming Signal. For this reason, when the Trailing Target is hit, the forward moving Signal has not yet reached the Leading Target:

F 4.14

(2)

(1)

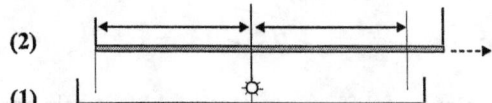

A brief Moment later, however, the forward moving Signal finally hits the Leading Target but this Event happens only **AFTER** the rearward Signal has already hit its Target and part of the Signal has traveled beyond it.

F4.15

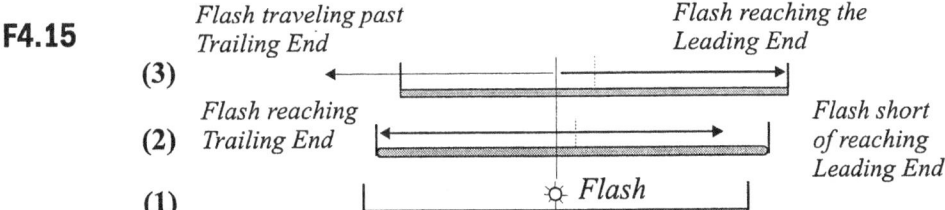

Mapping the above experiment

The illustration above *(Figure 4.15)* shows that a relativistic influence of Motion makes the Ends of <B>'s Stick ***non-simultaneous to*** <A>. The arrival of the rearward Signal at the Trailing End is visualized by <A> as occurring **BEFORE** the arrival of the forward Signal at the Leading End, the Time Segment between these arrivals being $T''>0$. But this is not the way <B> views the whole thing. For her, the Light Signal is hitting both End-Targets ***simultaneously.*** And within **HER** Home Frame, $t''= 0$. To let it all sink in, let us say ***again*** that <B>'s Home Frame is viewed by <A> as being in his (<A>'s) Other Frame and **WHAT to** <B> **is SIMULTANEOUS is NOT SO to** <A>. The **ST** Maps of <B>'s own Home Frame *(Fig. 4.16)* and <A>'s Other Frame *(also in Fig. 4.16)* illustrate the difference. The duration, T'' as well as t'' is the Time Segment between the Light Signal hitting the Trailing and Leading Ends of the Stick *($t''= 0$ and $T'' >0$). **The Time Segment, T'' (and also t'') is, therefore, a test of Simultaneity relative to the Observer whose Time Segment equals zero.** A basic Spacetime Map of the above Simultaneity experiment is shown next:

F4.16

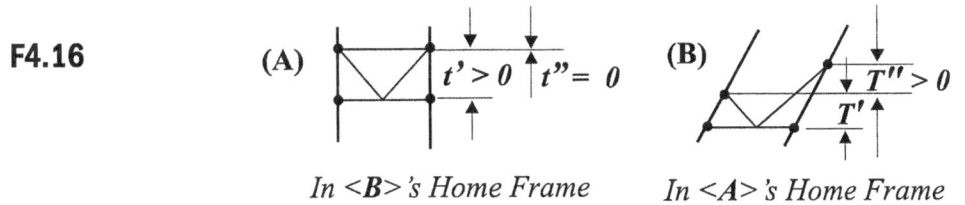

In <B>'s Home Frame   *In <A>'s Home Frame*

Before you continue, stop and read the last paragraph again to make sure its message is clear.

The source of confusion here is our common-sense reluctance to realize that what is true in <B>'s Home Frame is not so in <A>'s *(and our)* Home Frame. Therefore, note again that the moving Stick shown on the left side in Fig. 4.16 is observed in <B>'s Home Frame where she and the Stick are not moving. ***Observer***, <A>, however, ***sees*** <B>'s ***stationary HOME FRAME as the moving OTHER FRAME*** on the right side in Fig. 4.16 and all <B>'s Events there are depicted on <A>'s **ST** Map according to <A>'s point of view. Specific Events in two different Frames of Reference can be accurately viewed only ***from two separate and different observational points of view, one for each Observer!*** Remember: ***Observer Specificity*** is the monkey wrench in our intuition's gears! Once you understand it you are ready to tackle Lorentz Contraction, a complication many "thanks" again to Observer-Specificity which, in turn, is due to Relativity of Simultaneity.

Reading Time Order directly from the Map

The involvement of Time in the measuring of Lengths *(something we could hide by making T equal zero)* is still active by governing the **Time Order of Events**. This is a matter we need to review again before getting into **Lorentz Contraction** *(produced by Motion)* which itself is an ever-present feature of Special Relativity not well explained in most textbooks if explained at all.

There are three Time Order relationships. One of two Events can occur either **(a)** *before*, **(b)** *simultaneously with* or **(c)** *after* the other Event. We'll also find that a moving Stick has *NO Length* as such *except in reference to* its own *Worldbundle* and to a specific *Frame of Reference!!!* All this is contained in a well-drawn *ST* Map. In other words, the Stick is *located simply SOMEWHERE within its own invariant Worldbundle.* But looking at the Map is a luxury denied to us in real life. That's because *we cannot see simultaneous Events simultaneously (at the same Moment) in our real world*. Differing Time Order relationships now push us against - intuition blocks we haven't met before. To negotiate around them, we need to remind ourselves of a few essentials from past lessons:

(1) The *ITL*s of stationary Observers are horizontal, those of moving Observers are tilted.

(2) An *ITL* connects all Events *(or ST locations)* that are simultaneous to just one specific Observer *(whose ITL it is)*.

(3) The *ITL*s of all Observers who share the same anchoring Event, *O* are found only *within the Domain of the Present and the Events on those ITLs are always Space-connected* to all Observers regardless of their Velocity relative to each other *(as discussed in Chapter. III, Mapping Spacetime)* and *(their)* **Time Order is NOT directly observable!!!**

(4) *The Time Order of two Space-connected Events is never fixed*, is always relative and Observer-specific as these Events can be **(a)** simultaneous, **(b)** in one Time Order or **(c)** in Reverse Time Order, all according to the *ITL* positions of specific Observers.

(5) Different parts of the Stick such as its opposite Ends mapped at the same Moment *(simultaneously)* in its own Home Frame are *non-simultaneous* in another Observer's *(different)* Home Frame in relative Motion. This variable influence is exactly what determines their different *(measured)* Lengths for the different Observers.

(6) An Event positioned *lower* on the Map is always occurring earlier than the one located *above* it. By the same reasoning, an Event positioned higher on the Map is of later occurrence than the one located below. *BUT...*

(7) What is below or above on the Map, however, depends entirely on the *ITL* of that specific Observer under consideration, a complication due to Relativity of Simultaneity.

A Map Reader for help

A practiced eye can quickly determine *(simultaneous and non-simultaneous)* Time Order relationships. For less experienced students, a simple home-made **Map Reader** should help out.

A piece of blank paper serves well as a **Map Reader**. Just *keep its lower straight edge parallel to <A>'s ITL and place it over the Map*. Now slide it slowly upward until its lower border uncovers an Event *(or a position on the Map)* of interest. Everything uncovered thereafter by moving the Map Reader upward can be considered as having occurred *after* the first uncovered Event. Two *(or more)* Events seen just at the lower edge of the Map Reader can be considered to be simultaneous *(to <A>)*.

A really elegant Map Reader is made out of a piece of transparent, colored plastic by cutting a long, narrow horizontal slit across its center so that the upper and lower halves remain joined to each other. Now, everything seen inside this narrow slit *(kept in horizontal position)* can be considered as being simultaneous to <*A*>; other Events ready to occur in different Time Order are always kept visible under the transparent part of the Map Reader thereby allowing us to remain better oriented to all Events on the Map.

Sliding the Map Reader slowly upward reveals all Events on the Map in the correct *(<A>'s Time)* Order of occurrence and all Events at all Distances from <*A*> seen through the slit are always simultaneous *(to <A>)*. To analyze Events from <*B*>'s perspective, simply ***tilt the Map Reader into a position parallel to her ITL*** and slide it slowly upward while keeping the slit *(or the lower edge of the blank piece of paper)* parallel to <*B*>'s ***ITL***. This method always keeps you straight and helps you resist the urge to read the Map ***wrong*** from top ***down*** without paying attention to the different ***positions of the ITL***s of those Observers whose perspective you want to consider. A little practice with the Map Reader is all that most folks need. Now we are ready to analyze the Map of a moving Stick.

The Worldbundle of a Stick

As shown in the next illustration: **(1)** ***the Leading Side of the Stick's Worldbundle*** is *(always)* located on the side of the tilt which points in the direction of Motion, **(2)** the ***Leading End*** of the Stick is always touching the *(Leading)* Side of the Stick's tilted Worldbundle; and **(3)** the ***Trailing End*** of the moving Stick always touches the Trailing Side of the Stick's Worldbundle:

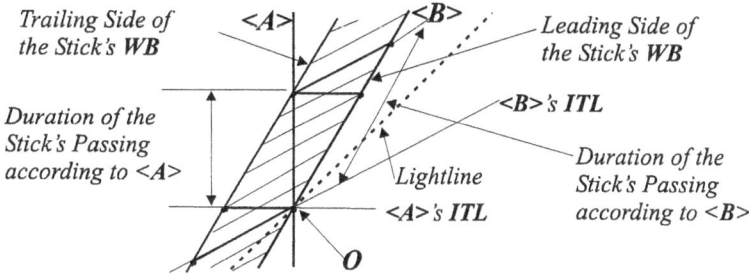

F4.17

Please, locate the following key components on Fig. 4.17:

(1) <*A*>'s ***WL*** *(vertical)*,

(2) <*B*>'s ***WL*** *(tilted)*, let's say at the Leading End of the Stick's ***WB***,

(3) The Stick's ***WB itself***,

(4) The Stick in <*B*>'s Home Frame, <*A*>'s Other Frame, at different Moments.

(5) The Moment when the Leading End of the Stick just begins to pass <*A*>,

(6) The Moment when the Trailing End of the Stick completes passing <*A*>,

(7) The total duration *(from beginning to completion)* of the Stick's passing of <*A*>,

(8) The total duration *(according to <B>)* of <*A*> passing <*B*> and her Stick *(as <B>* considers the Stick in her Home Frame and stationary!), and

(9) the Lightlines placed at 45 degree angle as to stationary <*A*>'s ***WL***. By the way, the Lightline pointing downward from Event, ***O*** is <*A*>'s and <*B*>'s ***Sightlines***, the only line-of-sight channels through which both Observers obtain their information.

By means of the Map Reader ***we can now see how <A>'s ITL cuts horizontally across the Stick's WB thereby indicating which Ends of the Stick in <B>'s Home Frame would <A> consider simultaneous.*** We also have to realize that ***while we get our information by just looking at the Map, such instantaneous observations are not possible in real life and all Observers simply have to depend on their Radar Probes for that information, a considerably Time-consuming, laborious procedure, indeed.***

This last statement is very important. The inability to actually see simultaneous Events is due to the fact that Light needs more Time to travel from the more distant End of the Stick to reach our eyes *(or camera)* compared to the Time needed for the Light originating from its nearer End. ***The separate Light Signals reaching us at any single Moment are, therefore, always of non-simultaneous origin!!!*** Having the Map conveniently in front of us places us into a position that appears deceptively real. With the Map in plain sight, we can instantly appreciate what is simultaneous to any participating Observer but note again that this is a luxury not duplicated in real life. In our study, the Map is an extremely helpful device giving us a lot of non-simultaneously obtainable information, thus helping us formulate a number of ***retrospective conclusions*** about Events already of record. These retrospective conclusions are never instantly available at any single Moment in real life.

Relativity of Simultaneity rules, placing <*A*> and <*B*> out of sync

Looking at the Map of F 4.17, we could immediately see, and pay close attention to this, that ***according to <A>, the Leading End of the Stick at an earlier Moment on <B>'s clock is simultaneous with the Trailing End of the same Stick at a later Moment on <B>'s clock!!!*** Let's repeat this last statement to pound it firmly into our minds:

According to <A>, the Leading End of the Stick at an earlier Moment on <B>'s clock is simultaneous with the Trailing End of the same Stick at a later Moment on <B>'s clock !!!

Please, keep reading the above sentence until it is becomes crystal clear. By moving the Map reader upward, you can also see ***WHEN*** the Leading End of the Stick just starts to pass <*A*> and ***WHEN*** the Trailing End just finishes passing <*A*>. Also note that the same two Events are seen differently by <*B*>. To her, it is <*A*> who is approaching then passing one of the Ends of her Stick, then the other End *(seen as trailing by <A>)* of her *(same)* Stick. After experimenting with the Map Reader held horizontally, try also moving the Map Reader upward with the slit *(or lower edge)* parallel to <*B*>'s ***ITL***. Remember: Observer Specificity and Relativity of Simultaneity are hard at work here just to confuse you*!*

Tackling Lorentz Contraction. first the Ends

Let's now examine this Simultaneity of the Stick's own non-simultaneous Ends once more but at two different Velocities, first *(Fig. 4.18A)* at a merely fast Velocity then at close to the Velocity of Light *(Fig. 4. 18B)*. You should be able to see that **(1)** *what is on the Map is a "frozen" record* of the Stick in Motion, that **(2)** *the TWO OPPOSITE ENDS of the same Stick at two different <B>'s Moments (in Fig. 4.18A) are also positioned CLOSER TOGETHER on our (and <A>'s) Isotemporal Line* and **(3)** at a higher Velocity *(in Fig. 4.18 B)*, the Stick's Worldbundle *(in <A>'s Other Frame)* becomes even more tilted and that the Endpoint Worldlines of the ***WB*** are even closer to each other *(on <A>'s ITL):*

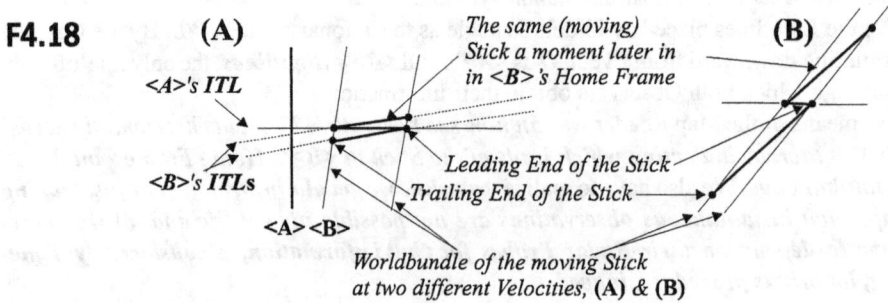

F4.18   **(A)**   *The same (moving)*   **(B)**
                              Stick a moment later in
<*A*>'s ITL   *in <B>'s Home Frame*

<*B*>'s ITLs   *Leading End of the Stick*
                             Trailing End of the Stick
<*A*> <*B*>

Worldbundle of the moving Stick
at two different Velocities, **(A)** & **(B)**

If you think all this is very odd, it certainly is but get used to it! There are even stranger things to come. At *(almost)* the Velocity of Light, the **Leading END** of an *"earlier Stick"* AND the **Trailing END** of *a "later Stick"* would *(almost)* *"touch"* each other on <A>'s **ITL** arriving at <A>'s location in Lineland at *(almost)* the same instant*!!!*. Let's repeat: **at *(almost)* the speed of Light, both Ends of the Stick arrive at <A>'s location at** *(almost)* **the same Moment** regardless of its Resting Length *(in <B>'s Home Frame!)*.

Odd, odd, odd *!!!* True. But remember, these simultaneous *(to <A>)* Ends belong to the **same Stick at different <B>'s Moments !!!** Let's repeat all this again: **the opposite Ends of the Stick as viewed by <A> belong to the same Stick but at two different <B>'s Moments.** At the impossible Velocity of Light, both Ends of the Stick at two different <B>'s Moments would arrive at, and hit us *(or <A>)*, at exactly the same Moment *!!!*

Let's take a pause, cool off and think. Let's construct a Map that shows the Stick at separate *(<B>'s)* Moments in her *(<B>'s)* own Home Frame and its opposite Ends connected by the *(single)* <A>'s Stick on <A>'s **ITL**. The **WB** of the Stick *(now considered non-moving)* is placed **vertically** while the now moving <A>'s **WL** is shown at a slant. In this **roles-reversed setting**, <B> views <A> as moving and the Stick at the two different <B>'s Moments are located one above the other in before-and-after Time Order. The two Ends of <A>'s Stick still touch the opposite Ends of the two different <B>'s Sticks at the same *(<A>'s)* Moment but these separate Ends *(of <B>'s Stick at two different <B>'s Moments)* are simultaneous only for <A>. You should now be able to see, or guess, that **<B>'s measurement of the Stick in this case is LONGER than what <A> would get.** In other words, the moving Stick **APPEARS** now **SHORTER** to <A> than it appears to <B>. The mathematical relationship between them is given by the Lorentz Multiplier as defined and explained in Chapter II.

F4.19

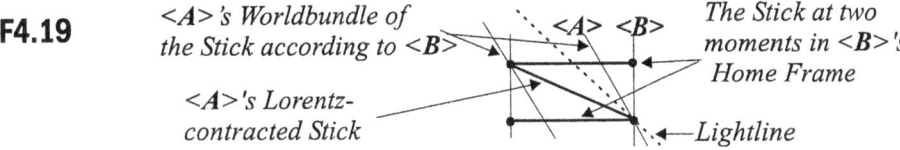

<A>'s Worldbundle of the Stick according to <B>   <A> <B>   *The Stick at two moments in <B>'s Home Frame*

<A>'s Lorentz-contracted Stick   *Lightline*

Filling in the middle

Now a new question: **What is there in the MIDDLE between the two Ends?** Answer: **The middle part of the same Stick at an INTERMEDIATE Moment in <B>'s Home Frame:**

F4.20

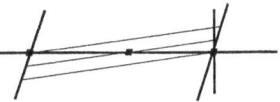

You should see that the midpoint of the Stick happens to be isotemporal or simultaneous to <A> at the midpoint of the Stick located between the two *(<B>'s)* Moments shown previously.

Now, if we draw several intermediate positions of the <B>'s Stick, we'll find that at different <B>'s Moments, different parts of the same <B>'s Stick are placed on <A>'s **ITL** between the Endpoint Worldlines that make up its Invariant Worldbundle:

F4.21

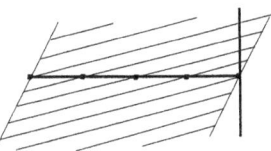

The same situation can be presented again from <*B*>'s point of view, showing the series of <*B*>'s Sticks cut *(or sectioned)* by <*A*>'s *(slanted)* Isotemporal Line:

F4.22

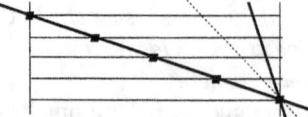

Finally, Lorentz Contraction defined

After all this verbal and visual exhortation, you should be able to understand that *A moving Stick for <A> is a MOSAIC of virtually infinite number of PARTS of <B>'s STICK that happen to be on his (<A>'s) ISOTEMPORAL LINE but at DIFFERENT MOMENTS in <B>'s Home Frame. The Leading and Trailing Ends of the Stick are closer to each other according to <A> thereby making the Stick seem "CONTRACTED" or shorter than the same Stick in <B>'s (stationary) Home Frame.*

What <*A*> is actually doing *(when measuring the moving Stick)* is reconstructing the Stick out of the many different parts of <*B*>'s Stick taken at different Moments in <*B*>'s Home Frame so that all the separate, mosaic parts are strung along on his *(<A>'s)* shorter Isotemporal Line *(in Fig. 4.20 & 4.21).* As a result, *the Ends of <B>'s Stick are placed closer on <A>'s ITL and its measured (more correctly, calculated) Length (of the Stick) is shorter than it is for <B> in proportion to Lorentz Multiplier.* That's *LORENTZ CONTRACTION* for you *!!!*

By the way, all the observations throughout the exposition were always verbally, repeatedly and tediously *TAGGED* as to *WHOSE* observations they represented. In practice, this tagging has to be done mentally, automatically, always routinely by habit and without fail.

Let's be reminded again that all these *(different)* Moments are simultaneous *(occurring at the same Moment)* only to <*A*> in his Home Frame and that <*B*>'s Home Frame was more insightfully visualized by showing it in <*A*>'s Other Frame *(Fig. 4.17-21)* Remember that <*B*>'s Sticks are always slanted on <*A*>'s *ST* Map due to being *isotemporal* to <*B*> in her own Home Frame. The Ends of <*B*>'s Stick and of <*A*>'s Stick are *"OUT OF SYNC !!!"*

A simpler way to visualize Lorentz Contraction

There is an idiotically simple way to visually illustrate the shortening *("Contraction")* of the moving Stick. Let us draw a picture of the moving Stick as seen from a position above at a given Moment and then again at a new position a short Time later, superimposed upon the "earlier" Stick. The double exposure gives us *the amount of overlap of the two pictures which corresponds to the Lorentz-contracted Length.*

F4.23

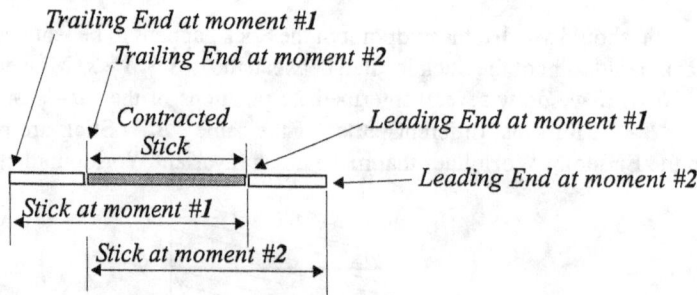

Trailing End at moment #1
Trailing End at moment #2
Contracted Stick
Leading End at moment #1
Leading End at moment #2
Stick at moment #1
Stick at moment #2

The order of Events in the last example is the same we saw in our earlier exercise: <*A*> considers the Leading End at <*B*>'s Moment #1 and the Trailing End at *(<B>'s)* Moment #2 as being simultaneous. There is no need to invoke anything other than the *fundamental disagreement between <A> and <B> as to what is simultaneous to both of them and it is the Relativity of Simultaneity that is producing this apparent "Contraction" of the Stick in <A>'s Home Frame of Reference, a Contraction that is NOT DIRECTLY VISIBLE !!!*

It is misleading to say that the moving Stick is somehow *"ROTATED"* as it is often described in some textbooks to rationalize the coming closer of the Leading and Trailing Ends from <A>'s point of view. According to the *(incorrect)* idea of "rotation," the Stick would then approach Observer, <A> hitting him broadside with its middle part, not with its Ends, thereby taking the Lorentz-contracted Stick out of the 1-D Lineland and into a 2-D Flatland. As you see, the *MOSAIC VIEW of Lorentz Contraction* makes much better sense and is really not any more difficult to grasp.

Using Radar Probe to measure a moving Length

With the Simultaneity problem now made crystal clear *(or is it still like mud?)*, let us finally proceed with the simple case of measuring a non-moving Stick's Length using a *single-Signal* Radar Probe in the usual 1-D setting.

Note that a single Signal hits the Reflecting Targets *(at both Ends of the Stick)* **sequentially** producing two *NON-simultaneous reflections* which are then returned back to <A> at the two *sequential R-Moments* as shown on the next Map. The Time Segment between these two reflections, R_1R_2 equals twice the Length of the non-moving Stick:

F4.24

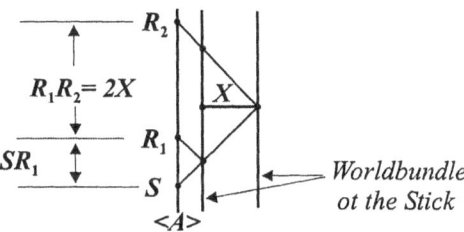

Doppler Shift raises its ugly head

In case the Stick is in Motion, the two reflected Signals received by <A> from the *approaching* Stick *(at R_1 and R_2)* are separated by a *shorter Time Segment* than those received from the *receding* Stick compared with the Time Segment obtained with a non-moving Stick:

F4.25

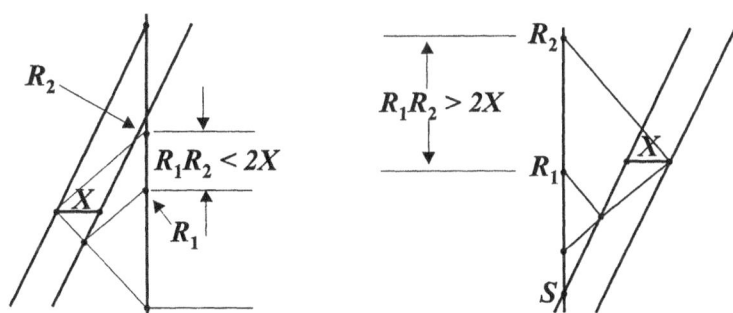

The R_1R_2 *information (about the moving Stick)* can be converted into *Length information* using the Doppler method. But at this stage of our study, we don't know how, not yet. So the Doppler method has to wait until Chapter VI, Doppler Optics. For now, we'll have to resort to the technically difficult procedure capable of producing visually clear-cut results by using *two separate Signals, S_1 and S_2 so precisely timed that, according to <A>, the two Signals hit the two Ends of the Stick simultaneously. The R_2 and R_3* (Fig. 4.26 below) *mark the ARRIVAL of simultaneous (to <A>!) REFLECTIONS from both Ends of the Stick and we can see from the Map that $R_2R_3 = S_1S_2$*. This *equality itself is a sign of success in having precisely timed the S_1 and S_2 Signals in order to obtain simultaneous Reflections off the two Ends of the Stick.* We can ignore the other two reflected signals *(R_1 and R_4)* that do not contain useful information.

F4.26

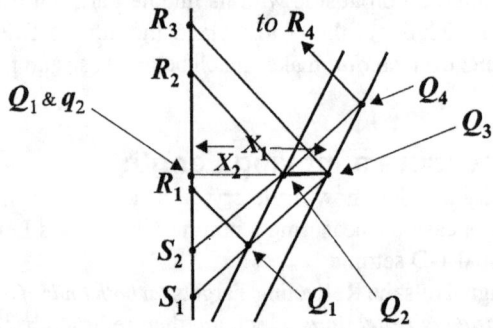

Let us re-map the experiment so that only those reflections we can use, renamed R_1 and R_2, are shown, Fig. 4.26. Now the Length of the Stick, $X=L$ is equal to R_1R_2 *(as already pointed out)* is equal to S_1S_2 and is also equal to the difference of the two Distances, X_1 and X_2 from <A> at the q-Moment. The Moment, q_1 coincides with q_2 and both are additionally simultaneous with Q_1 and Q_2. So if $S_1S_2 = R_1R_2$ then also $S_1S_2 = R_1R_2 = Q_1Q_2 = X$. Here X is the Lorentz-contracted Length whose Endpoints are located on <A>'s Isotemporal Line and the Distance between them *(the Length of the Stick for <A>)* is determined by <A> using the information provided by simultaneous reflections returned *(to <A>)* by the Radar Probe Signals at R_1 and R_2:

F4.27

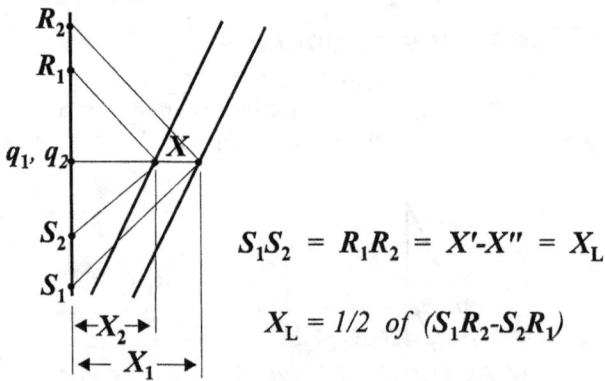

$$S_1S_2 = R_1R_2 = X'-X'' = X_L$$

$$X_L = 1/2 \ of \ (S_1R_2\text{-}S_2R_1)$$

Let's be reminded that Lorentz Contraction made <A>'s X shorter than <B>'s x? Let's spell out their mathematical relationships again. Note that now $X<x$, exactly the opposite of what we saw with Distances, Lorentz Multiplier still being the conversion factor:

$$X = x\sqrt{1-v^2} \ , \qquad x = X\frac{1}{\sqrt{1-v^2}}$$

Praise to *ST* Maps

The illustrations in this chapter have been indispensable for making all descriptions easy to follow. Without the *ST* Maps it would have been extremely difficult to point out all the complex relationships we had to dig through. Such maps are a great aid to intuitive understanding. So if there is a puzzle, any puzzle, map it out! Put all things in their correct places, look for identities and relationships. A good *ST* Map always helps to solve your puzzle. Doing the inevitable math is then greatly simplified and can be left to the very last.

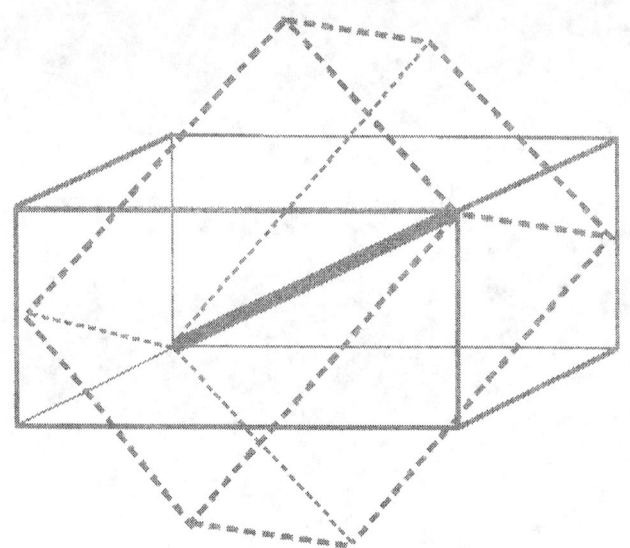

V - DOPPLER ACOUSTICS

Doppler Effect and its usefulness

Rapid technical advances in the 19th century placed a network of railroad tracks across most countries with railroad trains providing easy access to far-away places. A most spectacular sight was a fast-moving railroad train pulled by a noisy *steam powered engine* chugging along and blowing its *whistle* at railroad crossings. To someone a good distance away, there was nothing unusual about the whistle. Those standing close to the tracks, however, noticed that *the whistle sound suddenly dropped to a lower pitch at the exact Moment of Passing.* In our time, we hear the sudden pitch change in the noise produced by a variety of vehicles on the road. A cartoonist has sketched the audible experience of a passing truck as changing from **E** to **O**:

F5.1

.ₑₑEEEE**EEOO**Oooₒₒₒₒ.

This pitch change, known as the ***DOPPLER EFFECT*** was named in honor of Johannes Doppler who was the first *(in 1842)* to explain it and point out its potential usefulness in astronomy. Since then it has opened up numerous applications in science and industry. In ***astronomy***, Edwin Hubble *(in 1922)* measured the dim Light of distant galaxies and showed that the magnitude of Doppler shift increased with galactic Distances and concluded that the observable universe was expanding. In ***underwater sonar***, it is used to determine not only if the target is moving but also if it is approaching or receding and at what Velocity. It is employed in ***air traffic control*** and has become an indispensable tool in ***weather forecasting***. In ***medicine***, one of the many uses of Doppler effect is to measure blood flow in major arteries and by the changes in the Velocity of flow to detect narrowing of blood vessels.

In technical terms, ***Doppler Effect is a CHANGE in the observed FREQUENCY of a Signal due to Motion*** and Doppler Kinematics is the study of Motion as revealed by the Doppler Effect. In other words, Doppler Effect is produced b y M otion and can, in turn, be used as information about that Motion. ***Any Signal with a constant frequency such as SOUND with a definite pitch or LIGHT with a definite color can serve as a vehicle of Doppler information.*** Although Radar frequencies have no visible color, they are still capable of exhibiting the telltale frequency shift in case of Motion. The magnitude of the Doppler Effect depends upon the Velocities, first of the ***Signal SENDER, from now on designated as*** $<S>$, then also of the ***Signal RECEIVER***, $<R>$, the only quantities that can vary from zero to extremely high, the Velocity of Sound here ***NOT*** being the limit.

In Doppler studies, we'll use either a ***steady Signal,*** or ***pairs of ultra-short Signals*** with equal Time intervals between. Because we are most familiar with Sound passing through the air surrounding us, we will examine Sound propagation in air as preparation for studying the more important Doppler behavior of Light. ***There are two areas of Acoustics that are related to our study of Special Relativity***: (1) ***Certain features of elementary acoustics of Sound propagation in air are pertinent to the attempts in the late 19th century to detect the existence of ether,*** the hypothetical and still unobserved medium considered necessary for the propagation of Light. **(2)** ***Doppler behavior of Sound in air shows important parallels with Doppler behavior of Light and Radar Signals.*** Despite marked dissimilarities between Sound and Light, the really difficult Doppler Optics will come to a clearer focus if we work with the simpler Doppler behavior of Sound first.

A brief survey of elementary acoustics

The Velocity of Sound *(from hereon indicated by the letter, s* measured in air is 330 meters or approximately 1,000 feet per second. Lucky for us. We need this convenient, round figure for easy calculations. For instance, whenever you see a flash of lightning, you can estimate its Distance by counting the seconds from visible strike to audible thunder, allowing a 1000-foot Distance for each Second counted. Besides air, Sound is transmitted by a variety of media in gaseous, liquid and solid states. *The Velocity of Sound is always specific for different conducting substances.* A medium for the conduction of Sound is an absolute necessity: no medium, no conduction, no detectable Sound and no Doppler Effect.

One can legitimately ask, and this is *a very important question, whether the Motion of <S> or of <R> can influence the measured Velocity of Sound in air?* The first part of the question *(Velocity of Sound in air)* is determined by placing two Sound detectors a fixed Distance apart on the ground and moving the Sound Source at various speeds towards and away from the *(lined-up)* Sound detectors connected to a timing and recording equipment. Let us accept as established that while the Motion of the Sound Source does change the pitch of the Sound *(to the trackside Observer), the Velocity of the Sound itself in the medium (the air), once produced and set free to travel through it, remains uninfluenced by the Motion of its Source* provided that *no significant movement is imparted to the conducting medium (air) surrounding the Source (train) by the Motion of the Source through that medium.* Numerous examples of the independence of Sound Velocity from that of its Source can be listed. Perhaps the best illustration is the jet aircraft's ability to accelerate to supersonic speeds, then travel ahead of its noise. But even in this extreme case, the air speed of the noise still remains the same and is uninfluenced by the aircraft's own Velocity. The presence or absence of this influence by the Motion of <R> is another matter but we'll look into it soon enough.

To begin investigating elementary acoustics, let's do a virtual experiment by first assigning to you, the reader, the role of <R> then positioning both <S> and yourself (<R>) a fixed Distance of 1000 feet apart from each other in a perfectly windless day. Incidentally, all our scenarios in this chapter are set up on windless days so that the ground Velocity and the air Velocity of Sound remain equal. Now let <S> be a large bell and let a friend strike it with a hammer every 1 second. For convenience, electronic equipment can be used to perform and record all events planned. Using the familiar method of Spacetime Mapping *(I!I)* but standardized here to record Time in seconds and Distance in "Soundseconds," we can show the entire sequence of Events as follows:

F5.2

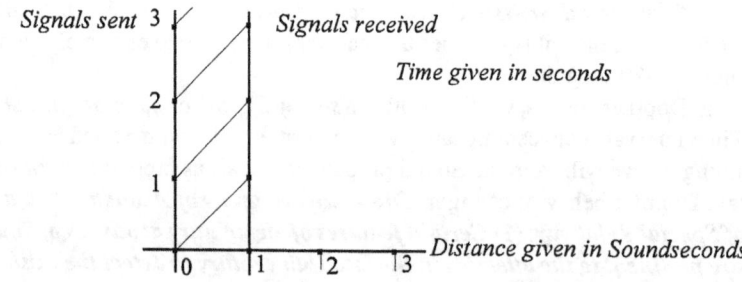

Considering the 1.0-Soundsecond <S>-to-<R> Distance, you should realize that the Moment your friend hits the bell for the first time, you hear nothing at all, not yet. You'll have to allow a full 1 second delay for the Sound to travel the 1000 ft intervening Distance to reach you. Only then would you start hearing the bell, each Sound 1 second apart, the interval equal to that between the strikes.

<S> and <R> a fixed Distance apart, moving, <S> leading <R>

For the next virtual experiment, *let the bell (<S>) and you (<R>) both 1000 ft apart move at the same Velocity, v in the same direction.* This can be accomplished by mounting both the bell *(<S>)* and you *(<R>)* with instruments 1000 ft apart on opposite ends of a long train and record exactly when the bell is struck and when its sound is heard. With constant Distance between <S> and <R> moving at the same Velocity, <S> leading <R>, *the previously stationary air is now observed blowing as an AIR WIND against and past <S> and <R>.* When a steady Velocity of 100 foot per second is established, the bell is struck at an agreed-upon moment. While the Sound is now traveling at velocity, *s* towards you, you are also moving at Velocity, *v* through the air toward the oncoming Sound. Therefore, the measurable, or *Effective, Velocity of Sound* is faster and *the Sound will reach you with less than the previously observed 1 second delay.* Let's designate the Time needed for the Sound to reach you (<R>) at the fixed 1000 ft Distance from <S> as *T'* and set up an equation to calculate its duration. But first, let's visualize the experiment as described:

F5.3

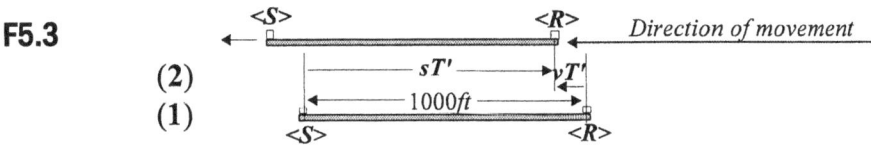

Now, the *T'* can be derived:

$$sT' + vT' = (s+v)\ T' = 1000 \text{ ft}$$

$$T' = 1000\ \frac{1}{s+v}\ ft$$

With Velocity values as given, we can calculate the numerical result in *(fractions of)* seconds for *T'* :

$$T' = 1000\ \frac{1}{1000+100} = \frac{1000}{1100}$$

$$T' = 0.818 \text{ seconds}$$

As you see, it now takes less than a second from the Moment the bell is struck to when the first Sound is heard.

To derive the new apparent or *"effective" Velocity, s'* of Sound as it travels through the 1000 ft Distance from <S > to <R >, we need to divide the fixed, moving 1000 ft Distance by the Time, *T'* it takes for the sound to get from <S > to <R > which comes out to be the sum of the Velocities, *s* and *v, (s + v)*:

$$s' = \frac{1000ft}{1000ft\ \dfrac{1}{s+v}} = s+v \qquad \text{Restated } s' = s + v$$

Summing it up: The Velocity of Sound observed by <R> when both <S> and <R> are moving at a constant Velocity through the Sound-conducting medium, <S> leading <R>, *the Effective Velocity comes out to be a sum of the two Velocities, s and v.*

\<S\> and \<R\> moving, 1000ft apart, \<R\> leading \<S\>

The same experiment can be repeated but with the train moving at 100 feet per second backward, \<R\> leading \<S\>. Again, the experiment can be first visualized:

F5.4   (2)   \<S\>   \<R\>   *Direction of movement*

(1) sT'
                   —1000ft— vT'
            \<S\>   \<R\>

$$sT' = 1000 \text{ ft} + vT'$$

And T' can be calculated $sT' - vT' = 1000 \text{ ft} = (s - v)T' = 1000 \text{ ft}$

$$T' = 1000 \text{ ft} /(s - v) = 1000/(1000\text{-}100)\, ft = 1000/900 \, ft$$

$$= 1.111 \text{ seconds}$$

This time, the first Sound is heard after a longer delay.

The *effective Velocity, s'* of the Sound traveling through the 1000 ft *distance that is fixed and moving through the air, \<R\> leading \<S\>*, comes out to be the difference of *s* and *v*, (*s-v*):

$$s' = \frac{1000 ft}{1000 ft \dfrac{1}{1-v}} = s - v = 900\, ft/sec \qquad \text{Restated} \quad s' = s - v$$

If we repeat our initial two experiments with both \<S\> and \<R\> on the moving train and the bell is struck repeatedly every 1 second, each Sound will be arriving at \<R\> with the same *(shortened or prolonged)* delay but all the Sounds will still heard with the 1 second intervals between them, evidence that *even with changed "EFFECTIVE" VELOCITY of the Signal but without a CHANGE in the DISTANCE between \<S\> and \<R\>, the intervals between the Sound Signals received remain unchanged and there is NO Doppler Effect.*
The important new feature influencing the effective Velocity of Sound created by the moving train experiment is the *production of the AIR WIND (or wind made up of the signal-conducting medium) around and through the entire setup!* Here the effective \<S \>-to-\<R \> propagation of Sound is either speeded up or slowed down by the Air Wind so all the Signals seem to be propagated at increased or reduced *Effective Velocity*. But as pointed out already, with uninterrupted sequence of Signals and constant \<S\>-to-\<R\> Distance, we *(at \<R\>)* would not notice anything different at all. If, however, single bell strikes are used and the Time from each bell strike to the arrival of each Sound at \<R\> is determined, we would see that the same *(increased or decreased)* amount of delay is observed with each individual bell Sound. So, *the Velocity of Sound conduction over a moving 1000 ft Distance from the bell to the Observer is definitely influenced by the Air Wind* even if *NO change is noticed in the Time intervals between the Signals received.* With forward movement of the train, *\<S\>* leading *\<R\>*, the *Signal transit Time, as calculated above is reduced by a factor of 1/(1+v).* Similarly, with backward movement of the train at the same Velocity through the air, *\<R\>* leading *\<S\>*, the *Signal transit time is increased by a factor of 1/(1-v).* And *the effective Signal Velocity, s' through the fixed space between \<S\> and \<R\> is the inverse of the transit Time, T'.*

In conclusion, **(1)** *With continuously repeated Signals, the influence of Motion on the effective Sound Velocity remains HIDDEN* unless the transit Time is checked for each individual Signal in the series; **(2)** *With constant, fixed <S><R> Distance, there is NO Doppler Effect.* The effect of Distance change will be taken up shortly.

The virtual experiments described above may seem tediously overworked, but *the Sound conduction in the moving train experiment is nothing other than the sonic equivalent setup of the Michelson-Morley Experiment (MME).* This was the historic experiment that could not be explained except by invoking Special Relativity. To create a more obvious parallel to it in acoustics, let us just suppose that *we (or a genie working for us) can make the air around the train experiment totally undetectable* while keeping its Sound conducting properties unchanged.

The question that begs to be asked then is this: *how can we be sure that the air actually exists?* A search for the only possible solution leads us directly to the sonic equivalent of the *MME.* Never mind that *the "acoustic theory" of sound requires its existence.* Intending to prove that the optical equivalent of air, the ether, actually existed, Michelson and Morley designed and performed the technically difficult experiment. Remember that with a continuous series of Signals, the observed intervals between them *(analogous to the wavelengths)* remain the same and that there would be *NO* Doppler Effect. So what can be used to prove conclusively that the air actually exists? Aha*!* we have just become familiar with all the tricks we need. Let's use them:

> ### *If air cannot be observed directly, its existence can be proven through the influence that the Air Wind has upon the "effective" Velocity of Sound measured over a fixed moving Distance.*

Although no scientist at that time had serious doubts that ether existed, it was something that had not been demonstrated experimentally. Advances in technology toward the end of the 19th century made it possible to set up the experiment and when it was finally performed, it produced totally unexpected results. The details of the experiment will be taken up in Chapter X, Michelson-Morley Experiment.

Just one more thing: If the moving-in-air experiment is performed within a fully enclosed train so that Sound conduction takes place through the *air trapped inside* and traveling with the train at the same Velocity, then no Air Wind would exist *inside* the train *(all wagons interconnected!)* and the Velocity of Sound conduction observable to *<R>* would not be changed by Motion. The same situation would exist if the air outside the train would remain wrapped around it like a *padding* so that Observer, *<R>* riding along and placed *outside* the train, would *NOT* be able to detect the Air Wind due to absence of its influence on Sound conduction.

Using Sonar to measure Distances

Before getting into the equally important topic of the Doppler behavior of Sound, let's briefly see how we can measure Distances with Sound waves. *Sonar Probes* work pretty much the same as Radar Probes. Let the stationary Observer *<A>* send out a Sound Signal toward *<B>*. That Signal would bounce off *<B>* and return to *<A>* who then would record the Time from *S* to *R* and calculate the Distance of *<B>* at moment, *q* just as we did with the Radar Probe. Note that Time on the Time Axis is recorded in seconds and Distance on Space Axis is recorded in Soundseconds:

F5.5

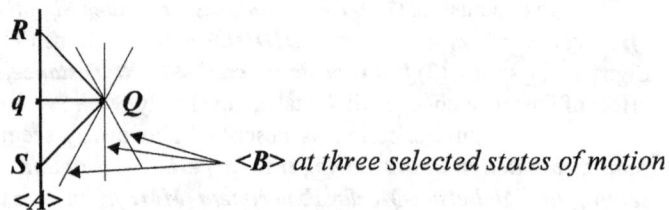

<B> at three selected states of motion

Whether *<B>* is moving or stationary, the Sound Signal round trip *(from S to Q to R)* makes up the Time, **SR** which is twice the Soundtime Distance, **qQ**.

If *<B>* is a bat flying about in a cave, the flight Velocity of the bat amounts to a higher percentage of the Velocity, *s* of Sound than Earth's Velocity compared with the Velocity of Light. Half of the **SR** duration is still a very good measure of the **qQ**-Distance, a situation not only similar to, but also different from, what we encountered with Radar Probes:

F5.6

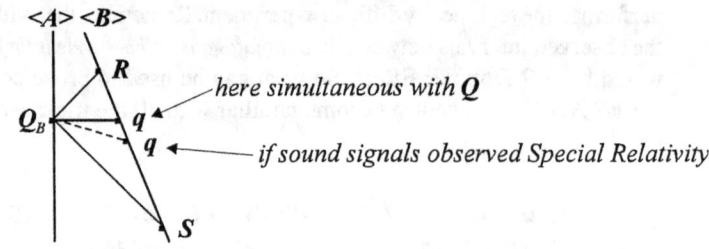

here simultaneous with Q

if sound signals observed Special Relativity

Supersonic Motion

Let us next map the path of supersonic aircraft noise both in approaching and receding modes. Note that the noise Signals, here in form of brief noise bursts generated one second apart, are received more closely together from the approaching aircraft and more widely apart after passing. Moreover, *in approaching mode, the arrival order of individual bursts is REVERSED* while the originating order is retained after passing. This, by the way, must not be mistaken for evidence that Time Order itself has become reversed *(in approaching mode),* as it is often alleged when featuring examples of Reversed Order of Reception of Light Signals coming from objects in superluminal Motion, all cases hypothetical, of course. The question of Time Arrow as demonstrated with Sound in the present chapter comes up in a different disguise with the topic "Faster Than Light." in Chapter IX. Now the supersonic Map:

F5.7

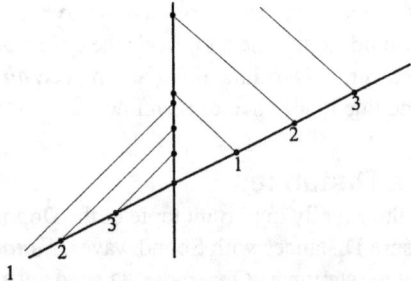

At exactly the velocity of Sound, and with near-collision Passing Distance, the noise produced in approaching mode would arrive at the Observer all at once, producing the "best" **SONIC BOOM** ever heard with *(almost)* no temporal spread of the crack-of-the-whip-like Sound blast:

F5.8

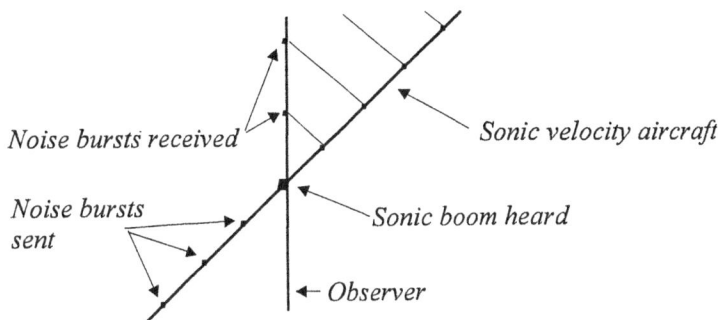

Noise bursts received

Sonic velocity aircraft

Noise bursts
sent

Sonic boom heard

← Observer

It would also be useful to visualize the relationship between the Velocity of Sound and that of the Source by showing the positions of sequentially spreading Signals in 2-D Space produced by <*S*> moving at different Velocities, first at $v = 0$, and at $v > 0$, then also $v = s$ and, finally, $v > s$:

F5.9

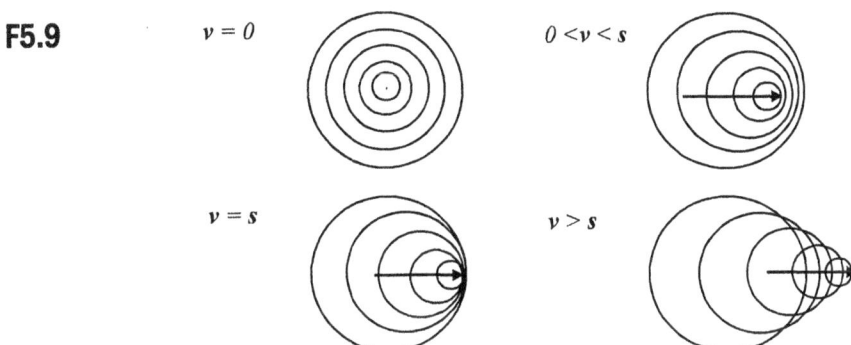

$v = 0$ $0 < v < s$

$v = s$ $v > s$

A particular kind of supersonic Motion is of special interest. Let a futuristic aircraft pass us at near-collision proximity and at a Velocity that *(at least acustically)* could be considered as "absolute." Here a half-luminal Velocity would be an example good enough for being an ***"Absolute-Velocity"*** Motion compared with sonic Signals. As we cannot hear the approaching aircraft, the audible experience would give the impression that two separate and identical copies of the aircraft materialized in a sonic blast at the Passing Distance and then flew away in opposite directions at the Velocity of Sound *(in air)*:

F5.10

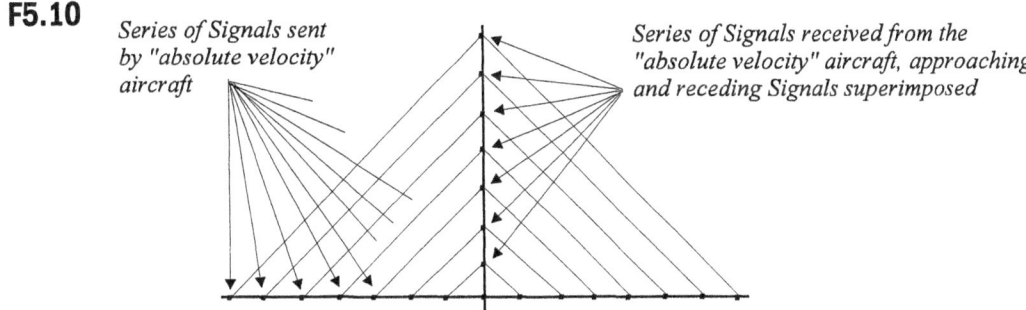

*Series of Signals sent
by "absolute velocity"
aircraft*

*Series of Signals received from the
"absolute velocity" aircraft, approaching
and receding Signals superimposed*

Generating Doppler Effect

The virtual experiments described so far in this chapter should have made us able to correctly interpret a somewhat different experiment where *either <S> or <R> is moving at a velocity of 100 ft/sec toward, or away from, the other participant thereby changing the Distance between them at a constant rate.* One may wonder if it makes any difference whether <S> or <R> is moving. In other words, is it only the relative Motion between them that counts *(just what we saw in "Axioms And Operations" with Radar Signals)?* Let's find out from specific examples:

With <S>to<R> Distance changing, *each Sound is heard with either progressively shorter or longer arrival Times at <R> depending on whether the intervening Distance is decreasing or increasing.* Note that those <S>-to-<R> signal transit Times do not remain the same in duration as it was in the previous "moving train" series. So *if the <S>-to<R> Distance is changing (decreasing or increasing) constantly at the same rate, the Signal arrival delays are also changing at the same rate with the result that all intervals between the Signals received are still equal but now differ in duration from those (intervals) between the Signals sent.* Thus,

> *Doppler Effect is created ONLY when the <S>to<R> Distance keeps changing.*

The general statement about Doppler kinematics on the first page of this chapter indicated that Doppler Effect is produced by Motion. But we have now determined that Motion by itself does not cause Doppler Effect. While it is the <S>to<R> Distance change that does it but the *AIR WIND,* while modifying the individual <S>to<R> transit Times, does not actually produce Doppler Effect. We'll see also that *The Air Wind itself does not influence the magnitude of Doppler Effect even if <R> is moving through the air causing (<R> to register) a change in the Effective Velocity of the Signal.* This change *(in the Effective Velocity of the Signal)* is not seen if it is only <S> who is moving through the air. In other words, *Motion of the Signal Source, as we we have seen, does not influence the Velocity of the Signal observed by <R>.*

To cement everything into place, let us summarize what we have learned piecemeal in this chapter. To keep everything as simple as possible, we'll assume that either <S> or <R> is moving *(but not both)* and that the days are absolutely windless.

For Doppler Effect to occur, a continuous change in the <S>-to-<R> Distance is necessary. In that case*:*

(1) *In APPROACHING mode, the Signal travel Time becomes progressively shorter. In RECEDING mode, the Signal travel Time becomes progressively longer. Without change in the <S>'to-<R> Distance, the intervals between Signals received remain constant but compared with originating intervals are shorter in approaching mode and longer in receding mode.*

(2) *It is the movement of <R> through the air that produces Air Wind Effect thereby additionally modifying the magnitude of the Doppler Effect. The Motion of <S> through the air does not produce it (Air Wind). Without the <S>-to-<R> Distance change, the Air Wind itself does not produce Doppler Effect.*

(3) *The Air Wind Effect is reduced or eliminated altogether if Sound conduction takes place through the air that is "dragged" along with equal-Distance <S>to<R> setup. This Air Drag may be incomplete if the air outside the train is only partially put into Motion or may be complete if the air forms a thick "air padding" around the train (the train carrying both <S> and <R>). It may also be complete if the air is completely enclosed and interconnected inside of the train with the Sound conduction occurring through the fully enclosed air-space.*

Now, with this much made clear, we can go ahead and derive representative examples of the Doppler Effect using actual numbers to make the argument more explicit. The examples illustrated are as follows:

(1) <*R*> stationary, <*S*> first approaching then receding.

(2) <*S*> stationary, <*R*> first approaching then receding.

As usual, everything is easier to follow if fully visualized. So, let's use Sound Velocity of 0.1*s* which equals 100 ft/sec. The derivations that follow are presented only visually and by numbers without any verbal persuasion and all this should be easy enough to trace out on the Maps provided.

<*R*> stationary, <*S*> approaching

F5.11

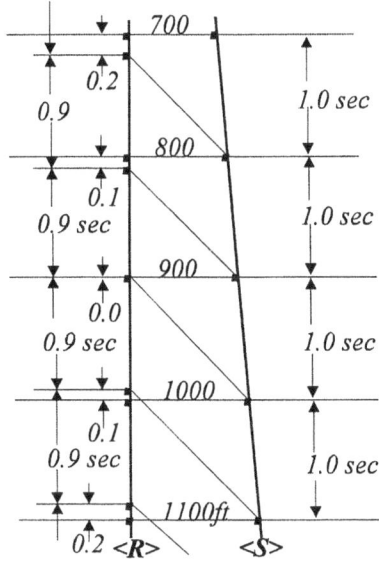

That was easy! Let's do the same with algebraic symbols, *s* = 1 and *X* in Sound-seconds:

F5.12

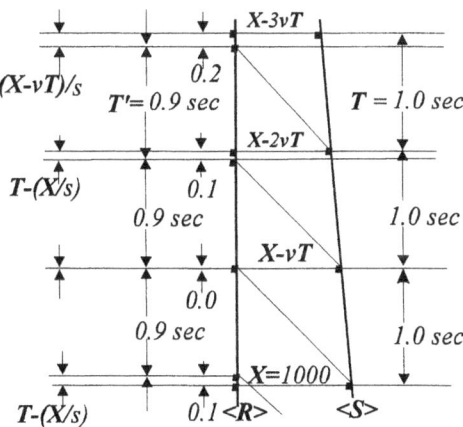

And the derivation of T' *(the Time interval between received Signals)* in the above example:

$$T' = \left(\frac{X - vT}{s} \right) + \left(T - \frac{X}{s} \right) = \frac{(X - vT) + (sT - X)}{s}$$

As the $+X$ and $-X$ cancel out, we are left with: $\dfrac{(sT - vT)}{s} = \dfrac{(s-v)T}{s}$, so

$T' = (1-v)T$, v given as a fraction of s.

<R> stationary, <S> receding

F5.13

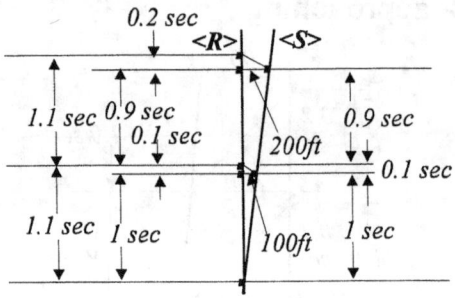

And the equation is $T' = (1+v)T$

<S> stationary, <R> approaching

From General Acoustics, we learned that the Air Wind Effect amounted to $1/(1+v)$ in approaching mode, both <S> and <R> *(at a fixed Distance from each other)* moving, <S> leading <R>. This same quantity also happens to be the Changing Distance Effect *(same as Doppler Effect)* but here only <R> moving in approaching mode:

F5.14

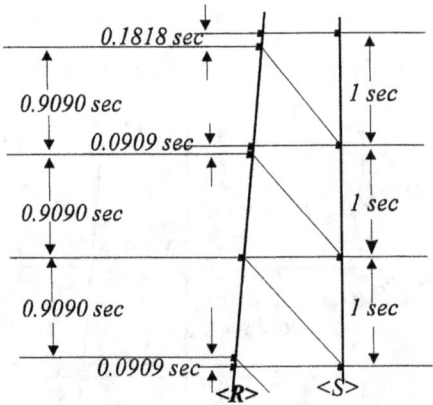

So the result of the Air Wind Effect in approaching mode is: $T' = \dfrac{1}{1+v}T$

and the Effective Velocity is: $s' = 1+v$.

<S> stationary, *<R>* receding

F5.15

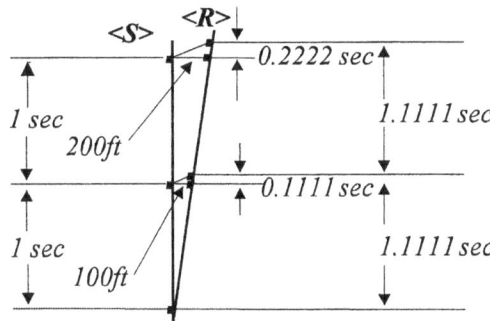

And the Air Wind Effect is: $T' = \dfrac{1}{1-v}\,T$ and the Effective Sound velocity, $s' = 1-v$

    To make the differences between the various T' values more obvious visually, let's plot the results also at $v = 0.5s$ first with *<R>* then *<S>* stationary.

<S> moving at *v* = 0.5s, first approaching then receding

F5.16

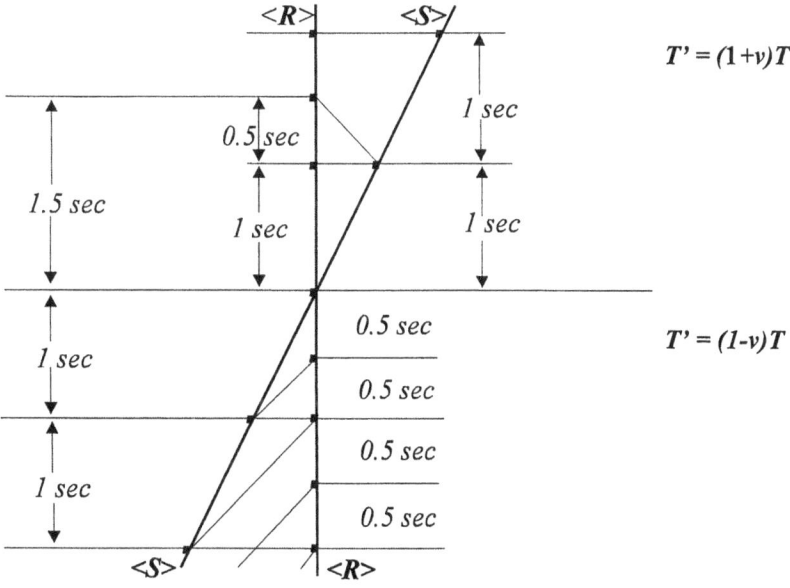

<R> moving at *v* = 0.5*s*, first approaching then receding

F5.17

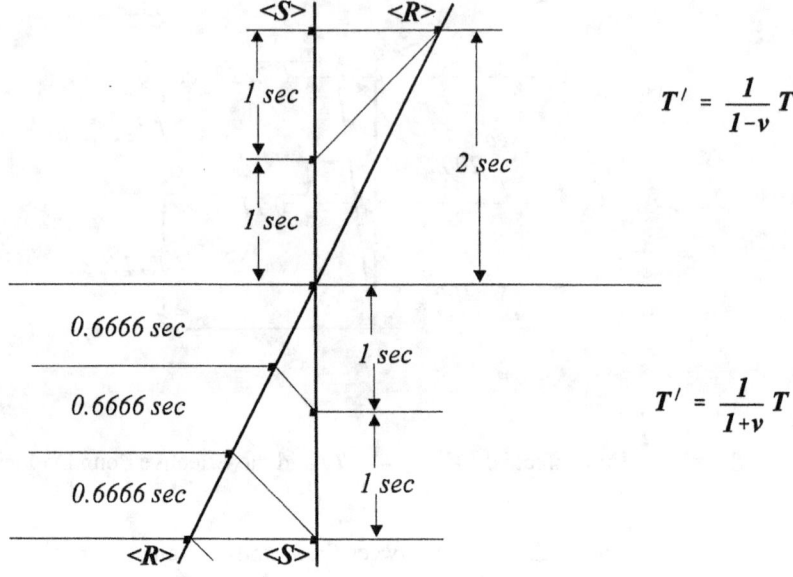

$$T^{\prime} = \frac{1}{1-v} T$$

$$T^{\prime} = \frac{1}{1+v} T$$

To compare all ourDoppler results with their derivations of *T'* listed separately, let's show them on a suitable, combined map *(with intervening Distance changing)*:

F5.18

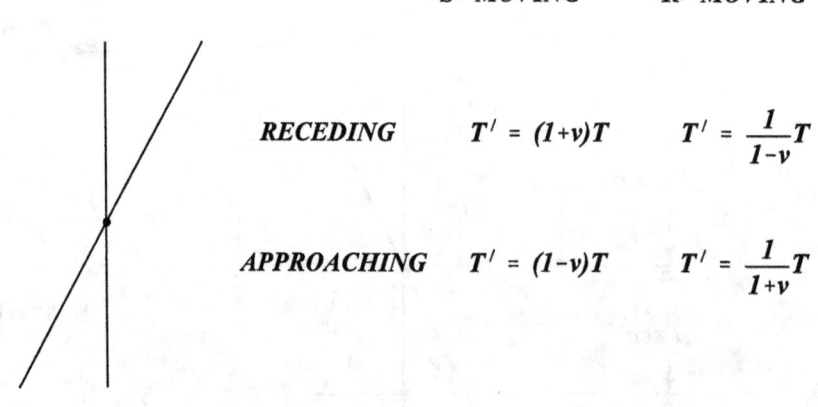

	<S> MOVING	*<R> MOVING*
RECEDING	$T^{\prime} = (1+v)T$	$T^{\prime} = \frac{1}{1-v}T$
APPROACHING	$T^{\prime} = (1-v)T$	$T^{\prime} = \frac{1}{1+v}T$

No two results are identical. What is missing here is something we can call, for lack of a better expression, ***SYMMETRY !*** Let's blame it on the additional element of the Air Wind. In the next chapter, Doppler Optics, we'll see that in contrast with Acoustics, there is no Ether Wind and the Doppler Effect there works out "symmetrically" with Light as well as Radar Signals.

Summing it up:

(A) *If <S>-to-<R> DISTANCE REMAINS CONSTANT and BOTH <S> and <R> ARE MOVING, NO Doppler Effect is generated.* Moreover:

(1) *If <S> is leading <R>,* the Sound *Signal travel Time (from <S> to <R>) is shortened* as the observed *Signal Velocity is influenced by the Air Wind* which *makes the Effective Signal Velocity into a sum of the individual Velocities: s' = s + v (same as 1+v).* Note that *the Signal travel Time is <S>-to-<R> Distance divided by s+v* while *the Time intervals between Signals received still remain the same.* But:

(2) *If <R> is leading <S>,* the Sound *Signal travel Time (from <S> to <R>) is lengthened by the Air Wind making the effective Signal Velocity into a difference of the two Velocities: s' = s - v (same as 1-v). The Signal travel Time is <S>-to-<R> Distance divided by s-v, the Effective Signal Velocity* while *the Time intervals between the Signals received still remain the same.*

(B) *If <S>to<R> DISTANCE is CONTINUALLY CHANGING and only one participant is moving then Doppler Effect IS observed because the intervals between Signals received will be different from original intervals sent.* Also:

(1) *If only <R> is moving and experiencing Air Wind, <S>* remaining stationary, *the intervals between Signals received in APPROACHING mode will equal the original intervals at <S> divided by s+v (same as 1+v). In RECEDING mode,* the *intervals between Signals received will equal the original intervals at <S> divided by s-v (same as 1-v).* But:

(2) *If only <S> is moving ,<R>* stationary, there is *NO Air Wind to influence results, the intervals between Signals received in APPROACHING mode will be equal to the original intervals at <S> multiplied by s-v* (same as *1-v). In RECEDING mode,* the *intervals at <R> equal the original intervals at <S> multiplied by s+v (same as 1+v).*

Next, for the Doppler Effect, **DE**, we'll use the *ratio of the interval between Signals received to the intervals between Signals sent,* (*T'/T*) to specify the magnitude of the Doppler Effect.

(1) *First,* <S> and <R> Distance diminishing, only <R> moving:

$$T' = \frac{1}{1+v}T$$

and the Doppler Effect, **DE:**

$$DE = \frac{T'}{T} = \frac{1}{1+v}T\frac{1}{T} \quad \text{or} \quad DE = \frac{1}{1+v}$$

(2) *Next,* <S> and <R> Distance increasing, only <R> moving:

$$T' = \frac{1}{1-v}$$

and the Doppler Effect, **DE:**

$$DE = \frac{T'}{T} = \frac{1}{1-v}T\frac{1}{T} \quad \text{or} \quad DE = \frac{1}{1-v}$$

(3) *First again*, <S>-to-<R> Distance diminishing, only <S> moving:

$$T' = (1-v)\,T$$

and the Doppler Effect, *DE:*

$$DE = \frac{T'}{T} = \frac{(1-v)T}{T} \quad \text{or} \quad DE = 1-v$$

(4) *Next*, <S>-to-<R> Distance increasing, only <S> moving:

$$T' = (1+v)T$$

and the Doppler Effect, *DE:*

$$DE = \frac{T'}{T} = \frac{(1+v)T}{T} \quad \text{or} \quad DE = 1+v$$

At this time, we don't need to take up the additional question as to what constitutes the Doppler constant, *k*. This we'll see in the next chapter, Doppler Optics.

VI - DOPPLER OPTICS

Get set, ready, go

The chapter on Doppler Acoustics provided a brief introduction to Doppler Kinematics along with the consideration of a medium believed necessary for the transmission of all wave-like phenomena that, of course, included also Light. Doppler Acoustics was introduced as background information against which the more important Doppler Effect with Light can now stand out in better contrast. A general familiarity with the Doppler behavior of Sound is recommended before proceeding with Doppler Optics.

Switching from Sound to Light

Doppler behavior of Light is in many ways similar to that of Sound. By habit, we consider them on equal terms except for their markedly unequal Velocities of propagation. But *the relativistic character of Light,* not seen with Sound, *is best explained by a single feature unique to Light which happens to be the absence of a medium (or the non-dependence on it) for transmission thereby explaining also the ABSENCE of the Wind Effect.* The question about the existence of such a medium was settled with finality by the historic Michaelson-Morley experiment *(Chapter X).*

In Doppler Optics, the *absence of the Wind Effect* makes the Motion of the Signal Source and of the Receiver through the intervening Space individually irrelevant. Only the relative Motion, the simple intervening Distance change between them, is all that matters. So *the Light (or Radar) Signal is unique not only by the* **(1)** *absence of a known medium for transmission but also by* **(2)** *possessing the same Velocity in all Inertial Systems. In addition, the Signal is* **(3)** *the only modality that can reach across the wide open spaces and interrogate its dimensions that are calibrated only by the Velocity of Light as to Distance and are totally inaccessible to direct measurement.*

The uniqueness of Light as an exclusive yardstick further underscores the impossibility of performing classical measurements in Space. As we already *(should)* know, one-dimensional Distance, Length and Time measurements do not yield the same values to all Inertial Observers in Motion relative to each other. Remember, the new *Absolute* is the *INVARIANT which has the same value to all Observers.* And it's all due to *RELATIVITY OF SIMULTANEITY,* itself a consequence of the unique properties of Light.

Doppler elements created

Let's take two astronauts, $<A>$ and $<B>$ and keep their Space vehicles at a constant Distance from each other. Now let $<A>$ send a series of Light or Radar Signals to $<B>$, each Signal 1 msec apart. Are both Observers standing still or are they moving at the same Velocity on parallel paths? The question is irrelevant. Only the change of Distance between them is all that matters. In any case, $<B>$ receives $<A>$'s Signal 1 msec apart as shown in Fig. 6.1:

F6.1

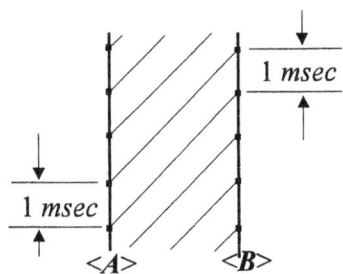

Next, let's place Astronaut < *C*> between <*A*> and <*B*> with <*C*> moving away from <*A*> and toward <*B*>. Let <*C*> send Signals 1 msec apart to both <*B*> ahead and <*A*> behind. The **ST** Map shows that <*A*> will receive <*C*>'s Signals at intervals that are longer than <*C*>'s and <*B*> will receive them at intervals shorter than <*C*>'s:

F6.2

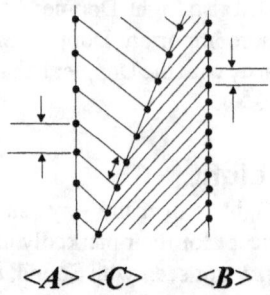

<*A*> <*C*>   <*B*>

Let's now use the same arrangement of <*A*>, <*B*> and <*C*> but let <*A*> send a series of Signals 1 msec apart toward <*B*>. As those Signals pass <*C*>, <*C*>'s equipment immediately repeats them even anticipating their timing so that <*C*>'s Signals are triggered exactly when <*A*>'s Signals pass him. Both Signals (<*A*>'s and <*C*>'s) combine and travel together towards <*B*>. Those of <*C*> also travel back to <*A*>. Next, let <*C*>'s Velocity relative to both <*A*> and <*B*> be sufficiently great so that his clock registers <*A*>'s Signals 1.2 msec apart. Still, the timing of the Signals at <*B*> would be 1 msec apart, same as those originating at < *A*>. The Time Segments between Signals reflected from<*C*> back to <*A*> sustain the same amount of lengthening as did the Time Segments between Signals coming to <*C*> from <*A*> *(1.2x1.2=1.44)*. To avoid clutter, the path of only two Signals from <*A*> to <*C*> and to <*B*> and back to <*A*> are shown:

F6.3

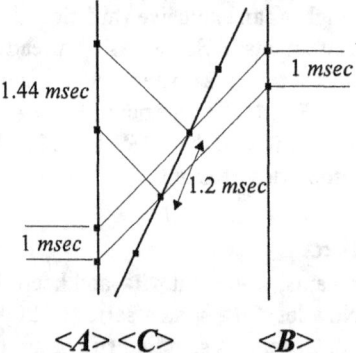

1.44 *msec*

1 *msec*

1.2 *msec*

1 *msec*

<*A*> <*C*>   <*B*>

The Doppler Optics Map above provides a clear visual demonstration of the most fundamental Doppler relationships between the timing of Light (and Radar) Signals at <A>, <B> and <C>. All Doppler optical phenomena can be derived from these very basic features. Their simplicity is deceptive. Examine the depicted Time Segments very carefully. A solid understanding of this particular Map will save you a great deal of puzzlement later.

Doppler Effect and Constant defined and calculated

Now, let's calculate *the magnitude of the Doppler Effect* from the available information expressed as *the ratio of Time interval received divided by Time interval sent.* Thus:

The Doppler Effect of <*A*>-to-<*C*> equals 1.2/1.0 = 1.2
The Doppler Effect of <*A*>-to-<*B*> equals 1.0/1.0 = 1.0
The Doppler Effect of <*C*>-to-<*B*> equals 1.0/1.2 = 0.833
The Doppler Effect of <*C*>-to-<*A*> equals 1.44/1.2 = 1.2
The Doppler Effect of <*A*>(-to-<*C*>)-to<*A*> equals 1/(1.2)(1.2) = 1/1.44

Now the Signal-to-Signal intervals: $t_C = 1.2T_A$, and $T_B = 0.833t_C$. But T_B is equal to T_A, thus $T_B = T_A$. Note that Time, T measurements by <*A*> and <*B*> are given in upper-case letters (T) and for <*C*> in lower-case letter (t) in recognition that <*C*>'s clock is located in a moving system and his Time measured is, therefore, different from those of both <*A*> and <*B*>.

Finally, the ***DOPPLER CONSTANT, k*** is ***the Doppler Effect in the Receding Mode.*** The Doppler relationships shown on Fig. 6.3 between <*A*>, <*B*> and <*C*> are next *(algebraically)* derived and labeled on the next Map, Fig. 6.4 according to the explanation given above. Study it again with great care and be certain that you understand it:

F6.4

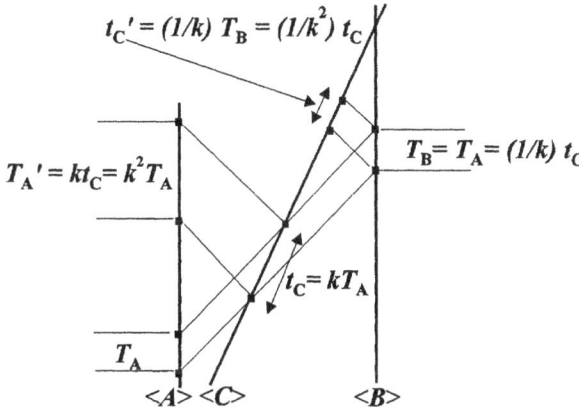

$$t_C' = (1/k)\, T_B = (1/k^2)\, t_C$$

$$T_A' = kt_C = k^2 T_A$$

$$T_B = T_A = (1/k)\, t_C$$

$$t_C = kT_A$$

$$T_A$$

<*A*> <*C*>   <*B*>

In receding motion, the multiplier that applies is k, in approaching Motion it is 1/k, the INVERSE of k. That's it *!!!* If you compare the Doppler effect of Light with that of Sound *(Chapter V)*, you can see that with Sound, the inverse relationship was ***NOT*** observed.

So far, we have not determined ***HOW*** k and v are related and ***WHAT*** v would produce ***WHAT*** k values. However, there are a few things we can reasonably expect:

(1) A certain value of k should stand for a certain Velocity.
(2) If $v = 0$, then $k = 1$.
(3) The greater the Velocity, the greater the magnitude of k.
(4) We should be able to convert v values to k values and vice versa.
(5) The k values should allow us to convert <*A*>'s Time measurements to <*C*>'s Time measurements and vice versa. In fact, if we know any two of the three values: k, t_C and T_A *(or T_B as in Fig. 6.3)*, we should be able to derive the third.

From Doppler Effect to Doppler analysis of Motion

Now we are finally ready to tackle Doppler analysis using our Radar Probe. We'll assign <*A*> to be the Stationary Observer in his Home Frame and place < *B*> into his *(<A>'s)* Other Frame *(in Motion relative to <A>).* The Passing Event at *O*-Moment is common to both <*A*> and <*B*> and serves to synchronize their clocks. At that *O*-Moment, the first of two Radar Probe

Signals is sent by $<A>$ to $<B>$ who receives it *(almost)* exactly at the same O-Moment. The second Signal is sent by $<A>$ at the S-Moment and is received by $<B>$ at the Q-Moment. Actually, the Radar Signal sent at O-Event duplicates the timing action of the Passing Event and can, therefore, be omitted on the Map:

F6.5

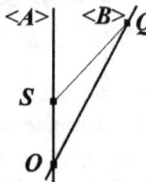

The information we can use consists of the Time Segment from O to S measured by $<A>$, and from O to Q measured by $<B>$. The Time, OS is T_S *(S for "sent")* and OQ is t_R *(R for "received")*. We can now state that:

$$k \text{ is } OQ/OS = t_R/T_S. \text{ Also, } t_R = kT_S \text{ and } T_S = (1/k)t_R$$

After the Radar Signal bounces off $<B>$, it returns back to $<A>$ at moment, R:

F6.6

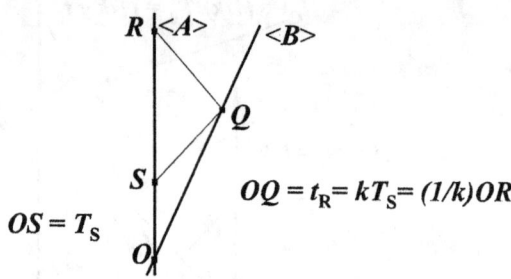

$$OS = T_S \qquad OQ = t_R = kT_S = (1/k)OR$$

A treasure trove of Doppler relations can be found here and is detailed below. Look at them carefully and practice reproducing them with pen and paper so that what you have seen will become part of your intuition:

$$OQ \text{ on } <B>\text{'s } WL \text{ equals } t_B = kT_S$$
$$OR \text{ on } <A>\text{'s } WL \text{ equals } kt_R = kkT_S = k^2T_S$$
$$t_R = (1/k)OR. \text{ Also } T_S = (1/k)t_R = (1/k^2)OR \text{ and } T_S + SR = kt_R = k^2T_S$$

Let us re-map all the information above with the Time intervals now labeled as Doppler derivatives containing k, *(Fig. 6.7)*:

F6.7

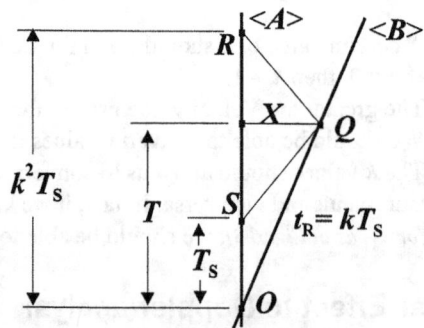

Next let's identify the components of the Radar Probe in terms of the basic Coordinate *(T and X)* projection values. Please, match the corresponding identities in Figures 6.7 and 6.8.

$$OS = T\text{-}X, \qquad OR = T\text{+}X, \qquad X = \tfrac{1}{2}(OR\text{-}OS), \qquad T = \tfrac{1}{2}(OR\text{+}OS).$$

F6.8

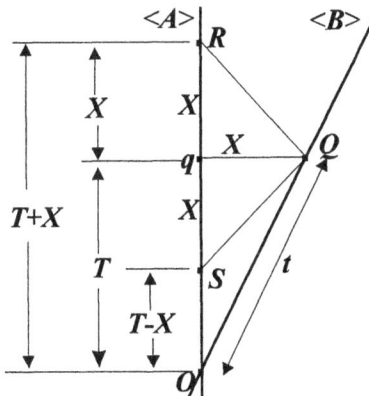

Next, let us combine the two Maps into a Doppler Map and label all the *T* and *X* values *(in Home Frame)* and *t* and *x* values *(in the Other Frame)* as Doppler data. Again, study it carefully and make it part of your intuition. Understanding the Doppler relationships now will make the rest of this chapter a piece of cake *(chocolate or vanilla?)*:

F6.9

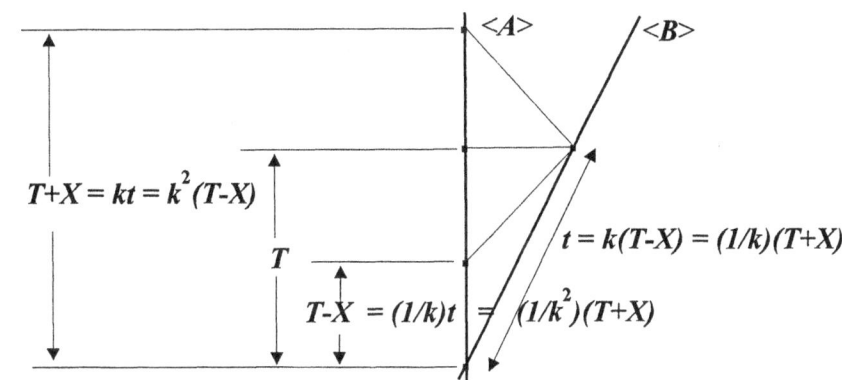

Doppler Constant from Velocity info

Using our basic Time and Distance data, *v* is *(as we already know)*: $v = X/T$

To make *k* and *v* mutually convertible, let us again map $<A>$'s T_S which is $T-X$ and show $<B>$'s t_R as Doppler-derived from $T-X$ which comes out as $t_R = k(T-X)$:

F6.10

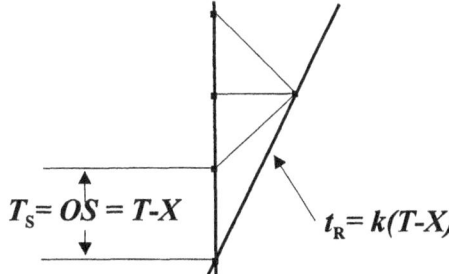

Now $k = t_R/T_S$, but knowing that in terms of <A>'s data, $t_R = T\sqrt{1-v^2}$ and $T_S = T - X$, therefore:

$$k = \frac{t_R}{T_S} \quad \text{and so} \quad k = \frac{T\sqrt{1-v^2}}{T-X}$$

By dividing both the Numerator and the Denominator in the last expression by T, we eliminate the T in the Numerator. In the Denominator, T is converted to 1 and X to X/T which is nothing other than v. Thus, we get:

$$k = \frac{\dfrac{T\sqrt{1-v^2}}{T}}{\dfrac{T-X}{T}} = \frac{\sqrt{1-v^2}}{1-v}$$

But $\sqrt{1-v^2} = (\sqrt{1-v})(\sqrt{1+v})$ and $1-v = (\sqrt{1-v})(\sqrt{1-v})$.

This allows us to rewrite the last equation as follows:

$$k = \frac{(\sqrt{1-v})(\sqrt{1+v})}{(\sqrt{1-v})(\sqrt{1-v})}$$

After canceling out identical square roots of $(1-v)$ in the Numerator and Denominator, we are left with the Doppler Constant, k derived from v:

$$k = \frac{\sqrt{1+v}}{\sqrt{1-v}} \quad \text{rewritten as} \quad k = \sqrt{\frac{1+v}{1-v}} \quad \text{so that} \quad k^2 = \frac{1+v}{1-v}$$

Deriving v from k

Now that we have k, we need to turn the tables around and express v in terms of k. For this, we'll take the k^2 equation above, develop it and rearrange everything so that v is finally brought to one side of the equation:

$$k^2(1-v) = 1+v$$

$$k^2 - k^2v = 1 + v$$

$$k^2 - 1 = k^2v + v$$

$$k^2 - 1 = v(k^2 + 1)$$

and so:

$$v = \frac{k^2 - 1}{k^2 + 1}$$

which is v obtained from k as ordered.

Deriving *v* from Doppler-converted coordinate data

If *v* could be derived from *k* using Minkowskian Coordinate values, we would have evidence that Doppler Optics is compatible with Minkowski's geometry. For this, we'll take *v* from our **ST** Map data and derive *k* from data converted to show Doppler relationships.

The Radar Probe data is always Minkowskian. For a review, let us show the basic Radar Probe again:

F6.11

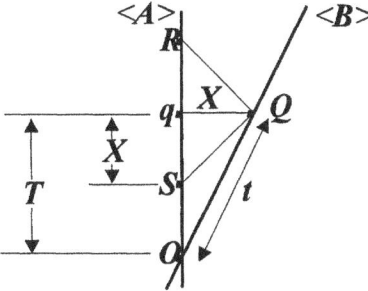

We start out with $v = X/T$ but knowing that $X = \frac{1}{2}$ *(OR - OS)* and $T = \frac{1}{2}$ *(OR + OS)*, we can restate the magnitude of *v* as:

$$v = \frac{\frac{1}{2}(OR - OS)}{\frac{1}{2}(OR + OS)}$$

After canceling the ½ in the Numerator and Denominator and restating the above equation completely in terms of *T* and *X* (as **OR** = T+X and **OS** = T- X), we get:

$$v = \frac{OR-OS}{OR+OS} = \frac{(T+X)-(T-X)}{(T+X)+(T-X)} = \frac{T+X-T+X}{T+X+T-X} = \frac{2X}{2T}$$

which gives us our familiar: $$v = \frac{X}{T}$$

Next, using Doppler-converted values for **OR** and **OS** *(which are Radar-Probe derived)*: $OR = k^2(T\text{-}R)$ and $OS = (T\text{-}S)$, we can re-write *X* and *T* in terms of Doppler data. First the *X:*

$$X = (OR - OS)$$

$$= k^2(T - X) - (T - X)$$

$$= (k^2T - k^2X - T + X)$$

$$= (k^2T - T) - (k^2X - X)$$

$$= T(k^2 - 1) - X(k^2 - 1)$$

$$= (T - X)(k^2 - 1)$$

Similarly, we can do the same for T:

$$T = (OR + OS)$$

$$= k^2(T - X) + (T - X)$$

$$= (k^2 T - k^2 X + T - X)$$

$$= (k^2 T + T) - (k^2 X + X)$$

$$= T(k^2 + 1) - X(k^2 + 1)$$

$$= (T - X)(k^2 + 1)$$

Combining the Doppler-converted X and T allows us to complete our Doppler detour and obtain v:

$$v = \frac{X}{T} = \frac{\frac{1}{2}(OR - OS)}{\frac{1}{2}(OR + OS)} = \frac{(T-X)(k^2-1)}{(T-X)(k^2+1)} \qquad \text{and finally } v = \frac{k^2 - 1}{k^2 + 1}$$

And so, from Minkowskian data converted to Doppler data, we have obtained v in terms of k.

Again checking Doppler data against Minkowski

There is another way we can show that Doppler Optics is mathematically in agreement with Minkowski's ST geometry. Let's set up a problem in concrete numerical values that can be solved both by Doppler method and by using the Minkowski's ST equation. If both give the same result, their equivalence *(of Map and Doppler methods)* would again be confirmed.

In the ensuing discussion, please, identify on the Figure 6.12 each item mentioned to keep yourself firmly oriented to the task at hand. To start, let us confirm that <A> and <B> are stationary to each other and 1 Lightsecond apart. Next, let <C> first move past <A> then also past <B> as mapped in Figure 6.12. Let <C>'s $v_C = 0.5$ by letting him travel 1 Lightsecond Distance in 2 seconds *(<A> and <B> stand 1 Lightsecond apart)*. At the exact Moment when <C> passes <A>, <C> sends a Signal toward <B>. That Signal reaches <B> exactly 1 second after the O-Moment according to <A>'s and <B>'s commonly shared Home Frame. At $v = 0.5c$, it takes 2 seconds for <C> to travel from <A> to <B>, so at the exact Moment when <C> passes <B>, <C> again sends a Signal this time back toward <A>. That Signal reaches <A> in 1 second *(after <C> passes <B>)*. On <A>'s WL, the Time from Event, O to the receipt of <C>'s Radar Signal, amounts to 3 seconds. For <B>, the Time from the receipt of the first <C>'s Signal to the Moment when <C> passes her (<B>) is 1 second. Thus $T_A = 3$ sec and $T_B = 1$ sec. The only thing we don't know is how much Time was shown on <C>'s clock from the Moment of <C> passing <A> to <C> passing <B>. This unknown Time interval on <C>'s clock is t_C which is to be calculated from information contained in the paragraph just concluded. Now, the ST Map:

F6.12

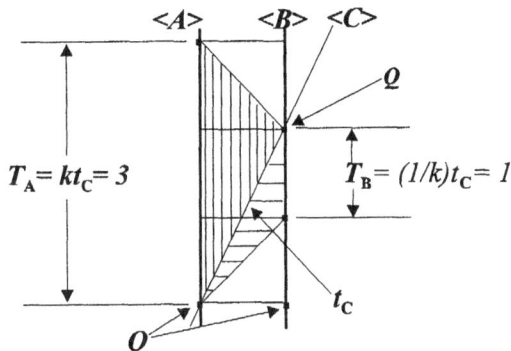

The Doppler Constant can be obtained in two different ways:

(1) As $k = \dfrac{T_A}{t_C}$ and also $= \dfrac{t_C}{T_B}$, so $\dfrac{T_A}{t_C} = \dfrac{t_C}{T_B}$, $t_C^2 = T_A T_B$ and, finally $t_C = \sqrt{T_A T_B}$

As $T_A = 3$ and $T_B = 1$, therefore: $t_C = \sqrt{3} = \mathbf{1.732}$

(2) Using the Minkowski's *ST* geography *(and equation)* we can also calculate t_C from *<B>*'s *OQ* counted from *<C>* passing *<A>* to passing *<B>*, altogether 2 seconds. With $v = 0.5$, $T_{OQ} = 2$ seconds and using the Lorentz Multiplier, we'll get:

$$t_C = T_{OQ}\sqrt{1-v^2} = 2\sqrt{1-0.5^2} = 2\sqrt{1-0.25} = 2\sqrt{0.75} = (\sqrt{4})(\sqrt{0.75} = \sqrt{4x0.75} = \sqrt{3}$$

Restated: $t_C = \sqrt{3} = \mathbf{1.732}$

which is exactly the same as calculated using the Doppler Constant. This proves that the two methods are equivalent.

Let us set up an identical Doppler Optics experiment *(as in Fig. 6.12)* with *<A>*, *<B>* and *<C>* but with *<C>* traveling at Velocity $v = 0.25$. Let the more adventurous readers calculate the time t_C from the mapped data using the Doppler method first, then by Minkowski's geometry:

F6.13

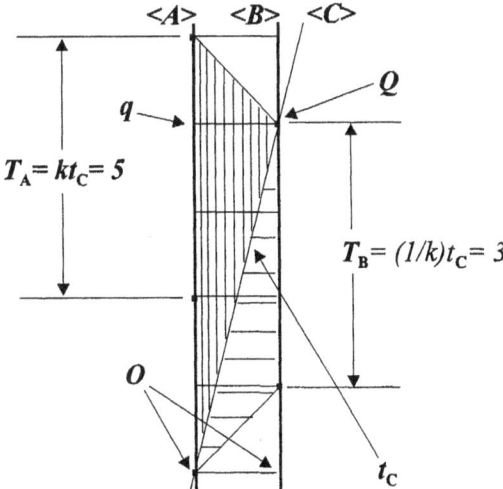

Doppler Optics and Time "Dilation"

For Time Dilation, we'll use the Twin Paradox Experiment of Chapter II, p.21, 22. In this, one of the twins is sent off to a high-speed trip. After traveling a given distance, he turns around and returns, all the time carrying an accurate clock. Let us ignore the technical difficulties with acceleration-deceleration and map his trip as accomplished, curved lines showing the accelerations and decelerations during the trip:

F6.14

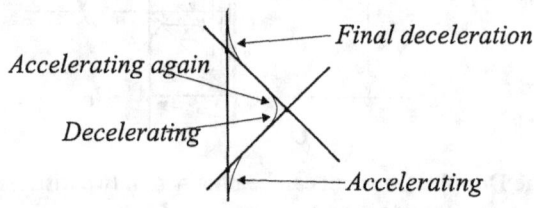

The trouble with the above plan is that there are no exact Moments when one Velocity changes instantly into another. The need to calculate the Time spent accelerating and decelerating makes our task very difficult. To eliminate the changing velocities from the Map, let's hire two astronauts, *<B>* and *<C>* to run a time-relay round trip each moving at equal, constant velocities relative to *<A>* but in opposite directions as shown on the next illustration. Note that there are three Events *(Moments)* when the astronauts pass each other in Space : Moment, **BA** when *<B>* passes *<A>*. Moment, **BC** when *<B>* passes *<C>* and Moment, **CA** when *<C>* passes *<A>*. Each Astronaut records exactly how much Time passed between crossings:

F6.15

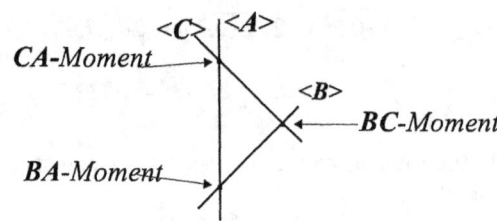

Now the results:

(1) Because the Time measured by *<B>* from Moment **BA** to **BC** is equal to the Time measured by *<C>* from **BC** to **CA**, those Time segments *(of <B> and <C>)* are equal: $t_B = t_C$.

(2) The Moment, **BC** occurs at half-way point on *A>*'s Time clock between Moments, **BA** and **CA** as established by Radar Probe. The entire experiment can now be mapped as follows:

F6.16

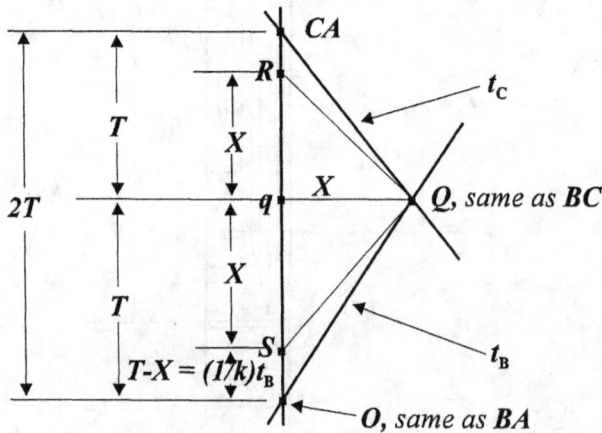

(3) Using Radar Probe terminology and Doppler Optics math, t of <*B*> equals t of <*C*> and can be calculated as follows:

$$t_B = t_C = k(OS) = k(T\text{-}X)$$

(4) The time interval for <*A*> between Moments, *BA* and *CA* equals $2T$.
(5) As $t_B = t_C$, and $t_B + t_C = 2t$ (*t without subscript, applies to both* <*B*> *and* <*C*>).
(6) The value of $2T$ can now be calculated from t and k:

As $2T = OR + R$-to-*CA*, and R-to-*CA* $= OS$, so $2T = OR + OS$. But, again, as $OR = kt$ and $OS = (1/k)\,t$, so

$$2T = kt + \frac{1}{k}t = t\left(k + \frac{1}{k}\right) \quad \text{Restated:} \quad 2T = t\left(k + \frac{1}{k}\right) \quad \text{from which we can}$$

develop $2t$ *(round-trip by* <*B*> *and* <*C*>*)* as "dilated" Time to match 2 *T*: $2T = t\dfrac{k^2+1}{k}$ and

$t = 2T\dfrac{k}{k^2+1}$ from which we get $2t$ *(t of* <*B*> *plus t of* <*C*>*)*: $2t = \dfrac{4kT}{k^2+1}$, the Time Dilation we have seen already demonstrated visually.

Now that we have the needed equation, *let 2T be 1 hour* and see the results for $2t$ given in numbers at different velocities expressed by k:

(1) With $k = 1.0$ *(v = 0 as there is no Motion)*, $2t = 2T = 1$ hour.
(2) With $k = 1.5$ *(v = 0.62)*, The Time <*A*> spent between Moments *AB* and *CA* while <*B*> and <*C*> clocked a total of *0.92 h.*
(3) With $k = 2$ *(v=0.77)*, $2t = 0.8$ h.
(4) With $k = 3$ *(v=0.89)*, $2t = 0.6$ h.
(5) With $k = 10$ *(v=0.99)*, $2t = 0.2$ h.
(6) With $k = 100$ *(v=0.9999)*, $2t = 0.02$ h.
(7) With $k = 1000$ *(v=0.99999999)*, $2t = 0.002$ h.

Of course, we can reverse the roles by fixing $2t$ to be 1 hour so that $2T = 1$ hr without Motion and calculate the progressively increasing values of T at progressively increasing Velocities, calculations left for you to do:

This exercise shows that the Velocities have to be considerable, much more than the $0.5c$ to make a good deal of difference in what <*A*> and <*B*> measure on their clocks.

Incidentally, the seemingly low value of $k = 1.0001$ represents the Velocity of the Earth in orbit around the Sun, a Velocity about 30 km per second or 0.01% of c, or $0.000,01c$, actually a rather slow Velocity, indeed.

Doppler Optics And Lorentz Multiplier
Next, let us derive our handy Lorentz Multiplier using Doppler relationships between x and X, both obtained from Radar Probe data but expressed as Doppler data.

F6.17

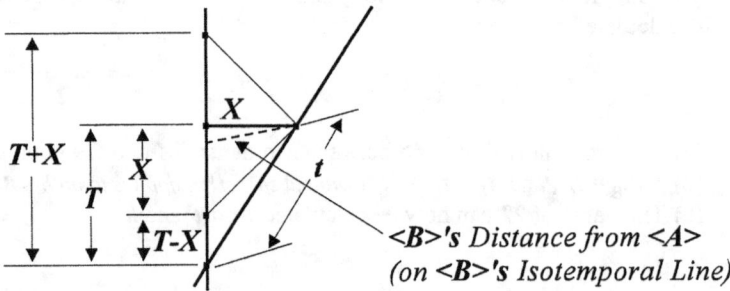

<B>'s Distance from <A>
(on <B>'s Isotemporal Line)

As shown above, in Fig. 6.17, $t = k(T-X) = (1/k)(T+X)$. As $v = X/T$ and also $v = x/t$, we can say that $X/T = x/t$, t here given as Doppler information.

Again, as $t = k(T-X)$ and we also know that $\dfrac{x}{t} = \dfrac{X}{T}$, we can state that *(with k replaced with its Doppler derivative from v):*

$$x = \frac{X}{T}t \quad \text{and also} \quad x = \frac{X}{T}k(T-X), \quad \text{and} \quad x = \frac{X}{T}k(T-X) = \frac{X}{T}\left(\sqrt{\frac{1+v}{1-v}}\right)(T-X)$$

To develop the last expression further, let us multiply the right side of the last equation *above (which is a **PRODUCT**, not a sum !!!)* first by T then by $1/T$, thus not changing the total value of the expression at all. In the development of the expression, the *(X/T)T*-part becomes X and the *(T-X) /T*-part becomes $1-v$:

$$x = \frac{X}{T}(T)\left(\sqrt{\frac{1+v}{1-v}}\right)(T-X)\left(\frac{1}{T}\right) = X\sqrt{\frac{1+v}{1-v}}(1-v)$$

As $1-v$ equals: $(\sqrt{1-v})(\sqrt{1-v})$, the initial result obtained can, again, be developed further:

$$x = X\left(\sqrt{\frac{1+v}{1-v}}\right)(1-v) = X\frac{(\sqrt{1+v})(\sqrt{1-v})(\sqrt{1-v})}{\sqrt{1-v}}$$

Canceling out identical expressions in the Numerator and the Denominator, we get:

$$x = X(\sqrt{1+v})(\sqrt{1-v}) \quad \text{which simplifies into:} \quad x = X\sqrt{1-v^2}$$

and that is what we have previously obtained through Minkowski's geometry.

A Doppler application in astronomy

Until now we have been studying Doppler Effect by the change in the Time interval between Signals. When dealing with a steady Signal, the duration of one wave *(or wave length)* is equivalent to the Time interval, T between the two separate, individual Signals. The math we worked out can be converted into frequency data of any steady Signal. Take notice, however, that there is an inverse relationship between wave length, λ and frequency, f :

$$f = \frac{1}{\lambda} \quad \text{and also} \quad \lambda = \frac{1}{f}$$

The Time Segment between two Signal bursts is related to wave length so that $\lambda = cT$, with $c = 1.0$, $\lambda = T$ and $f = 1/T = 1/\lambda$, restated: $f = 1/\lambda$.

In astronomy, Doppler change in the frequency of emitted or reflected Signal can be used as information capable of telling us:

(1) if the Source is stationary or moving
(2) what is the velocity in case of Motion
(3) is the Motion toward or away from us

Doppler Constant is the ratio of **received** intervals, t to original s **ent** intervals, T in receding mode. In astronomy, practically all Signal Sources are receding. If a steady Signal is sent and bounced back from a distant object, we'll have all the data we need to calculate the Velocity of the object. If, on the other hand, the received Signal is generated by a distant object then all we have is only the frequency of the received signal *(the original frequency or wave length is unknown)*, we are **NOT** in a position to calculate the Doppler Constant. It is fortunate, however, that the Light from distant stars contains patterns of frequency bands each pattern identified as having been emitted by known chemical elements. Their frequencies are known to us from accurate laboratory measurements *(without Motion of the Source)*. This information enables us to calculate k by dividing laboratory-determined frequency by the frequency of received signal. Changing the wave length information to frequency information gives us an "upside down" equation because of their inverse relationship:

$$k = \frac{\lambda_{received}}{\lambda_{known}} \quad \text{and also} \quad k = \frac{\dfrac{1}{f_{received}}}{\dfrac{1}{f_{known}}} = \frac{f_{known}}{f_{received}}$$

To demonstrate this Doppler application in actual numbers, let us take a band of frequencies with known lab value of, say, $f = 1,220,000$ and the received slower frequencies, $1,200,000$ from a celestial object *(using rounded out numbers for easier calculation)*. This gives us:

$$k = 1,220,000 / 1,200,000 = 1.0167$$

As $k > 1$, the Light Source must be moving away from us, and the recession Velocity is:

$$v = \frac{k^2 - 1}{k^2 + 1} = \frac{1.0167^2 - 1}{1.0167^2 + 1} = \frac{0.03368}{2.03368} = 0.01656c$$

which is $0.01656 \times 300,000 = 4,968$ km/sec, a mighty fast Velocity indeed but not unexpected for a star or galaxy located far away.

Doppler for measuring Lengths

In Chapter IV, we used a pair of Radar Signals to measure the Length of a moving Stick. The Signals had to be so finely timed that the first Signal, S_1 hit the distant End and the second Signal, S_2 hit the closer End of the Stick exactly at the same Moment, or simultaneously, reflecting both Signals back to <A> at R_1 and R_2 . We had to do this by a method that is technically very difficult because we hardly knew then how to use Doppler. To make measuring of Lengths possible in the first place, we had to attach to both Ends of the Stick reflecting targets because *the more distant End of the one-dimensional object, such as the Stick, could otherwise not be viewed or measured in Lineland.* Providing Radar access to the far End of the Stick and the solution of the measuring procedure were demonstrated in Chapter IV, Figure 4.6.

Now we are finally ready to see how the Length of a Stick can be measured with just one Radar Signal.

Finally, measuring a stationary Stick by Doppler

Let's place a Stick *(equipped with reflecting Targets)* first motionless at a given Distance from <A> *(one End pointing directly toward <A>)* as shown in Figure 6.18. Our *ST* Map includes the Worldline of <A> and the two Endpoint Worldlines of the Stick. *These Endpoint Worldlines are separated in Space by a Distance that defines the Length of the Stick.* Because the End Targets are at unequal Distances from the Observer, they *(the two attached End Targets)* are hit sequentially *(one after the other)* by a single Signal. *The Time Segment between the two returned reflections contains the Doppler-coded information about the Stick's Length.*

On the next Map, the Lightlines of the single Radar Signal and its two reflections are recorded by <A> at Moments, R_1 and R_2:

F6.18

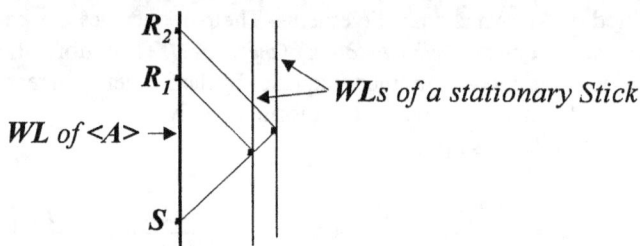

The upper part of the Map is redrawn below but zoomed in for more detail. Note that the stationary object's Length, X equals half of the Time between R_1 and R_2. The R_1 is produced by returning Signal reflected from the nearer End of the Stick and R_2 by the Signal reflected from the more distant End *(pointed out again just in case you missed it)*. Note that with non-moving Stick, the Doppler Constant is 1.0 and *NO* Doppler Effect is created:

F6.19

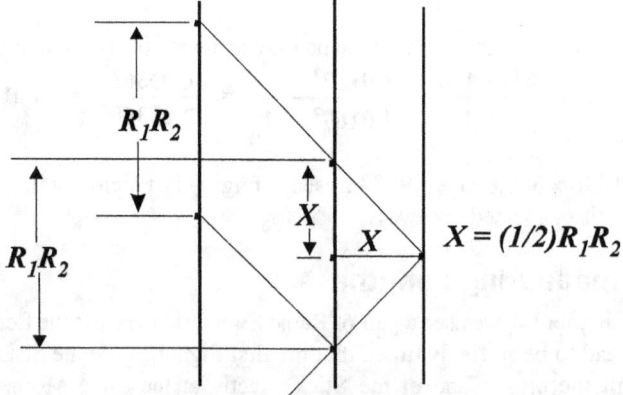

$$X = (1/2)R_1R_2$$

From the above Map, you should be able to see that X, the Length of the Stick, equals half of the R_1R_2 Time Segment. So there is no need to send two difficult-to-time Signals to hit both Ends of the Stick at exactly the same Moment as we had to do in Chapter IV *(Measuring Lengths)*. So, with a stationary Stick, $X = \frac{1}{2}(R_1R_2)$.

Radar-probed Stick in receding mode

But what about measuring a *MOVING* Stick? To see the problem, let us look at the Map of a Radar-probed moving Stick in *receding* mode:

F6.20

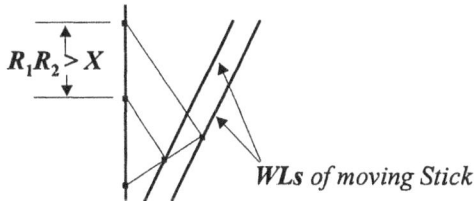

WLs of moving Stick

Here *the R_1R_2 Time interval is lengthened and is always longer than 2X: $R_1R_2 > 2X$.*

As you see, the lengthening of the R_1R_2 interval due to Doppler Effect is analogous to the Doppler Shift that lengthened or shortened the Time Segments between the two received Signals. *The Time Segment, therefore, contains the "Doppler-coded" Length information.* If we calculate the Lorentz-contracted Length, the "Dopplerized" Length information gets automatically decoded thereby hiding the Doppler Effect. In other words, when the invariant Rest Length, *x* or the Lorentz-contracted Simultaneous Length, *X* is calculated, our "mathemagic" makes them also *symmetrical*, exactly the same in approaching and receding Motion.

To see how to apply Doppler to moving Lengths, let us again map the last experiment in greater detail and label the key components:

F6.21

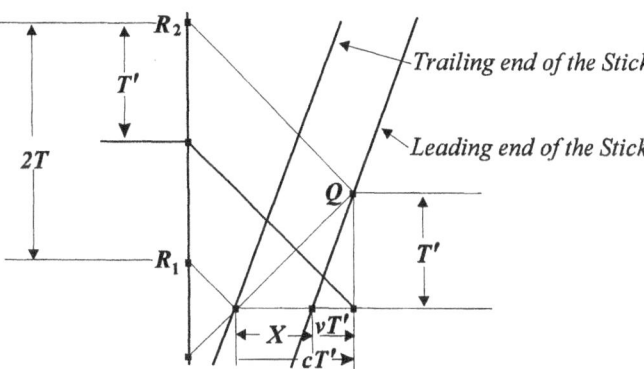

Trailing end of the Stick

Leading end of the Stick

You can see that the Tail End of the receding Stick is closer to the Observer and reflects the Radar Signal first, producing R_1. After this, the Signal travels not only the entire Length, X of the Stick but also the additional Distance needed to catch up, at Q, with the Nose End that has been moving away from the oncoming Signal. This additional Distance equals vT' which is the Stick's travel Time, T' times the Velocity, v. Now, look carefully at the Map. You should see that $T' = \frac{1}{2}(R_1R_2)$. This is a very important point as $2T'$ is something that <A> can measure firsthand. Now vT' added to X *(the Length of the Stick)* is the Distance traveled by the Radar Signal from Tail End to Nose End *(of the moving Stick)*:

$$cT' = X + vT'$$

The above information can be used to derive the value of X in terms of T':

$$cT' = X + vT' \quad \text{so}$$

$$X = cT' - vT' = T'(c - v). \text{ Restated: } X = (1-v)T'$$

But note that the X under consideration *is the Lorentz-contracted Length of the Stick and NOT its Resting Length, x (here not shown on the Map)*. Knowing the quantitative relationship between X and x and also how X was derived from T' we can say that:

$$X = x\sqrt{1-v^2} \text{ and also that } X = (1-v)\ T'$$

We can now equate both values of X shown above: $x\sqrt{1-v^2} = T'(1-v)$, so

$$x = T'\frac{\sqrt{1-v^2}}{1-v}$$

Again, by knowing that:

$$\sqrt{1-v^2} = (\sqrt{1-v})(\sqrt{1+v}) \text{ and that } (1-v) = (\sqrt{1-v})(\sqrt{1-v}),$$

we can calculate the Rest Length, x starting with:

$$x(\sqrt{1-v})(\sqrt{1+v}) = T'(\sqrt{1-v})(\sqrt{1-v})$$

and develop it further:

$$x = \frac{T'(1-v)}{\sqrt{1-v^2}} = \frac{T'(\sqrt{1-v})(\sqrt{1-v})}{(\sqrt{1-v})(\sqrt{1+v})}$$

By canceling the identical expressions, $(\sqrt{1-v})$ in the Numerator and Denominator, we finally obtain the value of x:

$$x = T'\sqrt{\frac{1-v}{1+v}}$$

But what do we have here? A big surprise!!! Recognizing that the Multiplier of T' is the inverse of k, the $1/k$, our result resolves beautifully into:

$$x = T'\frac{1}{\sqrt{\frac{1+v}{1-v}}} \quad \text{or} \quad x = \frac{1}{k}\ T'$$

meaning that x and T' are really Doppler-related.

Moreover, as $x = \frac{1}{k}T'$ so, interestingly, $T' = kx$.

Because we started out with the Length *(of the Stick)* that was Lorentz-contracted, we also need to acknowledge again that *Lorentz Contraction hides the Doppler Effect which is concealed (in-obviously contained) in the ST-geometric measurement of X.*

For finding the Doppler derivation of X, let's first recall that (1) $X = (\sqrt{1-v^2})\ x$, also $x = \frac{X}{\sqrt{1-v^2}}$, and that (2) $T' = \frac{1}{2} R_1 R_2$.

As T' was equal to kx *(that is $T' = kx$)* see page VI-16, and $x = \dfrac{X}{\sqrt{1-v^2}}$, so:

$$T' = k\frac{1}{\sqrt{1-v^2}}X \quad \text{and} \quad X = \frac{1}{k}(\sqrt{1-v^2})\,T' \quad \text{and} \quad X = \frac{1}{2k}(\sqrt{1-v^2})\,R_1R_2$$

In the same way, combining $1/k$ and $\sqrt{1-v^2}$, we get $X = (1-v)T'$

There is one more thing here that must be pointed out. In receding mode, the Single-Signal Radar Probe information, T' $(=\frac{1}{2}\,R_1R_2)$ becomes progressively longer with increasing Velocities. But as v approaches c, the Multiplier, $1-v$ becomes progressively smaller, thereby counteracting the resulting progressive R_1R_2 lengthening. So Lorentz Contraction still remains supreme and the Stick *(with its Ends simultaneous to $<A>$)* becomes progressively shortened by Lorentz Contraction. At Light Velocity, the calculated Length becomes zero because *the Multiplier, 1-v in the equation reduces to zero: 1-v = 1-c = 1-1 = 0 !* That *all simultaneous Lengths at Light Velocity contract to zero* we have already learned *(in chapter IV, Measuring Lengths)*. That *Lorentz Contraction is not modified by Doppler Effect* is new to us and that *Lorentz Contraction always wins,* is now clearly pointed out.

By the way, at Light Velocity, R_1R_2 becomes infinitely long which, when multiplied with zero *(1-v = 1-c = 0)*, gives, well, if not an unqualified zero *(infinity times zero)* for the calculated X value then at least an indefinite one. Practically speaking, anything like R_1R_2 at Velocity c simply cannot be measured.

Radar-probed Stick in approaching mode

As the Radar Signal travels toward the approaching Stick, it reaches its Nose End first and is reflected back to the Observer bringing information about the location of its Nose End at that particular Moment. The rest of the Signal continues traveling toward the Tail End which in the meantime keeps coming toward the oncoming Signal striking it just a Moment later. The Distance in Space between the two strikes *(at both Ends of the Stick)* is now *LESS than the (Lorentz-contracted) Length* of the Stick thereby shortening the R_1R_2 Time Segment between them. This remarkable result is shown next on the *ST* Map:

F6.22

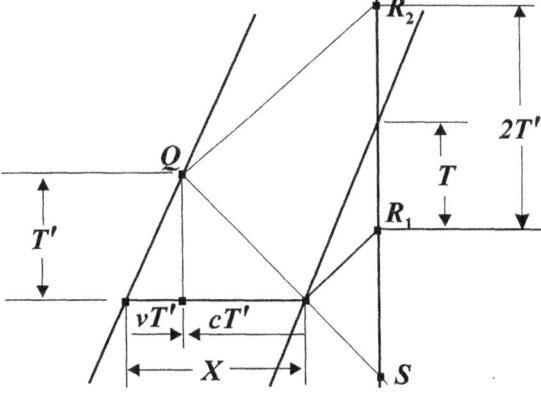

Here the *(Lorentz-contracted)* Length, $X = vT' + cT' = T'\,(v+c) = T'(1+v)$. Following the same procedure as used in the receding mode and adapted to the approaching mode, the Doppler-expressed Length of x comes out as:

$$x = \frac{1}{2}k(R_1 R_2) = kT' \text{ and:}$$

$$X = \frac{1}{2}(R_1 R_2)k(\sqrt{1-v^2}) = \frac{1}{2}(R_1 R_2)\left(\frac{\sqrt{1+v}}{\sqrt{1-v}}\right)(\sqrt{1-v})(\sqrt{1+v})$$

resolving into $X = \frac{1}{2}(R_1 R_2)(1+v) = (1+v)\,T'$

At almost the Velocity of Light, $1+v$ almost becomes $1+1 = 2$. But the Stick's Endpoint Worldlines also become closer and closer *(as observed by <A>)* with progressive Lorentz Contraction at increasing Velocities so that at the Velocity of Light, the $R_1 R_2$ Time Segment, T' becomes zero, keeping Lorentz Contraction again in full command at all Velocities.

Summing up single Signal Radar Probe results in Lineland

Doppler Constant can be used for measuring Lengths of moving objects *(such as the Stick)*. The primary data needed is **(1)** the $R_1 R_2$ Time Segment obtainable by means of a Single-Signal Radar Probe, and **(2)** the relative Velocity translatable into Doppler Constant. Here half of the $R_1 R_2$ Time segment equals T' *(that is $T'=(1/2)R_1 R_2$)*. From T' and v, we can calculate x and X *(as done above)*. The results feature the Doppler Constant, k. The values of X and x are now placed in full view for comparison:

In **RECEDING** mode:

$$x = \frac{1}{2k}(R_1 R_2) \text{ and also } = \frac{1}{k}T'$$

$$X = \frac{1}{2k}(R_1 R_2)\sqrt{1-v^2} \text{ and } = \frac{1}{k}(\sqrt{1-v^2})\,T' \text{ and also } = (1-v)T'$$

In **APPROACHING** mode:

$$x = \frac{1}{2}k(R_1 R_2) \text{ , and also } = kT'$$

$$X = \frac{1}{2}k(R_1 R_2)\sqrt{1-v^2} \text{ and } = k(\sqrt{1-v^2})\,T' \text{ and also } = (1+v)T'$$

By the way, the $R_1 R_2$ interval *(obtained by Radar Probe)* is an empiric, measured quantity and this in the receding mode becomes **lengthened.** In the approaching mode, it becomes **shortened.** When calculating the values of X and x, the Doppler influence on the $R_1 R_2$ interval becomes fully manifest by the k in the equation. Where X is calculated, the k combined with Lorentz Contraction becomes "$1-v$" in receding and "$1+v$" in approaching Motion. These Multipliers taking the place of k still counteract the Doppler Stretch and Compression of the $R_1 R_2$ Time Segments so that Lorentz Contraction always wins *(in the calculated results)*.

Doppler Effect (or is it Shift?) tansforms (deforms? distorts?)

As we next switch our attention from the calculated Doppler results to the ***direct results of seeing and recording the*** *(Dopplerized)* ***moving Stick photographically,*** we need to review the important lessons learned about the reach of our vision *(by Sightline!)* described in Chapter III, p.17 and 18 *(with Fig. 3.33)* restating our conclusions clearly before we start mapping the "new world" of ***Visual Doppler Effect:***

(1) ***Simultaneous happenings*** at different Distances ***can never be seen*** at the same Moment.

(2) What happens at ***different Distances*** can be seen at the ***same Moment*** *(by the Observer)* ***IF, and ONLY IF,*** these Events happen to be located on our Sightline and are ***sight-connected.***

Seeing the Stick approaching

Visual Doppler Effect is basically not different from the Doppler Effect which caused Time Segment *(interval)* changes between *(sequentially produced)* Signals received from a single moving Source. Visual Doppler Effect is obtained ***NOT*** by Radar Probe but ***by sampling ultra-short sections of continuous Light coming from the CONSTANTLY ILLUMINATED*** *moving* ***Stick in Lineland.*** With wide-open eyes *(or open camera shutter)*, the image received is blurred by superimposition of continuously overlapping images that make up what is best described as "Blue Streaks." But ***with an ultra-short shutter-speed camera, we can "freeze" the big long blur of the moving Stick into a single, sharp VISUAL IMAGE.*** This image, however, is now seen transformed by Motion in a way that is asymmetrical, different in approaching and receding modes within the 1-D Space that includes both the observed Stick and the Observer.

A sharp visual image of a constantly illuminated approaching Stick is formed by ***Light ARRIVING at exactly the same Moment*** to the camera's split-second-open shutter from both End–Targets of the Stick. But ***it takes longer for the Light from the more distant End to reach the camera compared with the Time for the Light arriving from the nearer End.*** Before the Light from the more distant End catches up with the nearer End of the *(approaching)* Stick, the latter keeps moving forward. Consequently, ***the Light Signal from the more distant End must have originated at an EARLIER MOMENT*** *(when it was located farther away)* compared with the Moment that produced the Light Signal from the nearer End. And this timing difference exists both in approaching as well as in receding directions of Motion in an asymmetrical fashion.

Because the information *(about the image)* of the Stick's Length is formed by the ***Light coming from its both Ends reaching the camera at the same Moment*** *(or simultaneously)*, so any *(fast)* Motion in Lineland will produce a change in the image ***making the Stick appear LONGER in approaching mode and SHORTER in receding mode.*** This result *(in Lineland)* is opposite to the result obtained by Radar Probe but the Doppler Visual Effect does ***NOT*** show the Lorentz-contracted *("measured")* Length, X that, incidentally, cannot be directly seen anyway under any condition and cannot be recorded by photographic means.

In other words, ***a Dopplerized visible-length image is formed simultaneously from non-simultaneously originated Signals*** while ***information about Lorentz-contracted, non-visible, simultaneous Length can be obtained*** *(only)* ***by calculation, not by photography.***

To illustrate the line-of-sight *(or Sightline)* argument, let us depict a fast moving Stick under constant illumination first in approaching mode. A side view shows more clearly how the visual image of its Length is "transformed" by Motion. The two successive positions of the Stick *(in the following illustration)* are Time-ordered and labeled accordingly **(1)** and **(2)**:

F6.23

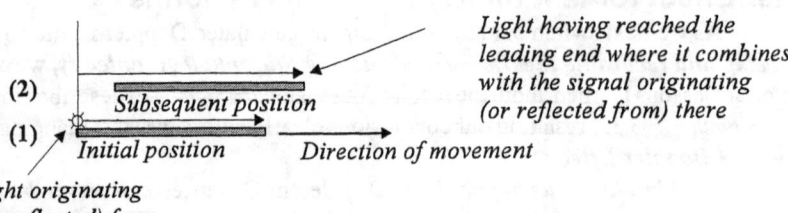

The total Path, cT' traveled by Light from the Tail End to the Nose End visualizes all the quantitative relationships:

F6.24

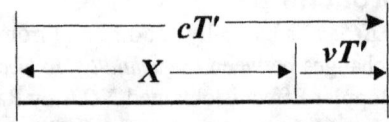

and can be written out mathematically as: $cT' = X + vT'$.

Here the Observer's position is in the direct path of the Stick. The right-pointing arrow-tipped lines of cT' and vT' *(in the above illustration)* mark the location where ***Light from the Tail End has caught up with the Nose End from where Light from both Ends travel together toward the Observer*** whose camera samples it during an extremely brief, single Moment thereby "freezing" a sharp image of the moving Stick on photographic film *(or in magnetic memory)* as its ***SIGHTLENGTH,*** $\mathscr{L}$ which ***appears longer than*** its *(non-visible)* Simultaneous *(Lorentz-contracted)* Length, X.

Now the ***ST*** Map of the above:

F6.25

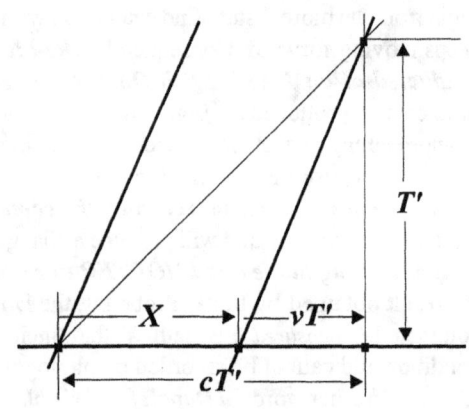

And the ***Sightlength,*** $\mathscr{L}$ of the Stick is:

$$\mathscr{L} = cT' = X + vT' \quad \text{where } cT' = T' \text{ (as } c = 1.0\text{)}.$$

Restated:

$$\mathscr{L} = T'$$

Solving for T':

$$T' - vT' = X \quad \text{and} \quad T'(1 - v) = X \quad \text{and also} \quad T' = X/(1-v)$$

So the Sightlength, $\mathscr{L} = T'$ and is also equal to $\dfrac{X}{1-v}$ in approaching mode.

Realizing that X on the Map is the Lorentz-contracted moving Length of x *(in <B>'s Home Frame)*, and that $X = x\sqrt{1-v^2}$, we can derive $\mathscr{L}$ in terms of the Proper *(or Resting)* Length

of the Stick:
$$\mathscr{L} = x\,\frac{\sqrt{1-v^2}}{1-v}$$

Developing the Multiplier of x :
$$\frac{\sqrt{1-v^2}}{1-v} = \frac{(\sqrt{1-v})(\sqrt{1+v})}{(\sqrt{1-v})(\sqrt{1-v})} = \frac{\sqrt{1+v}}{\sqrt{1-v}} = \sqrt{\frac{1+v}{1-v}}$$

Recalling that the above calculated multiplier of x is nothing other than the Doppler constant, k, we can state that:

$$\mathscr{L}_{\text{approaching}} = kx \quad \textit{(which is \textbf{Doppler-stretched} !)}$$

Thus the ***Doppler Effect transforms the Proper Length of the Stick into SIGHT-LENGTH*** by *(the amount of)* the Doppler constant proportional to the Velocity of Motion. And that's not all. Paradoxically *(so it seems)* **at the velocity of Light, the Stick's Sightlength becomes *infinite*** while its Lorentz-contracted *(but visually or photographically "unseen")*, **"measured" Length becomes *zero!!!*** To balance out the description developed, we need to add that the *(almost)* infinite Sightlength *(in 1-D Lineland)* could theoretically be seen *(or photographed)* only in approaching mode when the Stick travels *(almost)* at the Velocity of Light, both Ends of the Stick equipped with *(hopefully)* visible Targets.

Seeing the Stick receding

Next, let's look at the Visual Doppler Transformation in receding mode. Here, the Light from the more distant Nose End needs to travel less than the full Length of the Stick to reach the oncoming Tail End where it combines with the Light from it *(the Tail End)* on its way toward the Observer:

F6.26

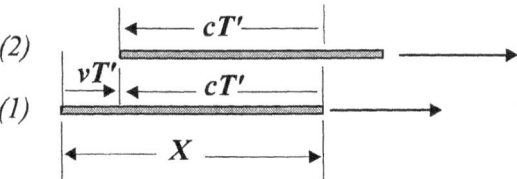

The Observer is positioned away from *(or behind)* the direction of Motion. The Light from the more distant Nose End moves toward the Tail End reaching it where the arrows of vT' and cT' in the illustration meet. There the Light from both Ends continue together in rearward direction and toward the Observer where it gets sampled by the super-fast-shutter-speed camera. The **Sightlength, $\mathscr{L}$** is now **shorter than** the Simultaneous *(Lorentz-contracted)* Length, X. The above scenario is mapped next:

F6.27

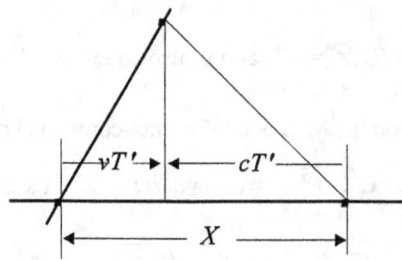

In mathematical terms:

$$X = vT' + cT'$$
$$= (c + v)T'$$
$$= (1 + v)T'$$

and so:

$$T' = \mathscr{L} = \frac{X}{1+v}$$

Changing X to its rest value, $x\sqrt{1-v^2}$ Also gives us the **Sightlength :**

$$\mathscr{L} = \frac{x\sqrt{1-v^2}}{1+v} = x\frac{(\sqrt{1-v})(\sqrt{1+v})}{(\sqrt{1+v})(\sqrt{1+v})} = x\sqrt{\frac{1-v}{1+v}} = \frac{1}{k}x$$

Restated: $\mathscr{L}_{receding} = \frac{1}{k}x$ **(Doppler-compressed !)**

In receding mode at the velocity of Light, the Sightlength becomes zero. Odd, odd, odd indeed*!!!* This is not because of Lorentz Contraction has somehow been increased. Sightlengths and Lorentz-contracted Lengths are entirely different things. *Lorentz-Contracted Lengths in Lineland CANNOT BE SEEN end-on (in line-of-sight) and are, therefore, invisible as such in 1-D Space.* Even the End-Targets of the Stick do not cooperate well with the best of intentions except, perhaps, in receding mode and at less than the Velocity of Light. More about it in Chapter XV, Sighting Blue Streaks.

Summing up Doppler Visual Transformation in Lineland

(1) In *constantly illuminated* scenarios, a camera with ultra-fast shutter speed gives us *SIGHTLENGTHS(Visual Lengths)* that are always *Doppler-stretched Proper Lengths (DOPPLER STRETCH) in approaching mode* and *Doppler-compressed Proper Lengths (DOPPLER COMPRESSION) in receding mode.*

(2) *Sharply delineated Visual Images in Lineland are obtainable* only through high-shutter-speed photography while *single-Signal Radar Probe never results in good visual images that can be seen directly.*

(3) *Lorentz Contraction* always neutralizes, or hides, Sightlengths by the *(Doppler-decoding)* math used but *its (Lorentz-contracted) results in Lineland are never visible as such or recordable by photography.*

(4) R_1R_2 obtained by a single-signal Radar Probe in receding mode *and* Sightlengths in approaching mode are *lengthened.*

(5) R_1R_2 obtained by Radar Probe in approaching mode *and* Sightlengths in receding mode are *shortened.*

Prelude to Sighting Blue Streaks

In Lineland, the Ends of a fast-moving Stick are separated from the Observer by Distances that are Unequal and this circumstance gives rise to the visible effects of Doppler Stretch or Doppler Compression. The Visual Doppler Effect and the other Doppler Effect hidden in the Radar Probe results are opposite in size both in approaching and receding modes of Motion *!!!* What we did not explore here was whether our moving Stick or any other object for that matter would produce a Doppler Effect at the exact Moment of Passing. This is a new topic for us because *at the exact Moment of of Passing without hitting the Observer, the Length of the object AND the Observer are no longer lined up, even by approximation, in our familiar 1-dimensional, line-of-sight direction of Motion.* At the Passing Moment, we are, therefore, no longer dealing with a 1-dimensional Space. The Space there is at least 2-dimensional and the Motion is best described as *Angular Motion.* The *Unequal Distances from the Observer are found not along the Length of the object but in line-of-sight direction across its THICKNESS.* The visual image there undergoes a change that is totally different from Doppler Stretch and Compression. In very-very fast Angular Motion, there are also other interesting kinematic effects that impact on visual images of all fast-moving objects and all these will be taken up in the last chapter of this book. Only then can we begin to appreciate the important sum-total of all the *Visual Transformations* that convert any familiar figure into a strange-indeed *UFO-like Blue Streak,* the proverbial super-fast object adopted for the title of this book.

VII - LORENTZ TRANSFORMATION

A Brief Definition

In Chapter II we used Lorentz Multiplier to calculate one Observer's measurements from those of another in situations where the Spacetime invariant "Distance" was marked by two Events: **(1)** the **O,** located at the crossing of their Worldlines and **(2)** the **Q,** located on one of the Observer's Worldline. In this chapter, the **Q-Event is no longer on anyone's Worldline** but is placed at various locations away from both Observers' **WL**s. With our earlier methods no longer adequate, the **Lorentz Transformation** we'll develop for new **Q**-Locations **enables us to derive X and T Segment measurements of one Observer** *(such as <A>)* **from those of another** *(such as <B> and vice versa)* **in a much wider variety of Spacetime situations** including those we have already mastered. In its capacity, **Lorentz Transformation can additionally reconcile unequal coordinate measurements of any Space or Time Segment** in a number of situations. The key component linking the two Systems such as <A>'s and <B>'s in our exercises is their shared relative Velocity, **v** along with the versatile Lorentz Multiplier, **β**. As usual, a **ST** Map gives us the best hint about how to proceed.

ST "Ladder" as a calculating guide

In the **ST** Map below *(Fig. 7.1A)*, <A> considers <B>'s Distance from him as X, while <B> considers <A>'s Distance from her as x. As you see, the X- and x-Maplines do not coincide. On <A>'s **WL,** <B>'s Q_B comes **before** <A>'s q_A and x is smaller than X, $(x < X)$. If, however *(Fig.7.1B)*, <B>'s Q_B is made to coincide with <A>'s q_A, the x' created in <A>'s Other Frame is longer than <A>'s X: $x' > X > x$. We can go up and down on both **WL**s adding Xs and xs between these Worldlines joining them together in tandem while producing an oddly shaped - Ladder that helps us find our mathematical steps in Spacetime. To bestow legitimacy to this odd **SPACETIME LADDER**, let us show it on the Map *(Fig. 7.1C)*:

F7.1

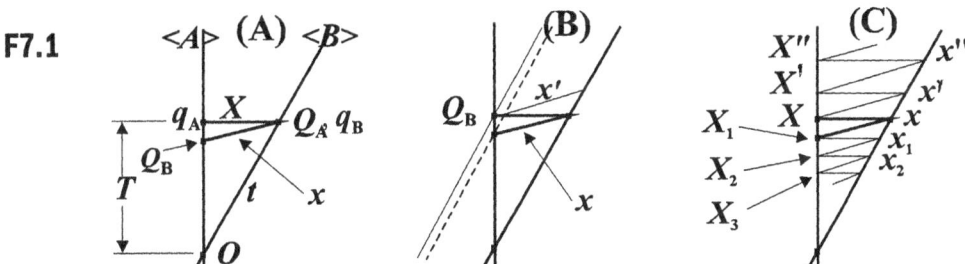

The initial two rungs of this Ladder are shown by heavier lines. The upward going X and x rungs are labeled with apostrophes and downward rungs with subscripts. The rungs get shorter and shorter going down and longer and longer going up thereby pointing to **a systematic change in the Distance between <A> and <B>.**

The Distance as seen by <A> and <B> *(and shown above on Figure 7.1A)* are well known to us from Chapter II. The X and x are connected end-to-end on <B>'s **WL**. But we can see another Distance, x' going upward, connected to X on <A>'s **WL** as shown on Figure 7.1B. The mathematical relationship between X and x' we have never explicitly explored. This we have to do now because we have to know how they are *(quantitatively)* related to each other before we can proceed confidently with the rest of this chapter.

While mathematical derivations are most instructive, we can also start out simply by **looking at the ST Maps.** This gives us intuitive orientation before tackling the math. So let just **(1)** draw the main outline of the Figure 7.1A first then **(2)** present the X and x totally from <B>'s point of view where the X is shown going upward, followed by (3) Fig. 7.1B again from <A>'s point of view but now with the new upward-going x' added:

F7.2

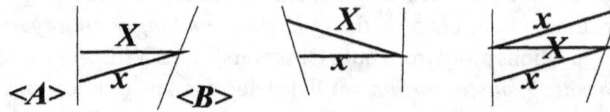

Recall, please, that the original x is smaller than X which in <B>'s Home Frame shows an upward slant exactly as x' on Fig. 7.1B. The quantitative difference between X and x is not altered in this different view and we know that it is given by the inverse Lorentz Multiplier: $X = x / \sqrt{1-v^2}$. We also know that $x = X\sqrt{1-v^2}$.

This visual demonstration of X and x makes it apparent that *x' and X must bear the same Qualitative relationships to each other as do X and x* thereby sparing us a more laborious, roundabout mathematical way of reaching the same conclusion:

$$x' = \frac{X}{\sqrt{1-v^2}} \quad \text{and} \quad X = x'\sqrt{1-v^2}$$

Thus we can say that adding another Distance going up *tandem* always increases the preceding Distance by the inverse of the Lorentz Multiplier and adding another tandem Distance going down we decrease the preceding Distance by the Lorentz Multiplier. In other words: *Each step up, Xs alternating with xs, increases their measurements by the inverse of the same Multiplier and each step down decreases their measurements by the same Multiplier.*

Unless you are used to working with inverse values, it is better to stick to the simpler *(not the inverse)* expression. That is why we'll be always using the "straight" expression, β as our working ***Lorentz Multiplier.*** Reminded of this switch again, we are finally ready to provide the ***ST*** Ladder with measurements using β and its inverse as conversion tools. To make the comparative values of the rungs also visually obvious, we'll show β with the appropriate powers added for <A> and for <B>:

F7.3

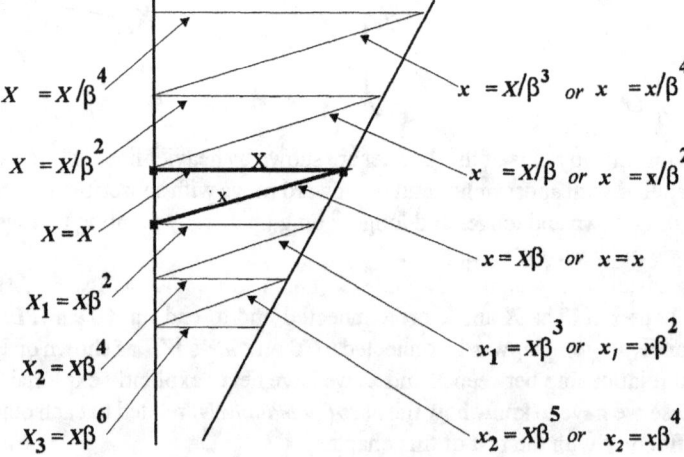

There's *v* in other *(unsuspected)* places

In our usual calculations, we have always multiplied Velocity, v with T or t, the Time Segment, to obtain the Distance, X traveled by <B> *(as observed by <A>)* or x traveled by <A>

(as observed by <B>) during the amount of Time indicated by *T* or *t*. Looking at the *ST* Ladder produced by the tandem additions of Distances, we can see that we have also divided <A>'s and <B>'s *WL*s into smaller Time Segments located between the tandem Distance attachments. For that reason, we need to know what these derivations *(of the small sections, T' and t')* amount to in order to add this knowledge to our Special Relativity tool kit for use in this and the next chapter. This addition may seem unimportant but we must not leave this matter wrapped in mystery.

　　　To show how these Time sub-Segments are derived, let's take a *ST* Ladder with just two rungs, X_1 and X_2 and one *x* in between. The small *T* Segment, *T'* equals $T_1\text{-}T_2$:

F7.4a

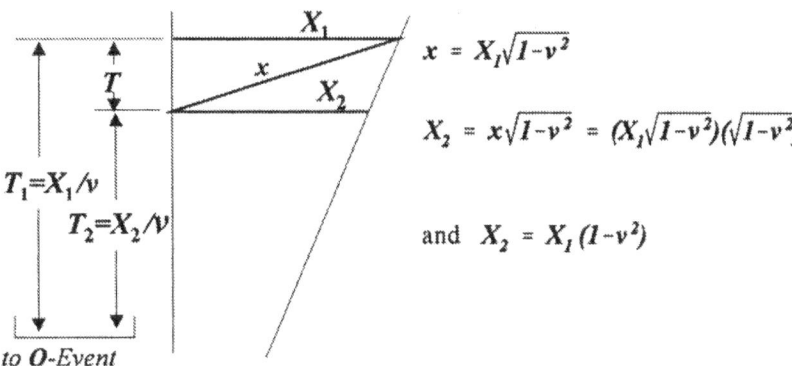

$$x = X_1\sqrt{1-v^2}$$

$$X_2 = x\sqrt{1-v^2} = (X_1\sqrt{1-v^2})(\sqrt{1-v^2})$$

$$\text{and} \quad X_2 = X_1(1-v^2)$$

The small *T* Segment, *T'* between the attachments of X_1 and X_2 to <A>'s *WL* can now be derived:

$$T' = T_1\text{-}T_2 = \frac{X_1}{v} - \frac{X_2}{v} = \frac{X_1 - X_1(1-v^2)}{v} .$$ Developing it further gives us:

$$T' = \frac{X_1 - X_1(1-v^2)}{v} = \frac{X_1-(X_1-v^2X_1)}{v} = \frac{X_1-X_1+v^2X_1}{v} = \frac{v^2X_1}{v} = \frac{vvX_1}{v} .$$

The final result can be restated: $\quad T' = vX_1$

By the same reasoning, the small *t* Segment on <B>'s *WL* can also be derived:

F7.4 b

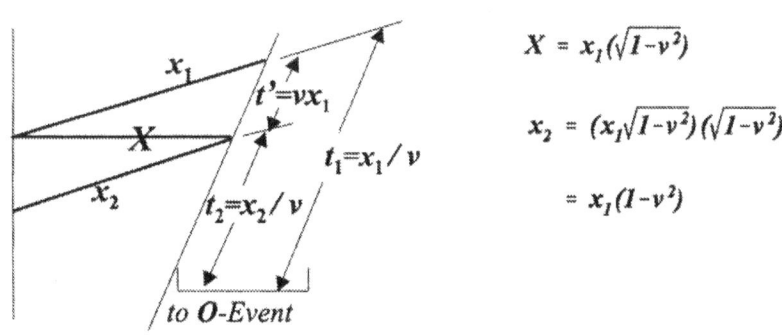

$$X = x_1(\sqrt{1-v^2})$$

$$x_2 = (x_1\sqrt{1-v^2})(\sqrt{1-v^2})$$

$$= x_1(1-v^2)$$

$$t' = t_1\text{-}t_2 = \frac{x_1}{v} - \frac{x_1(1-v^2)}{v} = vx_1 \quad \textit{(derivation process abbreviated)}$$

Now we have two ways v can link Observers $<A>$ and $<B>$. Not only does v represent the angle on the Map **(1)** where two **WL**s intersect *(at the O-Moment)* but now we have also added the angle **(2)** where the two corresponding **ITL**s *(X and x)* intersect. The first case is known to us but the second case, now added, needs additional comments:

(1) The size of the angle formed by intersecting **ITL**s is visually and quantitatively equal to the v formed by the intersecting **WL**s and is the same for both $<A>$ and $<B>$. The visual v does not become distorted by Motion as do the Map lines standing for x and t *(of $<B>$)* that always become **LONGITUDINALLY distorted** by Motion and do **NOT** remain true to scale as explained in Chapter III.

(2) The X, x with either T' or t' form a triangle analogous to that formed by T, t along with either X or x. These two triangles are, therefore, proportional to each other so that $T/t = X/x$.

(3) To obtain T' or t' we multiply X or x with v: $vX=T'$ and $vx=t'$. Each conversion has to be within the same Frame of Reference and the v has to be the angle between the Distances, X and x *(on the ITLs)*.

Now we can label the small Time Segments on the sides of the **ST** Ladder between the rung attachments:

F7.5

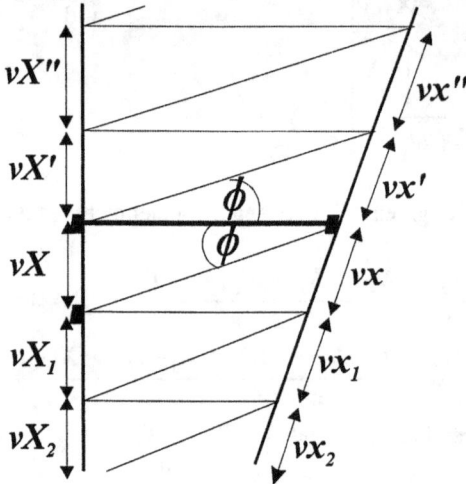

In standard situations, two Isotemporal Lines (**ITL**s) crossed by the two **WL**s mark off Distances *(X and x)* on those **ITL**s for both Observers. Each Distance, X or x when multiplied by β, the Lorentz Multiplier, gives us the Distance directly below, Xs alternating with xs. And all Distances between $<A>$ and $<B>$ bear a constant relationship to the those above and below. But in addition to calculating all rungs of the Ladder, we have also formalized the calculation of these small Time Segments on both **WL**s between the rungs *(as shown above on Fig. 7.5)*: $vX'' > vX' > vX >$ etc. and $vx'' > vx' > vx >$ etc..

Note again that wherever two **ITL**s cross, the angle formed between them is exactly the same as the one formed by the two **WL**s. So this angle, regardless of location, always stands for velocity, v between the two Observers. With v we can, therefore, derive any unknown side of a triangle such as Oq_AQ_A, Q_Bq_BO or $q_AQ_BQ_A$ *(see Fig. 7.1A)*, where the two sides, each in different Frames of Reference, form the angle, Φ. Two of the sides forming the triangle are **EITHER T** and **t OR X** and **x. All derivations can be carried out only for the third side within the same Frame,** Home or Other: $vT=X$, $vt=x$, $vX = T'$ and $vx = t'$ *(see above Fig. 7.5)*.

The derivation of a Time Segment from a Distance Segment is, therefore, new to us but it is a very handy trick for working out various examples of Lorentz Transformation in this chapter and we'll see that it has other applications as well.

Setting up sample cases of Lorentz Transformation

Let's pick an Event, Q positioned *AWAY from the WLs of both <A> and <B>*. The location shown here is quite different from those in all previous examples as *none of the Time Segments (from O to Q) here are Proper Times and cannot be measured directly by either Observer.* The Time *(T or t)* as well as the Distance *(X or x)* measurements from O to Q are now all *Coordinate Projections*. Each Observer can still calculate Coordinate Time *(O to Q)* and Distance *(WL to Q)* values from those obtained by the other Observer but the math is now more complicated. *The keys to conversion here are still v, the naked Velocity and β, the Lorentz Multiplier.* The three representative locations we need to consider are shown next:

F7.6

These unusual locations ordain that $x \neq (\sqrt{1-v^2})X$, $t \neq (\sqrt{1-v^2})T$, $x \neq vt$ and $X \neq vT$!!! Without knowing this, we can be easily thrown off the track. The Radar Probe lines will be omitted to reduce clutter and only X, x, T & t Map lines are shown below.

The *ST Map method* helps us visualize our task. Let's first draw the Map with Q_1 and mark off <A>'s and <B>'s Time *(T and t)* and Distance *(X and x)* Segments. Note that these *(Time and Distance)* Segments on the Map are *NOT* those we are used in previous exercises. Also their magnitudes no longer transform simply by means of the Lorentz Multiplier:

F7. 7

Case Q_1: x and t from X and T, that is <A>-to-<B> Transformation

Except for a general orientation, the initial Map, Fig. 7.7 above, does not really tell us how to proceed. The Coordinate Time Segments of <A> and <B> are located on their *WLs* and the Coordinate Space Segments are on those *ITLs* that join their *WLs* to Q, the main location of interest. In addition, the Time Segments are also shorter than Oq on both *WLs*. Some of the useful information on the Map can be visualized better by drawing parallel lines to the basic Space and Time Segments already mapped and labeling these Segments or their parallel equivalents with their derivations *(from measurements obtained by one or the other Observer)*. Note that all <A>'s lines and their parallels are square while <B>'s lines are tilted. The angles, ϕ standing for v are found between all crossings of their *WLs* as well as of their parallels, and between all crossings of their *ITLs* and of their parallels.

We'll be using vT, vt, the Lorentz's Multiplier, β and the odd vX and vx expressions to negotiate between the two Frames of Reference. All initial derivations are carried out in longhand. Subsequently, it may suffice to provide only starting positions and end results, leaving the in-between steps to those able to do equation crunching on their own.

On our first Case Map, the values of the useful line segments have already been derived *(with help from v or β)* from those that can be *(directly or indirectly)* measured firsthand and are provided also visually without verbal persuasion. Each derivation should be scrutinized well so that nothing about them remains a puzzle as everything later in this chapter depends on these basic steps. Note again that in this part of our study, $X \neq vT$ and $x \neq vt$ *!!!* and that both $<A>$'s and $<B>$'s *ITL*s go from their *WL*s to *Q:*

F7.8

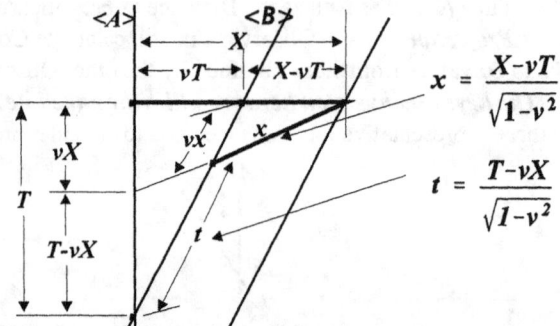

As Q here is not located on anyone's *WL*, both X and T are needed for the derivation of x *(from X and vT)* and of t *(from T and vX)*. Note that X in this first case is divided into two parts: vT and $X-vT$ *(the vT located between $<A>$'s WL and $<B>$'s WL)*. The $X - vT$ is what is left over from X after vT is subtracted from it. Also note, please, that X is not equal to vT and x is not equal to vt *!!!* That the remnant $(X-vT)$ is smaller than x is due to Lorentz Contraction. Now that all this is clear *(or is it?)*, both x and t have assembled themselves simply from the derived components already labeled:

$$x = \frac{X - vT}{\sqrt{1-v^2}} \qquad t = \frac{T - vX}{\sqrt{1-v^2}}$$

A Doppler detour, case Q_1: $<A>$-to-$<B>$ Transformation

It is instructive to see that Doppler method gives exactly the same results as obtained by the *ST* Map method. Again, let's look at the general approach shown on the Map below *(Fig. 7.9)*: $<A>$ sends a Radar Signal, S_A toward *(and past)* $<B>$ where it becomes also $<B>$'s own Radar Probe to Q. The Events, O and Q are the same to both $<A>$ and $<B>$. Events that are separate for $<A>$ and $<B>$ are q_A and q_B as well as R_A and R_B. The X, x, T and t needed for the conversions *(Transformations)* are labeled accordingly. Review the details of the Map below before proceeding:

F7.9

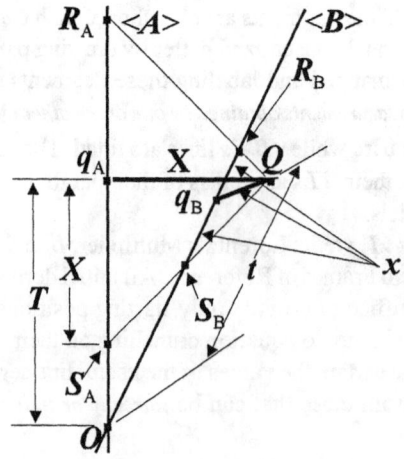

Next, let us define Time and Distance Segments for <A> and <B>. For <A>, X is $q_A Q$ and T is $O q_A$. For <B>, $x = q_B Q$ and $t = O q_B$.

Now we need to redraw the Map as a Doppler Map, leaving out some of the less useful Doppler derivations to reduce clutter but showing the Doppler relationships of $t+x$ to $T+X$ and $t-x$ to $T-X$:

F7.10

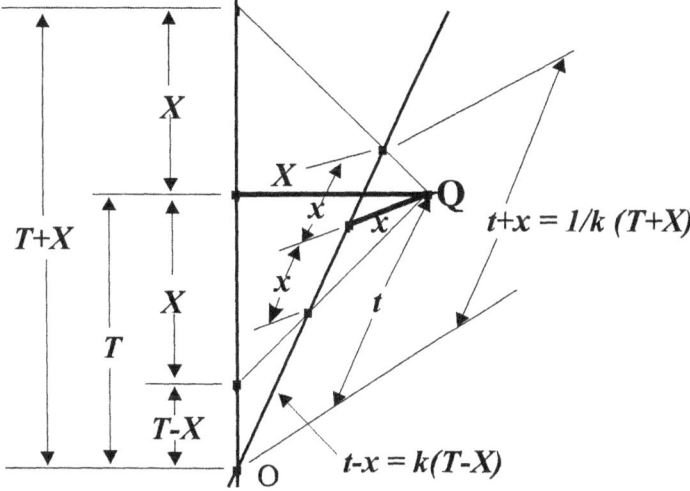

Because here we have two unknowns, x and t, we need two basic equations, (1) and (2) to provide the information needed for t and x derivations.

(1) $t - x = k(T - X)$; from this we get (a) $x = t - k(T-X)$ and (b) $t = k(T-X) + x$

(2) $t + x = (1/k)(T + X)$, and from this we develop: (a) $x = (1/k)(T+X) - t$ and (b) $t = (1/k)(T+X) - x$

Each equation can be used to derive both x and t. But these individually do not contain all the information needed. To bring in the extra information, let's take the derivation of x from the first equation in which the t is then substituted with its derivative from the second equation. In the combination, we still have a long and tedious equation-crunching of Doppler relationships between <A>'s and <B>'s measurements. To help you see the full reasoning process, all the steps will be done in longhand.

The initial equation for x from the above $(t-x)$ equation (#1 above) is:

$$x = t - k(T-X)$$

In it we'll substitute for the unknown t its derived value, $t = (1/k)(T+X) - x$ taken from the "$t+x$" equation (#2):

$$x = \left(\frac{1}{k}(T+X) - x \right) - k(T-X) = \frac{T+X}{k} - x - k(T-X)$$

Next, let's bring x from the right side of the equation over to the left side and then divide both sides by 2:

$$2x = \frac{1}{k}(T+X) - k(T-X), \text{ and } x = \frac{T+X}{2k} - \frac{k(T-X)}{2} = \frac{T+X-k^2(T-X)}{2k}$$

Substituting the v-derived values of k, $k = \sqrt{\dfrac{1+v}{1-v}}$, into the equation we get:

$$x = \frac{(T+X) - \left(\frac{\sqrt{1+v}}{\sqrt{1-v}}\right)\left(\frac{\sqrt{1+v}}{\sqrt{1-v}}\right)(T-X)}{2\left(\frac{\sqrt{1+v}}{\sqrt{1-v}}\right)}$$

The Numerator, N is a rather complex expression but it can be simplified by separating it from the Denominator:

$$N = (T+X) - \left(\frac{\sqrt{1+v}}{\sqrt{1-v}}\right)\left(\frac{\sqrt{1+v}}{\sqrt{1-v}}\right)(T-X)$$

and developing it further:

$$N = (T+X) - \left(\frac{1+v}{1-v}\right)(T-X) = \frac{(T+X)(1-v) - (1+v)(T-X)}{1-v}$$

and further:

$$N = \frac{T+X-vT-vX-(T-X+vT-vX)}{1-v} = \frac{(T-vT+X-vX-(T-X+vT-vX)}{1-v}$$
$$= \frac{T+X-vT-vX-T+X-vT+vX}{1-v} = \frac{T-vT+X-vX-T+X-vT+vX}{1-v}$$

The expressions, T and vX in the last equation come with both + and - signs which, therefore, cancel out, leaving us with something more manageable:

$$N = \frac{2X - 2vT}{1-v} = \frac{2(X-vT)}{1-v}$$

Combining the now simplified Numerator with the simple Denominator:

$$x = \frac{\dfrac{2(X-vT)}{1-v}}{2\dfrac{\sqrt{1+v}}{\sqrt{1-v}}}$$ and, again, developing it further:

$$x = \frac{2(X-vT)(\sqrt{1-v})}{(1-v)\ 2(\sqrt{1+v})} = \frac{(X-vT)(\sqrt{1-v})}{(\sqrt{1-v})(\sqrt{1-v})(\sqrt{1+v})}$$

In the last equation, we can cancel out the identical $\sqrt{1-v}$ expressions in the Numerator and Denominator and multiply the left-over $\sqrt{1-v}$ and $\sqrt{1+v}$ in the Denominator. The final result now pops out:

$$x = \frac{X - vT}{\sqrt{1-v^2}}$$

which is exactly the same we obtained earlier through the simpler *ST* Map method.

The value of t can be obtained by the Doppler method in a similarly laborious way. Let this be done by the more adventurous reader. What we have accomplished here is sufficient for our purpose: to show that Doppler method is equivalent to standard ST Map method.

Case Q_1: X and T from x and t, that is <B>-to-<A> Transformation

Let us first map the basic Spacetime landscape, provide it with parallel lines as done previously but label only those *(derived)* values we can use for the conversion:

F7.11

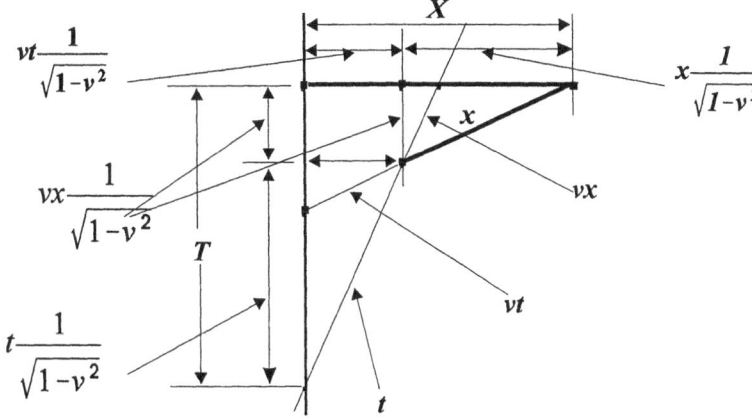

The component values can be taken right off the Map and the straightforward derivations of X and T can be simply written down:

$$T = t\frac{1}{\sqrt{1-v^2}} + vx\frac{1}{\sqrt{1-v^2}} = \frac{t + vx}{\sqrt{1-v^2}} \quad \text{and}$$

$$X = x\frac{1}{\sqrt{1-v^2}} + vt\frac{1}{\sqrt{1-v^2}} = \frac{x + vt}{\sqrt{1-v^2}}$$

Before proceeding with the Q in other positions, let us rewrite all the results obtained so far for the purpose of visually comparing them:

<A> to <B>	<B> to <A>
$x = \dfrac{X - vT}{\sqrt{1-v^2}}$	$X = \dfrac{x + vt}{\sqrt{1-v^2}}$
$t = \dfrac{T - vX}{\sqrt{1-v^2}}$	$T = \dfrac{t + vx}{\sqrt{1-v^2}}$

You can see that consistent with inequalities, $T > t$ and $X > x$, the Numerator in the <A>-to-<B> Transformation features a minus sign and the Numerator in the <B>-to-<A> Transformation features a plus sign. The equations are otherwise identical.

Case Q_3 solved by simply re-defining it

When considering the position of Event, Q_3 on <A>'s side rather than on <B>'s side on the Map, we should recognize right away that the situation is exactly like the Case Q_1 where Q_1 was on <B>'s side and T was greater than t, $T > t$. Everything with Q_3 works out exactly the same way as with Q_1 except that <A>'s and <B>'s positions are now **SWITCHED** and the inequalities are also switched: $T < t$, $X < x$.

<A>'s Home Frame   <B>'s Home Frame

F7.12

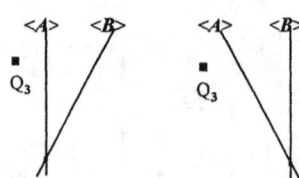

Note that in the final results of Q_3 Transformations, the + and - signs are reversed compared with those of Case Q_1:

$$x = \frac{X + vT}{\sqrt{1-v^2}} \qquad X = \frac{x - vt}{\sqrt{1-v^2}}$$

$$t = \frac{T + vX}{\sqrt{1-v^2}} \qquad T = \frac{t - vx}{\sqrt{1-v^2}}$$

Case Q_2: x and t from X and T, that is <A>-to-<B> Transformation

The Q in this case is located **BETWEEN** <A> and <B>, a situation quite different from what we had with Q_1 and Q_3. As usual, the ST Map places everything into full view.

To allow better display of important detail, we'll place Q fairly close to <A>'s **WL**. The <A>'s and <B>'s **ITL**s go from their own positions right to Q between them. On the right half of the illustration, the various components on the Map are labeled with their derivations:

F7.13

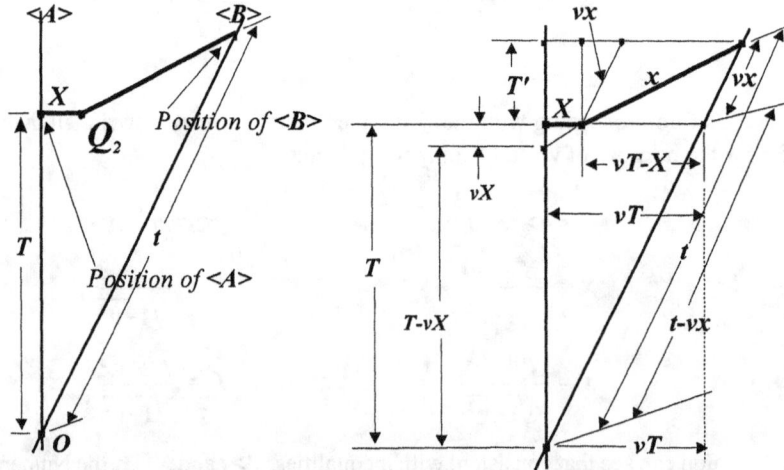

From the Map, you should be able to puzzle out directly that :

$$x = (vT - X)\frac{1}{\sqrt{1-v^2}} = \frac{vT-X}{\sqrt{1-v^2}}. \text{ Restated: } x = \frac{vT-X}{\sqrt{1-v^2}}$$

We can begin to build the equation for t using components already labeled:

$T = (t-vx)/\sqrt{1-v^2}$ from which we isolate t: $t-vx = T\sqrt{1-v^2}$, and $t = T\sqrt{1-v^2} + vx$

Next, we substitute x with its already derived value *(shown above)* and develop it:

$$t = T\sqrt{1-v^2} + v\left(\frac{vT-X}{\sqrt{1-v^2}}\right) = T\sqrt{1-v^2} + \frac{v^2T-vX}{\sqrt{1-v^2}}$$

$$= \frac{T(\sqrt{1-v^2})(\sqrt{1-v^2}) + v^2T-vX}{\sqrt{1-v^2}} = \frac{T(1-v^2) + v^2T-vX}{\sqrt{1-v^2}}$$

which can now be finalized after canceling out $-v^2T$ and $+v^2T$:

$$t = \frac{T-v^2T+v^2T-vX}{\sqrt{1-v^2}} = \frac{T-vX}{\sqrt{1-v^2}}.$$

Restated: $t = \dfrac{T-vX}{\sqrt{1-v^2}}$

Case Q$_2$: *X* and *T* from *x* and *t,* that is <*B*>-to-<*A*> Transformation

Starting again with the *ST* Map where useful parts on the Map are labeled with their derivations from x and t:

F7.14

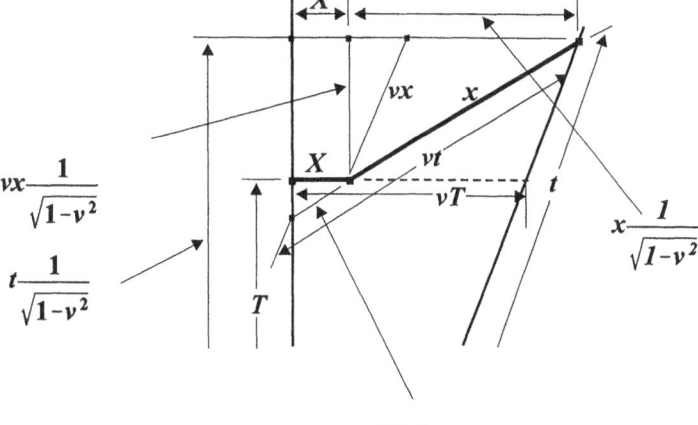

The *X* from *x* and *t*:

$$X + x\frac{1}{\sqrt{1-v^2}} = \frac{vt}{\sqrt{1-v^2}}, \text{ so } X = \frac{vt}{\sqrt{1-v^2}} - \frac{x}{\sqrt{1-v^2}}$$

which abbreviates into:

$$X = \frac{vt - x}{\sqrt{1-v^2}}$$

The T from t and x:

$$T = t\,\frac{1}{\sqrt{1-v^2}} - \frac{vx}{\sqrt{1-v^2}}$$ which abbreviates into: $$T = \frac{t - vx}{\sqrt{1-v^2}}$$

Let's rewrite all Case Q_2 derivations for comparison. Note that all Numerators feature the minus (-) sign and it is only in the X and x derivations where we have vT and vt instead of vX and vx:

$$x = \frac{vT - X}{\sqrt{1-v^2}} \qquad\qquad X = \frac{vt - x}{\sqrt{1-v^2}}$$

$$t = \frac{T - vX}{\sqrt{1-v^2}} \qquad\qquad T = \frac{t - vx}{\sqrt{1-v^2}}$$

Case Q_2: <A>-to-<B> Transformation by Doppler

The Doppler method again is complicated use but the procedure is similar to that of Case Q_1 we already solved, so let's repeat the same steps also here.

First, <A> sends a Radar Probe to Q where part of it is reflected back to <A> at R_A but the rest *(of this Signal)* continues toward <B>'s WL, arriving there at R_B. Next, <B> sends a Radar Probe to hit Q at exactly the at the same Moment when <A>'s Probe Signal arrives there from the other side *(a technically difficult but theoretically a feasible arrangement)*. From Q, <B>'s Radar Signal continues toward <A> along with reflected <A>'s Signal and is also reflected back to <B> arriving there at R_B along with <A>'s Signal. Look carefully at the Map and puzzle out the steps described:

F7.15

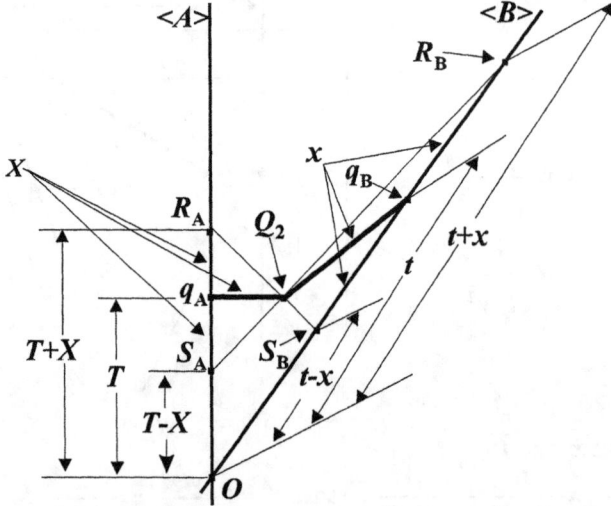

You should see that "$T - X$" goes with $t+x$ and "$T + X$" goes with t-x. The two starting equations are given first:

(1) $t + x = k(T - X)$ and
(2) $t - x = (1/k) T + X$

In this case, we'll solve for t first. Let's isolate it from the first equation:

$$t = k(T - X) - x$$

The x is extracted from the second equation:

$$x = t - (1/k)(T + X).$$

Substituting x derivation from the second equation into the t-derivation from the first and developing it, we get the complete value of t after a tedious march of steps:

$$t = k(T-X) - x = k(T-X) - \left(t - \frac{1}{k}(T+X)\right) = k(T-X) - t + \frac{1}{k}(T+X)$$

$$2t = k(T-X) + \frac{1}{k}(T+X) = \frac{k^2(T-X)}{k} + \frac{T+X}{k} = \frac{k^2(T-X) + (T+X)}{k}$$

restated: $2t = \dfrac{k^2(T-X) + (T+X)}{k}$ and $t = \dfrac{k^2(T-X) + (T+X)}{2k}$

Moreover, as $k = \sqrt{\dfrac{1+v}{1-v}}$, so $t = \dfrac{\left(\dfrac{1+v}{1-v}\right)(T-X) + (T+X)}{2\sqrt{\dfrac{1+v}{1-v}}}$

Developing the Numerator of the above equation:

$$N = \frac{(1+v)(T-X) + (1-v)(T+X)}{1-v}$$

$$= \frac{T-X+vT-vX+T+X-vT-vX}{1-v}$$

$$= \frac{2T - 2vX}{1-v}$$

$$= \frac{2(T-vX)}{1-v}$$

Combining Numerator with the simpler Denominator:

$$t = \left(\frac{2(T-vX)}{1-v}\right)\left(\frac{1}{2\sqrt{\dfrac{1+v}{1-v}}}\right) = \frac{2(T-vX)}{(1-v)\,2\,\dfrac{\sqrt{1+v}}{\sqrt{1-v}}} = \frac{2(T-vX)(\sqrt{1-v})}{(1-v)2\sqrt{1+v}}$$

$$= \frac{(T-vX)(\sqrt{1-v})}{(\sqrt{1-v})(\sqrt{1-v})(\sqrt{1+v})}$$

$$= \frac{T-vX}{(\sqrt{1-v})(\sqrt{1+v})}$$

With calculation completed and restated:

$$t = \frac{T - vX}{\sqrt{1 - v^2}}$$

To solve for x, we need to start out from the same two equations:

(1) $t + x = k(T - X)$
(2) $t - x = (1/k)\,T + X$

which gives us:

$x = k(T - X) - t$ from the first equation, and
$t = (1/k)(T + X) + x$ from the second equation.

Substituting the derivation of t taken from the second equation for the t in the first one, we get:

$$x = k(T-X) - \left(\frac{1}{k}(T+X) + x \right) = k(T-X) - \frac{1}{k}(T+X) - x$$

In the above development, the lone x needs to be brought to the left side of the (=) mark:

$$2x = k(T-X) - \frac{1}{k}(T+X)$$

Going through the development of this expression in the manner of the previous example *(but its laborious detail not shown)*, we get:

$$x = \frac{vT - X}{\sqrt{1 - v^2}}$$

Case Q$_2$: <B>-to-<A> Transformation by Doppler

We start out with exactly the same two Doppler equations given above for the <A>-to-<B> Transformation. From these the T and X can be obtained the same long-winded way. The final results are given below and are placed side-by-side along with the previous <A>-to-<B> Transformations in order to compare them:

$$t = \frac{T - vX}{\sqrt{1 - v^2}} \qquad T = \frac{t - vx}{\sqrt{1 - v^2}}$$

$$x = \frac{vT - X}{\sqrt{1 - v^2}} \qquad X = \frac{vt - x}{\sqrt{1 - v^2}}$$

Note that the equations in both <A>to<B> and <B>to<A> Transformations show the minus sign in all the Numerators just as we saw them in equations obtained by the Map method.

A sideline of Lorentz Transformation

It is interesting that Lorentz Transformation can be used also in the really simple case where the Q-Event is located in the familiar place on <B>'s *WL* and the information available consists of <A>'s coordinate measurements, X and T. Using Lorentz Transformation instead of Minkowski's *ST* equation, we'll be retracing our more familiar conversions of Distance and Time from one Frame of Reference to the other but doing it by this newly acquired Transformation method. Obtaining the same results will also prove then that the Lorentz Transformation is true and valid also in simpler, more elementary situations.

As usual, we'll start out with the basic *ST* Map for orientation and to point out that *OQ* is the *ST* "Distance" *(or the Invariant)* under consideration:

F7.16

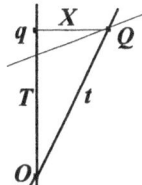

It is instructive to re-define the measured *(and calculated)* Space and Time Segments for both <*A*> and <*B*> in order to point out which of those are Proper and which are Coordinate measurements.

X_P - <*A*>'s own Proper Distance, Oq_A. This is a ***ZERO quantity by definition*** because <*A*> in his Home Frame is stationary and not going anywhere at all in Space. That *Oq* can also be a Space Segment was mentioned in Chapter II *(Axioms and Operations)* but this definition was not important enough then and was, therefore, not pointed out further.

X_C - <*A*>'s Coordinate Distance, $q_A Q$ which is **(1)** *Q*'s as well as **(2)** <*B*>'s Distance at the *Q*-Moment from him, *Q* being on <*B*>'s *WL*.

T_P - <*A*>'s Proper Time, Oq_A. This is what <*A*> is reading directly off his clock located in his Home Frame.

T_C - <*A*>'s Coordinate Time measurement, Oq_A of <*B*>'s Proper Time, *OQ* *(primarily <B>'s Proper Time, t in her Home Frame)* in <*A*>'s Other Frame.

The same terms in <*B*>'s Home Frame, x_P, x_C, t_P and t_C are defined in exactly the same way as Proper and Coordinate Segments in <*A*>'s Home Frame.

As *OQ* is located entirely within *B*>'s Home Frame and represents both Proper Space and Proper Time Segments for her, so <*A*>'s measurements have to be Projections of the same *OQ* on <*A*>'s Space and Time Coordinates, thus Coordinate measurements. These definitions again bring home the Observer-Specificity of all observations/measurements and the Minkowski's *ST* equation as further specified for $OQ : X_C^2 - T_C^2 = x_P^2 - t_P^2$. Now back to Transformations.

(A) x_P from X_C and T_C

The <*A*>-to-<*B*> Transformation of *OQ* from <*A*>'s Home Frame to <*B*>'s Home Frame *(<A>'s Other Frame)* is done by taking the Lorentz Transformation equation for *x* *(from coordinate values of X and T)* but in the Numerator, we'll either use *vT* in place of *X* or *X* in place of *vT*, both cases resulting in zero value for the Numerator: *vT-vT*=0, *X-X=0:*

$$x = \frac{X - vT}{\sqrt{1-v^2}} = \frac{X - X}{\sqrt{1-v^2}} \ , \ \text{giving us} \ x = 0, \text{ the } x \text{ is in} <B>\text{'s Home Frame,} <A>\text{'s}$$

Other Frame.

Restated: *x = 0*

If you think that *x* for *B*> should be <*A*>'s Distance from <*B*>, you are mistaken. The Invariant here is *OQ* with both *O* and *Q* on <*B*>'s *WL*. The *qQ* Map line produces <*B*>'s Distance from <A> but when OQ is considered within <B>'s own Home Frame *(as well as in <A>'s Other Frame)*, it becomes equal to **ZERO!** Remember, <B> views herself to be stationary, non-moving. Therefore, for her, going from O to Q is *not going anywhere at all in Space* so that she herself remains in the same place *(or location)* within her own Home Frame *!!!*

(B) t_P from T_C and X_C

To get<A>-*to*-<B> conversion of $OQ = t$, we can do it in two different ways: either **(a)** Replacing v in the vX in the Numerator with X/T, or **(b)** Replacing X in the vX with vT. We'll take up each of these choices individually:

(a) Replacing v with X/T:

$$t = \frac{T - vX}{\sqrt{1-v^2}} = \frac{T - \frac{X}{T}X}{\sqrt{1-v^2}} = \frac{\frac{T^2 - X^2}{T}}{\sqrt{1-v^2}} = \frac{T^2 - X^2}{T\sqrt{1-v^2}}$$

From Axioms an Operations, we learned that $\sqrt{T^2-X^2} = T\sqrt{1-v^2}$, and as

$$T^2 - X^2 = (T\sqrt{1-v^2})^2 = T^2(1-v^2), \quad \text{thus also:}$$

$$t = \frac{T^2(1-v^2)}{T\sqrt{1-v^2}} \quad \text{which is developed further into} \quad t = T\frac{(\sqrt{1-v^2})(\sqrt{1-v^2})}{\sqrt{1-v^2}}$$

resulting in: $t = T\sqrt{1-v^2}$

and that is what we expected in the first place*!!!*

(b) Replacing X with vT:

$$t = \frac{T - vX}{\sqrt{1-v^2}} = \frac{T - v(vT)}{\sqrt{1-v^2}} = \frac{T-v^2T}{\sqrt{1-v^2}} = \frac{T(1-v^2)}{\sqrt{1-v^2}} = \frac{T(\sqrt{1-v^2})(\sqrt{1-v^2})}{\sqrt{1-v^2}}$$

so: $t = T\sqrt{1-v^2}$, which is again what we already know

Bringing vX and vx into better focus where Q is on <B>'s WL

Now that we have seen vX and vx quite a few times in our equations, it would be useful to point out their roles in the different conversions *(or Transformatiions)* where these new derivations *(vX and vx)* can be properly applied. Recall that < A>-to-<B> conversion of X *(that is Coordinate measurement, $q_A Q$ in <A>'s Home Frame)* into x *(in <B>'s Home Frame, <A>'s Other Frame)* resulted in x *(that is equivqlent to x_P)* $= OQ = 0$, which is <B>'s own Space Segment *(x = 0)*. But OQ is also <B>'s *(Proper)* Time Segment, $t=T\sqrt{1-v^2}$. Next: OQ, if considered in <A>'s Home Frame, becomes both **(1)** his Coordinate Space measurement, $X_C=q_A Q$ and also **(2)** his Coordinate Time, $T_C=Oq_A$. These can be transformed to x_1 or t_1, or to x_2 and t_2. Both Trans-formations clearly illustrate both vX and vx:

F7.17

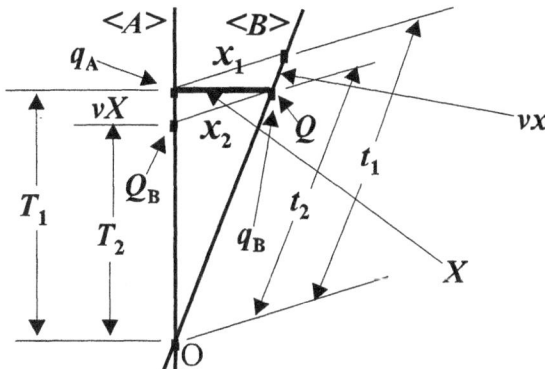

Again, *OQ* can be either *<B>*'s *(Proper)* Space Segment, *x=0* or *<B>*'s *(Proper)* Time Segment, *t>0*. In *<A>*'s Home Frame, however, it becomes by projection *(!) <A>*'s Coordinate Space Segment, *X (that is* $q_A Q$) and also his Coordinate Time Segment, *T (or* Oq_A). With all these components identified, the *vX* and *vx*. Segments have now been fully identified and labeled.

It is in these last examples that we can finally see that the odd *vX* and *vx* participate in the Lorentz Transformation by ***adding to or subtracting from the length of a given T or t (* Oq_A *or* B*) Time Segment.* Another look at the above Fig. 7.17 should tell you that (1) T_1-*vX* transforms into t_2, (2) T_2+*vX* transforms into t_1, (3) t_1-*vx* transforms into T_2 and (4) t_2+*vx* transforms into T_1. Using the plus (+) or minus (-) sign before the odd *vX* or *vx* in the Trans-formation equations merely indicates that the resulting Transformation is going up or down the Spacetime Ladder. So if the Numerator contains the (+) sign, the Transformation always gives us a greater, that is, ***up-on-the-Ladder*** result and the (-) sign gives us *(in the Other Frame of Reference)* a lower or ***down-on-the-Ladder*** result.

Internal consistency checkup of the Lorentz-Transformation-based derivations

Let us take the equation for *X* as initially derived:

$$X = \frac{x + vt}{\sqrt{1-v^2}}$$

and replace *x* and *t* with their Lorentz-derived values. Where does this lead us? Let's see:

$$X = \frac{\dfrac{X - vT}{\sqrt{1-v^2}} + v\dfrac{T - vX}{\sqrt{1-v^2}}}{\sqrt{1-v^2}}$$

Let's develop the above expression. If the Lorentz Transformation is consistent and valid, we should get an equality that is also true:

$$X = \frac{\left(\dfrac{X-vT}{\sqrt{1-v^2}}\right) + v\left(\dfrac{T-vX}{\sqrt{1-v^2}}\right)}{\sqrt{1-v^2}} = \frac{\dfrac{X-vT+vT-v^2X}{\sqrt{1-v^2}}}{\sqrt{1-v^2}} = \frac{X-vT+vT-v^2X}{(\sqrt{1-v^2})(\sqrt{1-v^2})} = \frac{X - v^2X}{1-v^2}$$

So: $X = \dfrac{X(1-v^2)}{1-v^2} = X$

Restated: with $X = X$, internal consistency has been demonstrated! Similar proof can be construed *(but is not developed here)* for T, giving us $T = T$.

While $X \neq x$ and $T \neq t$ by a small or wide margin, Lorentz Transformation merely changes one measurement into one obtainable in the Other Frame of Reference and vice versa, so that both can be considered *"equal under Lorentz Transformation."*

One more sideline of Lorentz Transformation

Now let's ask: do those mutually converted or transformed Space and Time Segments observe Minkowski's Spacetime Equation, $X^2 - T^2 = x^2 - t^2$ as they should? Let's see. We'll replace $x^2 - t^2$ with their Lorentz-transformed equivalents and develop the resulting equation:

$$
\begin{aligned}
X^2 - T^2 &= \left(\frac{X-vT}{\sqrt{1-v^2}}\right)^2 - \left(\frac{T-vX}{\sqrt{1-v^2}}\right)^2 \\[2mm]
&= \frac{X^2-2vT+v^2T^2-(T^2-2vX+v^2X^2)}{1-v^2} \\[2mm]
&= \frac{X^2-2vT+v^2T^2-T^2+2vX-v^2X^2}{1-v^2} \\[2mm]
&= \frac{(X^2-v^2X^2) - (T^2-v^2T^2)}{1-v^2} \\[2mm]
&= \frac{X^2(1-v^2) - T^2(1-v^2)}{1-v^2} \\[2mm]
&= X^2 - T^2
\end{aligned}
$$

which shows that Minkowski's fundamental Spacetime Invariant relationship is observed by Lorentz Transformation.

By the way, if in the Transformation equation, the Numerator contains a (+) sign instead of the (-) sign $(x + vt)$, the above result is still duplicated $(x^2 - t^2 = x^2 - t^2)$ because the minus sign before the t-derivation *(in the Numerator)* becomes inserted into that position in the final step of calculation. Try it out on your own this time.

A final comment

We are merely belaboring the obvious: *A single transformation equation does NOT apply to all the possible locations of Q.* Each Spacetime case demands individual analysis to develop the mathematical statement specific to that particular situation. Standard undergraduate physics textbooks are in the habit of giving just one solution thereby creating an erroneous impression that one equation is all we ever need. By working through a selection of different Spacetime examples, we have found out more truth about Lorentz Transformation than we perhaps wished but certainly more than found in all publications for the student and the forever curious lay reader.

Lorentz Transformation is applicable to a great many Spacetime situations. We'll be using it next when we add velocities. We have also seen that *ST* Mapping and Doppler methods give the same results. In discussions about relativistic physics, we often see the expression: such-and-such results are *"invariant under Lorentz transformation."* This means exactly what it says. Translated: *a given quantity can measure out differently in different Inertial Systems but as the Invariant* (calculated from d ata o btained i n a ny F rame of R eference) *is always the same in all Systems, two different* (coordinate or proper) *measurements* (obtained in different Inertial Systems) *can be in perfect agreement even when their numerical results differ* (by a narrow or wide margin as the case may be) *!!!*

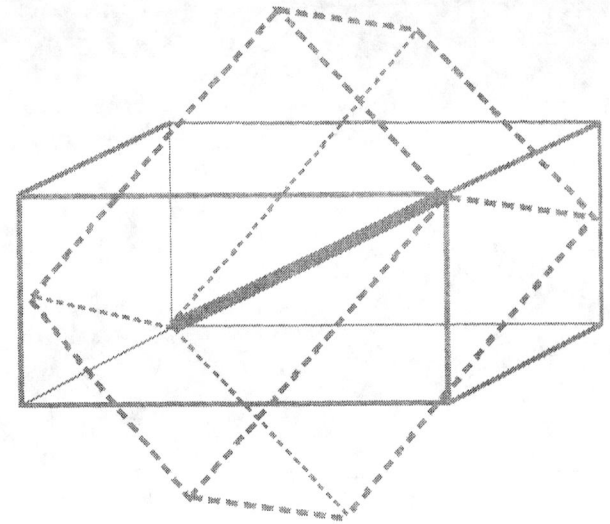

VIII - ADDING VELOCITIES

By Classical Mechanics

In classical mechanics the addition of Velocities is a simple arithmetical addition of the given Velocities. For example, if a railroad train is passing a bystander who is very close to the tracks and at the same time a stowaway on top of a railroad car is running in the forward direction, the Velocity of the stowaway relative to the bystander is simply the sum of the two Velocities: the ground Velocity of the train and the Velocity of the stowaway relative to the train.

Let v be the ground Velocity of the train and u the Velocity of a stowaway on top of the train. The sum of these Velocities, w would then be $w = v + u$. If $v = 50$ mi/hr and $u = 5$ mi/hr, then $w = v + u = 55$ mi/hr.

If, however, the stowaway was running towards the rear of the train at a Velocity that is $u = -5$ mi/hr *(minus sign indicating reverse direction)* then $w = v + u = 50 - 5 = 45$ mi/hr. In all cases *we either add or subtract Velocities* depending upon the directions involved.

Another example should further illustrate the adding and subtracting of Velocities. Let the stowaway be inside the train, leaning out of an open window and, armed with a bow and arrow, would shoot at the bystander while approaching him. Having missed his target, he would then turn around and shoot another arrow at the bystander after passing him.

Let the velocity of the arrow be u and let $u = 60$ mi/hr. In this case the first arrow would whizz past the bystander at $w = 50 + 60 = 110$ mi/hr and the second arrow at $w = -50 + 60 = 10$ mi/hr, (-) sign indicating opposite direction.

If, instead of shooting the second arrow, he would throw a rock at him at $u = -50$ mi/hr, the rock would then have a ground-based Velocity of $w = -50 + 50 = 0$ mi/hr. To the bystander, the rock would have *NO* horizontal Motion at all and would simply fall straight down from the hand that threw it*!*

Mapping Velocities to be added

We can plot the two different Velocities on a *ST* Map. The bystander would be <*A*>, the train <*B*> and the stowaway <*C*>. Because the actual Velocities plotted on the Map would not differ visually from non-Motion, <*B*>'s and <*C*>'s *WL*s would hardly deviate from <*A*>'s vertical *WL,* we need to employ really fast Velocities more like those employed in preceding chapters to demonstrate graphically the relativistic features of the addition process. Let us, therefore, map three astronauts: <*A*> standing still, <*B*> and <*C*> in real fast Motion, <*C*> faster than <*B*>. To review our mapping procedure, let us give each their turn being stationary:

F8.1

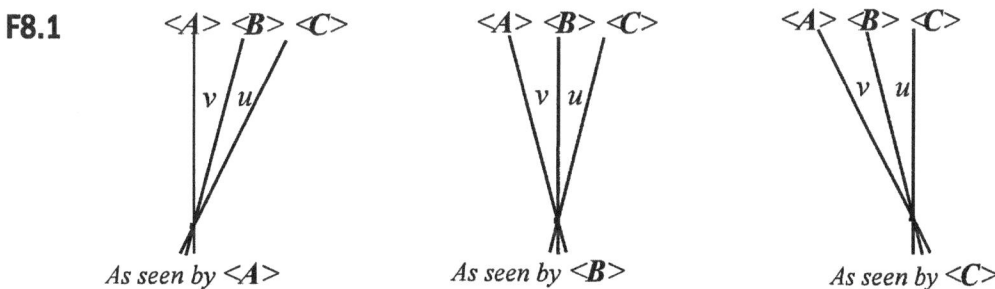

As seen by <*A*>   As seen by <*B*>   As seen by <*C*>

Here, the *AB* Velocity is v and can be measured directly by <*A*> as well as by <*B*>. The *BC* Velocity is u and can be measured directly by < *B*> and by <*C*> but, due to an accidental limitation of <*A*>'s equipment, it cannot be measured by <*A*>. Note that <*B*> can measure both *v* and *u* but anyone *(<A>, <B> or <C>)* with information of both *v* and *u* should be able to calculate *w*, the sum of *v* and *u*. The question is how.

Why simple addition wouldn't do ?

You may legitimately ask why on Earth *(or in Space)* would a simple arithmetical addition of Velocities not work*?* The impossibility is inherent in the requirement that in any Distance *(or Length)* measurement, the positions of both Ends of the Distance *(or Length)* must be taken *SIMULTANEOUSLY*, that is, *AT THE SAME MOMENT !* The trouble comes from the inability of two Observers in Motion to agree on which two Moments are simultaneous. This disagreement is shown by their non-coinciding Isotemporal Lines on the Map*!* And the only Moment they can regard as simultaneous *(to both of them)* is their near-collision Passing Moment *(the O-Event that lets them synchronize their clocks).*

Still, the *O* and *Q* are the two Spacetime Events that *(as usual)* mark off a Spacetime Invariant. And remember that the Invariant is an Absolute and of equal magnitude to all Observers. Unfortunately, the equality is obtained only in the calculated 4-D quantity that in practically all cases cannot be measured directly. What can be measured are its *PROJECTIONS* upon Space and Time Coordinates and it is in those measurements where differences show up if there is Motion relative to the participating Observers. By now, we should be sufficiently familiar with Lorentz Transformation and be amply prepared to add up *v* and *u* in the right, relativistic way.

Getting started

As usual, let us start by mapping the *WL*s of *<A>*, *<B>* and *<C>* but to better visualize the two Velocities, we need to map also their *ITL*s. What we need to add is the *AB* Velocity (*v*) and *BC* Velocity (*u*). As always, *<A>* will be the stationary Observer. But in order to show how all participants view what is going on, we'll let *<B>* and *<C>* also take their turns being stationary Observers:

F8.2

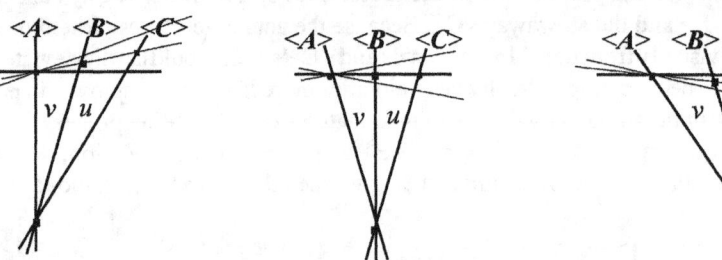

It would be helpful to recall that *<A>* can directly measure *v* but not having direct access to *<B>*to*<C>* Distance, he is not in position to measure *u*.

A question may now be asked: Considering that *<B>* has direct access to both *AB* and *BC* Velocities, why couldn't *<B>* measure both components of the *AC* Distance *(AB and BC)* simultaneously*?* With *exquisite timing of two Radar Probes*, both *AB* and *BC* Distances could be measured by *<B>* *on the same ITL*. The simultaneous *AC* Distance would then be a simple sum of *AB* + *BC*. The *AC* Velocity should be obtainable *(see Fig. 8.3)* by *<B>* *dividing the sum of simultaneous AB & BC Distances by* t_B the *single Time measurement* good for calculating both Velocities *(v and u)*:

$$\frac{AB}{t_B} + \frac{BC}{t_B} = \frac{AB + BC}{t_B}$$

That's right. This means we'll let <*A*> use <*B*>'s data about her *AB* and *BC* Distances along with the her single Time, $t_{(AB \text{ and } BC)}$ measurement. It is possible for <*A*> to measure the entire <*A*>to<*C*> Distance by a single Radar Probe and use the corresponding single *T* to get the AC Velocity but he cannot measure the <*B*>to<*C*> Distance or the corresponding Time Segment. So <*B*>'s data is essential for our exercise:

F8.3

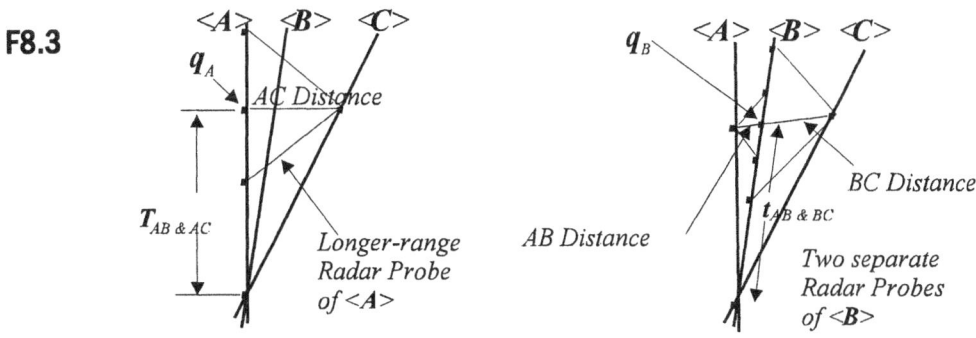

It may be technically difficult for <*B*> to get simultaneous <*A*>to<*B*> and <*B*>to<*C*> Distances with two separate Radar Probes but conceptually, simultaneous Distances are transparently simple and instructive once charted on the Map. To proceed, we'll assume that <*A*> obtained <*B*>'s two simultaneous *x* as well as her single *t* measurements *(even if he could and did actually measure Velocity, **v** himself)*. These are the only data used or needed. Note that none of <*C*>'s measurements are needed.

Let's start, however, with the straightforward Lorentz Transformation method that utilizes our newly acquired expertise from the preceding Chapter VII.

Velocity addition by Lorentz Transformation

Let's first map **(1)** the separate Space and Time components of *AB* and *BC* Velocities for <*A*> and <*B*> but with the Distance lines, *X* and *x* emphasized. After that, let's also draw **(2)** the Lorentz Transformation Map from the preceding chapter to demonstrate visually that adding velocities has a lot in common with the Lorentz Transformation, the most logical procedure to start with:

F8. 4

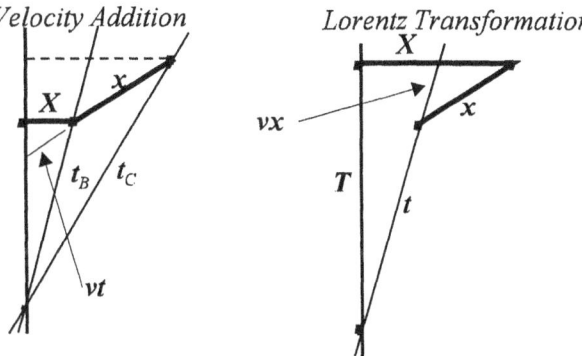

We'll next use the <*B*>to<*A*> Transformation to change the single *x* to *X* and *t* to *T* *(see Chapter VII)* then derive *w* by dividing the Lorentz-derived *X* by the Lorentz-derived *T* aided additionally by the single Velocity, *v* measured either by <*A*> or <*B*> :

$$As \quad X = \frac{x + vt}{\sqrt{1-v^2}} , \qquad T = \frac{t + vx}{\sqrt{1-v^2}} \quad so \quad w = \frac{x + vt/\sqrt{1-v^2}}{t + vx/\sqrt{1-v^2}} \quad and \quad w = \frac{x + vt}{t + vx}$$

To develop our near-complete result further, we need to divide both the Numerator and the Denominator by t::

$$w = \frac{\dfrac{x}{t} + \dfrac{vt}{t}}{\dfrac{t}{t} + \dfrac{vx}{t}}$$

Note that in the Lorentz Transformation, x is the \<B\>-to-\<C\> Distance *(location, Q being the same for \<A\> and \<B\>, see Fig. 7.5 and 8.4)*, thus $x/t = u$ (*NOT v !!!*) and $t/t = 1$. This will get us the final result without further difficulty:

$$w = \frac{u+v}{1+uv} \quad rearranged \ as \quad w = \frac{v+u}{1+vu}$$

What you see here is the classical addition of v and u *DIVIDED* by $1 +$ the product of v and u. This is the unique *SUM of two Velocities*. Take a good look at it!

Still another way to add

Our first stab at the addition was just too easy. To really understand the addition process, let's show the needed components on the Velocity addition Map below using x_1, x_2, t, v and u. We can figure out how to use \<B\>'s data about v *(also equal to \<A\>'s v)* and u *(unavailable to \<A\>)* to derive the values of X_1, X_2 and T_1, T_2 which will be the data needed to calculate w. We'll be using our newly acquired "*vx method*" to obtain T_2.

F8.5

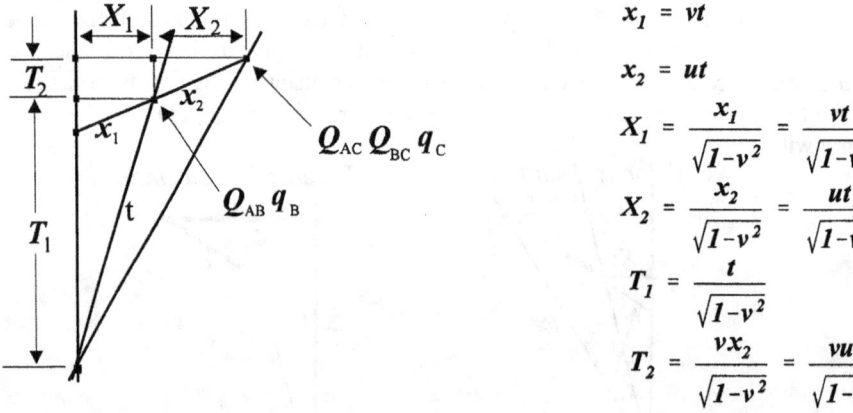

$$x_1 = vt$$

$$x_2 = ut$$

$$X_1 = \frac{x_1}{\sqrt{1-v^2}} = \frac{vt}{\sqrt{1-v^2}}$$

$$X_2 = \frac{x_2}{\sqrt{1-v^2}} = \frac{ut}{\sqrt{1-v^2}}$$

$$T_1 = \frac{t}{\sqrt{1-v^2}}$$

$$T_2 = \frac{vx_2}{\sqrt{1-v^2}} = \frac{vut}{\sqrt{1-v^2}}$$

The total of added Velocities is obtained dividing the sum of X_1 and X_2 by the sum of T_1 and T_2 :

$$w = \frac{X_{AC}}{T_{AC}} = \frac{X_1 + X_2}{T_1 + T_2} = \frac{vt/\sqrt{1-v^2} + ut/\sqrt{1-v^2}}{t/\sqrt{1-v^2} + vut/\sqrt{1-v^2}} = \frac{vt + ut}{t + vut} \qquad Restated: \ w = \frac{vt + ut}{t + vut}$$

In the last part of the equation, both the Nominator and the Denominator also need divided by t which step will give us:

$$w = \frac{v+u}{1+vu}$$ again what we obtained in our first derivation.

Let's proceed with another addition in which, instead of the sum of $T_1 + T_2$, we'll be using the sum of $T_3 + T_4$. We'll be again using the new *"vx-method"* in this case to obtain T_4. The appropriate Space and Time Segment derivations are placed on the right side on the *ST* Map:

F8.6

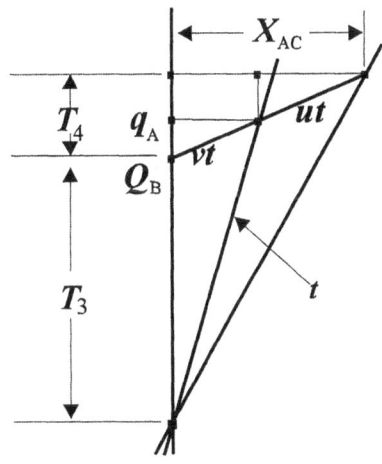

$$X_{AC} = \frac{x_1 + x_2}{\sqrt{1-v^2}}$$

$$= \frac{vt}{\sqrt{1-v^2}} + \frac{ut}{\sqrt{1-v^2}}$$

$$= \frac{vt + ut}{\sqrt{1-v^2}}$$

$$T_3 = t\sqrt{1-v^2}$$

$$T_4 = vX_{AC} = \frac{v(x_1 + x_2)}{\sqrt{1-v^2}}$$

$$= v\left(\frac{vt + ut}{\sqrt{1-v^2}}\right)$$

Now we can do the final assembly:

$$w = \frac{x_1 + x_2}{t} = \frac{X_{AC}}{T_3 + T_4}$$

$$= \frac{\dfrac{vt + ut}{\sqrt{1-v^2}}}{t\sqrt{1-v^2} + v\dfrac{vt + ut}{\sqrt{1-v^2}}}$$

Next, developing the more complex Denominator, $D = t\sqrt{1-v^2} + v\dfrac{vt+ut}{\sqrt{1-v^2}}$

$$D = \frac{t(\sqrt{1-v^2})(\sqrt{1-v^2}) + v(vt + ut)}{\sqrt{1-v^2}} = \frac{t(1-v^2) + v(vt + ut)}{\sqrt{1-v^2}}$$

$$= \frac{t - v^2 t + v^2 t + vut}{\sqrt{1-v^2}} = \frac{t + vut}{\sqrt{1-v^2}}$$

The $+v^2 t$ and the $-v^2 t$ in the above equation canceled out and left us with the much more manageable expression. Our Nominator and the simplified Denominator are, respectively:

$$N = \frac{t+ut}{\sqrt{1-v^2}} \qquad D = \frac{t+vut}{\sqrt{1-v^2}}$$

Combining next the Numerator with the Denominator:

$$w = \frac{\dfrac{vt + ut}{\sqrt{1-v^2}}}{\dfrac{t + vut}{\sqrt{1-v^2}}} = \frac{vt + ut}{t + vut}, \quad \text{restated:} \quad w = \frac{vt + ut}{t + vut}$$

Next, dividing both the Numerator and the Denominator by t, we get:

$$w = \frac{\dfrac{vt}{t} + \dfrac{ut}{t}}{\dfrac{t}{t} + \dfrac{vut}{t}} = \frac{v + u}{1 + vu} \quad \text{Rewritten:} \quad w = \frac{v + u}{1 + vu}$$

and what we have here is, again, the same result we got first but this time with more work.

Adding by Doppler method

Our lesson of Adding Velocities would not be complete unless we do it also by the Doppler method. As we have two Velocities, we have two Doppler constants: k_v and k_u which stand for v and u respectively. To refresh our memory, let us map the Doppler scenario. We do not need to know v or u. This information is already contained in the Doppler constants. First, the Map:

F8.7

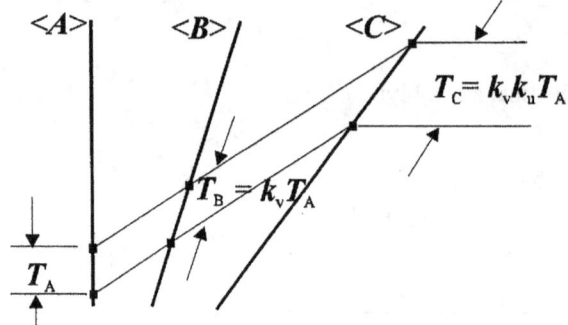

The Time span between the two signals received by $<B>$ is $k_v T_A$ and when these two signals reach $<C>$, they are *(for $<C>$)* $k_v k_u T_A$ Time Segment apart. The Doppler constant for velocity, w would then be:

$$k_w = k_v k_u$$

To convert the Doppler constant into Velocity, w we need to recall that:

$$v = \frac{k^2 - 1}{k^2 + 1}$$

and, therefore, it follows that:

$$w = \frac{k_w^2 - 1}{k_w^2 + 1} = \frac{(k_v k_u)^2 - 1}{(k_v k_u)^2 + 1} = \frac{k_v^2 k_u^2 - 1}{k_v^2 k_u^2 + 1}$$

We can develop the above expression by remembering that:

$$k_v = \sqrt{\frac{1+v}{1-v}} \quad \text{and} \quad k_u = \sqrt{\frac{1+u}{1-u}}$$

Therefore:

$$w = \frac{\left(\frac{\sqrt{1+v}}{\sqrt{1-v}}\right)^2 \left(\frac{\sqrt{1+u}}{\sqrt{1-u}}\right)^2 - 1}{\left(\frac{\sqrt{1+v}}{\sqrt{1-v}}\right)^2 \left(\frac{\sqrt{1+u}}{\sqrt{1-u}}\right)^2 + 1}$$

This bulky equation contains our solution for the Doppler Addition of Velocities. Let's start by squaring the bracketed separate Doppler constants:

$$w = \frac{\left(\frac{1+v}{1-v}\right) \left(\frac{1+u}{1-u}\right) - 1}{\left(\frac{1+v}{1-v}\right) \left(\frac{1+u}{1-u}\right) + 1}$$

Crunching the above equation, we'll get:

$$w = \frac{\frac{(1+v)(1+u) - (1-v)(1-u)}{(1-v)(1-u)}}{\frac{(1+v)(1+u) + (1-v)(1-u)}{(1-v)(1-u)}}$$

The identical Denominators cancel out, giving us the upper Numerator over the lower Numerator. We'll now multiply the members in the brackets, obtaining:

$$w = \frac{1+u+v+vu - (1-u-v+vu)}{1+u+v+vu + (1-u-v+vu)} = \frac{1+u+v+vu-1+u+v-vu}{1+u+v+vu+1-u-v+vu} = \frac{2u+2v}{2+2vu} = \frac{2(u+v)}{2(1+vu)}$$

and the final result can be rewritten as: $\quad w = \frac{v+u}{1+vu}$

A lot of work but all worth the effort. In this last example, we have seen again that Doppler gives us exactly the same result as obtained by the Lorentz*ST* Map methods used. And we again have the correct final equation: *The sum of two Velocities is the arithmetical sum of these velocities divided by the sum made up of their product added to one.*

A workout with numbers

To acquaint ourselves with the actual results, let us add together a few assorted Velocities, starting with the stowaway on the train described at the beginning of this chapter:

$$w = v + u = 50 + 5 = 55 \text{ mi/hr}$$

First, let us convert the above v and u into fractions of c, then add them together. Note that the Velocities are given this time in miles per hour and $c = 186,000$ mi/hr:

and:
$$v = \frac{50}{186,000 \times 60 \times 60} = 7.4671 \times 10^{-8}$$

$$u = \frac{5}{186,000 \times 60 \times 60} = 7.4671 \times 10^{-9}$$

The sum, w should turn out to be: $w = \dfrac{v + u}{1 + vu}$, let's see:

$$w = \frac{7.4671 \times 10^{-8} + 7.4671 \times 10^{-9}}{1 + 7.4671 \times 10^{-8} \times 7.4671 \times 10^{-9}}$$

$$= \frac{8.21381 \times 10^{-8}}{1.000,000,000,000,000,065,757,582}$$

$$w = 8.21381 \times 10^{-8}$$

We need to convert the above relativistic Velocity back to our everyday numbers:

$$w = 8.21381 \times 10^{-8} \times 186,000 \times 60 x 60 = 54.99967176$$

which is given in mi/hr, almost exactly 55 mi/hr, minor inaccuracies created by the low-power hand-held calculator, and is very close to the simple arithmetic sum, 50 + 5 or 60+(-5) of these Velocities.

The situation is not much different if we consider slightly faster Velocities such as possessed by the Earth in its full orbit around the Sun. The Velocity, approximately 30 km/sec, is 1/10,000th of c, written as 0.000,1. Let us now add together two speeds both 0.000,1:

$$w = \frac{0.0001 + 0.0001}{1 + 0.0001 x 0.0001} = 0.000,199,999,99$$

which is again almost exactly the arithmetical sum of 0.000,1 + 0.000,1 = 0.000,2. This example also succeeds pointing out that it takes Velocities much greater than what automobiles, airplanes, rifle bullets and Earth's orbit around the Sun can achieve to make the sum of Velocities noticeably different from the simple arithmetical addition that is satisfactory for all earthbound Velocities. So, let's play with some really fast Velocities.

Let's add two velocities, each 50% of c *(or 0.5)*:

$$w = \frac{0.5 + 0.5}{1 + 0.5 \times 0.5} = \frac{1.0}{1 + 0.25} = \frac{1}{1.25} = 0.8 = 80\%c$$

Adding two velocities, each 90% of c:

$$w = \frac{0.9 + 0.9}{1 + 0.9 \times 0.9} = \frac{1.8}{1 + 0.81} = \frac{1.8}{1.81} = 0.994475$$

Adding two Velocities, each 99% of *c*:

$$w = \frac{0.99 + 0.99}{1 + 0.99 \times 0.99} = \frac{1.98}{1.9801} = 0.999,949,5\%c$$

At last, let us add two Velocities, each equal to the speed of Light:

$$w = \frac{1+1}{1+1 \times 1} = \frac{2}{1+1} = \frac{2}{2} = 1.0 = 100\%c$$

It goes without saying that adding two Light Velocities or, perhaps, any Velocity to that of Light will not give us a Velocity exceeding that of Light. But let's also try 1 + 5 anyway:

$$w = \frac{1+5}{1+1 \times 5} = \frac{6}{6} = 1.0 = 100\%c$$

Adding any velocity to the Velocity of Light turns out surprisingly like adding water to an already full bucket. Regardless of your efforts, the bucket just cannot get any fuller than full. This we can understand, it agrees with our experience. One would only wish that things would work the same way in reverse. Wouldn't it be nice to take water out of a full bucket with the bucket nevertheless staying full. So what about subtracting a given Velocity (0.5) from that of Light? Let's find out. Actually, subtracting *u* from *v* is the same as adding *-u* to *+v*:

$$w = \frac{v-u}{1-vu} = \frac{1-0.5}{1-1 \times 0.5} = \frac{0.5}{1-0.5} = \frac{0.5}{0.5} = 1.0!$$

Let's take 0.999,999 from 1.0:

$$w = \frac{1+(-0.999,999)}{1+(-0.999,999)} = \frac{0.000,001}{0.000,001} = 1.0$$

The result again shows that subtracting a Velocity that is either substantially less or almost the same as that of Light from the full Light Velocity, we still remain with the full, undiminished Velocity of Light. So we can say that subtracting 0.9 from 1.0, we would still end up with 1.0, the full Velocity of Light.

But how about subtracting 1.0 from 0.9, that is the full Velocity of Light, from something less than that? Let's see:

$$w = \frac{0.9 + (-1.0)}{1.0 + 0.9(-1.0)} = \frac{-0.1}{0.1} = -1.0$$

We obtained the full Velocity of Light but this time with a negative sign. Instead of the unchanged +0.9, we obtained -1.0 that is the unchanged 1.0 with the minus sign indicating that it was now going in the direction opposite to that of the Velocity from which the Velocity of Light was subtracted. Interesting *!* It is the unchanged Velocity of Light we got.

Now 1.0 from 1.0 *(Light from Light):* $\quad w = \frac{1-1}{1-1} = \frac{0}{0} = ?$

The last subtraction gives an indefinite answer. It means that whenever we subtract a Velocity of Light from the Velocity of Light, the answer is zero divided by zero, a result that does not have a definite value or, perhaps, can have any value you would want.

In any case, the Velocity of Light is like an always full bucket of water after all. You can keep subtracting Velocities from it over and over again and you would still have the full speed of Light left over. This is quite unlike our experience with a bucket of water that can be emptied by repeatedly taking out small amounts. But subtracting a full Velocity of Light from one that is less than that gives us a result that is even more than odd: it gives us the subtracted Velocity of Light in all cases!

Adding the same Velocity repeatedly, starting from $v = 0$ is an important exercise because it reveals an inobvious trapdoor for errors when dealing with some of the cosmological problems such as using Hubble's law for calculating the age of the universe. Let's next go ahead and add a given Velocity over and over starting with $v = 0$:

$0 \Rightarrow 0.2 \Rightarrow 0.38 \Rightarrow 0.54 \Rightarrow 0.73 \Rightarrow 0.81 \Rightarrow 0.87 \Rightarrow 0.91 \Rightarrow 0.94 \Rightarrow 0.96 \Rightarrow 0.97$

$0 \Rightarrow 0.3 \Rightarrow 0.55 \Rightarrow 0.73 \Rightarrow 0.73 \Rightarrow 0.84 \Rightarrow 0.91 \Rightarrow 0.95 \Rightarrow 0.97 \Rightarrow 0.98 \Rightarrow 0.99$

$0 \Rightarrow 0.5 \Rightarrow 0.80 \Rightarrow 0.93 \Rightarrow 0.99$

$0 \Rightarrow 0.98 \Rightarrow 0.997 \Rightarrow 0.999$

What the above series of Velocities tells us is this: It may take a few additions to get to the Velocity, v = 0.9 but it takes at least the same number of additions and more to get from 0.9 to 0.999999999 and beyond. That's not all: it takes an almost limitless number *(actually an infinity)* of additions to get really close to $v = 1.0$ *!!!* The astronomical Distances, however, are often calculated according to the Hubble Constant, H so that $D = vH$ *!!!*

Now, that makes the quarter Distances from v to c: 0 to 0.25, 0.25 to 0.5, 0.5 to 0.75 and 0.75 to 1.00 *all equal* giving us the total age of the universe, of all things, 14 - 15 billion years. The actual numbers used in the calculation were a little different *(from those in the last paragraph)* and were conforming to the actual measurements obtained. But the simpler numerical demonstration used here gives a reasonably good sense of the concepts employed.

Summing it up

Our tedious experience with Adding Velocities can be summarized by saying that:

(1) for very small values of v and u, the Denominator of the expression is very-very close to 1 and dividing the arithmetic sum of the Velocities by a value that is very-very close to one gives us a result that is very-very close to the simple arithmetic sum of v and u.

(2) If, however, v and u are more substantial in value, the relativistic sum becomes a lot smaller than their simple, arithmetic sum.

(3) As all Velocities are given as fractions of c, their individual magnitudes are always smaller than 1 and their arithmetic sum in the Numerator is always smaller than 2. The Denominator of the equation is always greater than 1 but always smaller than 2 and the ratio can never become equal to or exceed 1.

(4) Even adding two Velocities of Light together gives us a sum that is no more than 1, just the single speed of Light.

(5) Relativistic addition of very fast Velocities tells us that the Velocity of Light is the absolute speed limit. This is the conclusion about the nature of Light that has never been experimentally disproved.

But. . .(a big but)

Suppose we want to split the speed of Light into two equal halves so that if you would subsequently add the halves together, you would get the original speed of Light back again. Can we do it? Let's start by recalling that when we added $c = 1.0$ and $c = 1.0$, we obtained w that was also equal to 1.0. Now let's express the result as follows:

$$w = \frac{c+c}{1+cc} = \frac{2c}{1+c^2} = \frac{2c}{1+1} = \frac{2c}{2} = c$$

Let's next solve for c but this time without reducing c or c^2 to 1.0. Working out the details, we get, of all things, a quadratic equation *(which doesn't change our general result)*:

$$w(1+c^2) = 2c$$

$$w + wc^2 - 2c = 0$$

$$wc^2 - 2c + w = 0$$

Let's recall that the general solution of a quadratic equation, $ax^2 + bx + c = 0$, is:

$$x = \frac{-b \pm \sqrt{b^2 - 4ac}}{2a}$$

As $a = w = 1$, $b = -2$, and $c = w = 1$, we can solve this general equation by making the necessary substitutions. And we get two solutions:

$$v = \frac{2+\sqrt{2^2 - 4x1x1}}{2} \qquad\qquad v = \frac{2-\sqrt{2^2 - 4x1x1}}{2}$$

The final calculated results are:

$$v = 1 \qquad\qquad v = 1$$

As you see, it makes no difference how we split the c. In all cases, splitting the Velocity of Light into two equal parts gets us two separate full Velocities of Light, $c = 1$ and $c = 1$. By now we have demonstrated that adding to the Velocity of Light or subtracting from it, splitting it or, most probably, performing any other manipulation, we simply cannot change the Velocity of Light *!!!* It is interesting how adding Velocities really works and how indestructible the Velocity of Light has proven to be.

If we cannot split the Velocity of Light, can we split one that is less than 1.0? Let us take $w = 0.8$ and split it into two equal parts *(recall that when we added 0.5 to 0.5, we obtained 0.8)*:

$$w = 0.8 = \frac{2v}{1+v^2}, \quad so \;\; 0.8(1+v^2) = 2v \quad \text{and also}$$

$$0.8v^2 - 2v + 0.8 = 0$$

and these:

$$v = \frac{2+\sqrt{(-2)^2 - 4x0.8x0.8}}{2x0.8} \qquad\qquad v = \frac{2-\sqrt{(-2)^2 - 4x0.8x0.8}}{2x0.8}$$

$$v = 2 \qquad\qquad v = 0.5$$

are our two solutions*!*

Of the two possible results, only the second one, obtained by using the minus sign in the equation, corresponds to what we would ordinarily be expecting: something less than 0.8 and certainly not anything more than 1.0 *!* We already know what we got from adding 0.5 and 0.5 together. But for more than mere curiosity, let us see what happens if we add 2.0 and 2.0 together.

$$w = \frac{2+2}{1+2x2} = \frac{4}{1+4} = \frac{4}{5} = 0.8$$

Now, isn't that strange! 2 + 2 = 0.8, back to the unsplit 0.8. Obviously, the equations somehow work well forwards and backwards. What we cannot tell is whether Velocities above 1.0 have any practical or theoretical significance, at least for now. In any case, those faster-than-Light Velocities really do not belong to a discussion of Special Relativity. We have seen on more than one occasion that the speed of Light is the absolute limit of all velocities. Or is it?

The question of superluminal Velocities takes us into an area that is just too complex for a short paragraph at the end of a difficult chapter. In order to do justice to this interesting topic, we should find out if "tachyonic pursuits" lead us to unexpected surprises. This will be examined in the next chapter.

Truth and consequences

The results of relativistic addition of Velocities can now be restated:

(1) The speed of Light can only be approximated, not fully reached or exceeded no matter how much we may boost a given subluminal Velocity. This restraint is imposed on all Motion and makes Light an effective limit to all naturally occurring Velocities.

(2) The speed of Light itself is unalterable, fixed, always remaining the same. It cannot be increased or decreased no matter how much we may add to or subtract from it. In this way, Light is unique among all natural phenomena, a true constant in nature, something that cannot exist at any other magnitude and never stays put at any location in Space.

The constant velocity of Light has been well established as an empirical fact. It was not always believed to be so. Einstein first guessed it as an act of insight and made it one of the cornerstones of Special Relativity. Here we have been forced to the same conclusion by a more circuitous route examining the relativistic addition of Velocities. It was not really a discovery but the uncovering of a consequence hidden in the equation.

IX - FASTER THAN LIGHT

Is it impossible?

Like a once popular televison series, this chapter can also be called "Mission Impossible." On TV, the challenges were always met with the most ingenious solutions. But what we are attempting here is absolutely the most impossible task: *describing Motion faster than Light.*

According to Special Relativity, nothing goes faster than Light and that's that. Agreed! Settled! Final!...Or is it? Well, one exception does suddenly come to mind. Do you know what *CHERENKOV RADIATION* is? And while we are on the subject, would anyone, please, tell us what professor Feynman meant when he said that *POSITRONS are backward electrons*, that is, "renegade" electrons going backward in Time. Also, why did professor Guth propose a *SUPERLUMINAL INFLATIONARY STAGE* during which the original Big Bang expanded thousands or, perhaps, more than millions or billions of Lightyears within an extremely small fraction of a second right after, or as part of, the Big Bang? Was he there to see it? And, finally, what about this *NONLOCALITY*, that "Spooky Action at a Distance," the thorn in Einstein's mind, the 30 year argument with Niels Bohr he lost? The topics listed do not belong to Special Relativity and will not be taken up here but the mere fact that these puzzling issues have been raised does suggest that *perhaps some things just might be going faster than Light* after all. And this possibility makes it worth our while to explore the *KINEMATICS of SUPERLUMINAL VELOCITIES* while leaving their actual existence undecided.

What is Cherenkov radiation?

Those who haven't heard of Cherenkov radiation may now be informed that *this radiation is the PHOTONIC equivalent of a SONIC BOOM*. Now, sonic boom is rather well known. It is heard when a supersonic aircraft flies overhead so that the noise created while approaching reaches us all at once at (*or immediately after*) the Moment of Passing. Similarly, *Cherenkov radiation is a bright flash of Light, a photonic boom, caused by the bunching up of Light created by a cosmic ray particle speeding through a transparent crystal going faster than the Light emitted in the process!* This remarkable phenomenon does not lose its odd luster when qualified by an explanation that the Velocity of Light in the crystal is less than what it is in the vacuum. So Special Relativity is not violated. But *how come that winning the race with Light cannot be repeated in the vacuum?* A good question. Those of you interested about the nature of this and all the other exceptions are on your own.

Bumping against "Light Barrier"

There has been a great deal of speculation about Velocities above that of Light, the commonly accepted upper limit. In the Chapter, "Adding Velocities," we learned that Velocities cannot be boosted to the exact speed of Light, not to mention going beyond that. Experimental work has never convincingly established Superluminal Motion as a possibility. What's more: there are currently no candidates among the menagerie of known waves and particles capable of breaking the cosmic speed record.

The impossibility of superluminal speeds must somehow be prohibited by the nature of four-dimensional Spacetime. After all, if Spacetime contains everything there is in the universe, it can set and enforce its own traffic rule which is: *There shall be NO crossing of the "LIGHT BARRIER!"* Superluminal Velocities would still be consistent with the Light Barrier concept but with a new twist: *the speed limit could be imposed strictly as usual but also in "REVERSE" direction* with the following "reverse" consequences: *Superluminal motion (1) cannot be slowed down to the Velocity of Light* and *(2) cannot be brought down past the Light Barrier into the subluminal range.* An interesting math (*taken up later in this book*) shows that all material objects increase their mass to infinity when accelerated to *c* thereby ramming against an impervious barrier that prevents anything from slipping past the speed of Light. But an even more interesting math (*not taken up in this book*) argues that *if an object is already going faster than Light, then its mass also increases to infinity when decelerated to c !!!* To understand the related $E = mc^2$, we have to understand **acceleration** first. This also comes up soon in this book.

Keeping s ubluminal a nd s uperluminal M otions s eparate a llows t he e xistence o f a n entirely unsuspected phantom domain populated exclusively by ghostly entities moving at superluminal speeds. This alien world may be so completely isolated from our normal, subluminal world that any interaction between them would be impossible. If, on the other hand, interactions do occur, these must then be so ordinary in appearance that they simply blend un-noticed into our familiar worldscape.

Monoverse + Monoverse = Duoverse

Recognizing two separate, independent worlds on each side of the Light Barrier makes it necessary to redefine our cosmos as being not really a *"UNIVERSE"* but a *"DUOVERSE"* made up of two separate *"MONOVERSES,"* one subluminal or *"TARDYONIC,"* the other superluminal or *"TACHYONIC."* The tardyonic half is easy to accept. It is the only one we know. No parts of the tachyonic half, on the other hand, have ever come to our attention. For the sake of argument let us simply assume that it exists and then proceed to figure out how Spacetime geometry can describe and map it.

The assumption itself does not demand proof and lets us have fun with *experimental imagination in Virtual Reality.* Indulging in unashamed phantasy has never been neglected. Think of all the science fiction stories written and those made into movies. Extrapolating from known physics to the superluminal has also been, so it seems, a recreational past-time for quite a few respectable scientists. They have been telling us among other things that if Velocities are pushed beyond that of Light, strange things begin happening: *everything there besides renegade electrons would be going BACKWARD in Time!*

When and where would Time Reversal happen?

Don't expect, however, that by crashing through the Light Barrier you can Time-travel into the Past, that once Time gets going in reverse direction, it keeps going that way as long as Velocity remains superluminal no matter what. Actually, Reverse Time concept is not that simple. That's because *Relativity of Simultaneity and Observer-Specificity are hard "at work" on both sides of the Light Barrier and what's more: Time Order is merely the other side of the (Relativity-of-) Simultaneity coin.* The whole thing, as you'll see, is really anti-intuitive. For that reason, the entire present chapter is dedicated to just this single topic. Fortunately, *we can explore Time Order not by pushing our luck in a Space vehicle but by topological manipulation of Spacetime Maps,* a method that lets us rely entirely on visual persuasion without entanglement with dangerous Space traffic or math that is beyond our grasp.

Now the big question: *Can there be such a thing as Reverse Time? Does Time in the tachyonic Monoverse really go into reverse?* To find out, let's send our experienced cosmonaut *<B>* on a superluminal trip to check it out. Her report from the tachyonic Monoverse should settle the question once and for all. But before such futuristic technology becomes available, we can reasonably suspect that *<B>*'s report from there will be a big *NO*, that *she will find nothing unusual going on "South of the border," that her Time progression never fails to slow down or turn around but that it always keeps going normally in the forward direction.* Then to really confound us good, she just might suddenly ask what the heck is wrong with us here going not only faster than Light ourselves but also backward in Time*!!!*

That Reverse Time is not observable inside the tachyonic realm should not be unexpected. It should follow the usual pattern shared by all relativistic effects of fast Motion such as Lorentz Contraction and Time Slowing *(Time Dilation)*. Remember that Lorentz Contraction and Time slowing are always observed *(actually obtained by calculation)* by **LOOKING IN on moving systems from the OUTSIDE** *(of that System)* **while within those Systems no such effects can be observed.** Just remember that if we ever observe someone else going at 0.9*c* and getting Lorentz-contracted, that someone else also would observe us going at 0.9*c* and getting Lorentz-contracted while **neither of us** would notice anything strange going on in our own Systems. So **Time Reverse, our new addition to that odd group of effects must always be taking place ELSEWHERE**. In other words, these effects are simply **OBSERVED to happen exclusively somewhere else and to someone else who would be totally unaware of them**. It is **a characteristic feature of all Inertial Systems wherein everything always appears just as good as normal. Natural laws operate exactly the same way in all Inertial Systems according to the all-inclusive Principle of Equivalence** *(same as the Principle of Relativity)* and as proposed by Galileo: **A state of uniform, linear Motion cannot be distinguished from a state of rest** *(that is, from non-Motion)* **by any** *(conceivable)* **test. As all inertial Observers can legitimately claim to be at rest, that is stationary** *(or non-moving)*, **so <B> in this strange new Monoverse would also be totally unaware of being in Motion or seeing anything unusual** within her own Monoverse. She would most likely point to us *(and to<A>)* as being superluminal and in Reverse Time Order. The Light Barrier must, therefore, be a side-reversing, **IMAGE-TRANSFORMING LENS** *(or mirror)* that produces **the APPEARANCE of Time Reversal which** *(on second thought)* **could not possibly be observed firsthand right then and there anywhere and/or by anyone !!!**

Let's repeat this important lesson: Observer-specificity just about demands that whenever we talk about <B>'s Time-Reverse, we mean that it is always **"AS VIEWED" by someone else like <A>, <C>, <D>, etc. who observe themselves to be in Normal Time Order as observed by all Inertial Observers wherever they are and Reverse Time is observed to happen only ELSEWHERE, that is in some OTHER Inertial System, in some OTHER Frame of Reference and is always OBSERVED ACROSS THE LIGHT BARRIER with Velocities faster than Light** *(relative to the Observer such as <A>)*. There is just one more thing to add: if <A> observes <B> going faster than Light then the relative Velocity is the same for both and <B> in turn would observe <A> as going faster than Light. Whether Reverse Time Order is always mutually observed is another question we'll tackle later on in this chapter.

Relativity of Simultaneity = key to Time Reverse

Haven't you noticed that reversal of Time Order is definitely not intuition-friendly. All our **cues about Time Order can come to us only by Signals from Tachyonic Systems under scrutiny.** We are fortunate that **Time Order in all Systems is always identical with the LOCAL order of Signal GENERATION.** This shifts the **burden of evidence** from the **order of Signal RECEPTION** to the **order of its GENERATION.** Unfortunately, most evidence for Time Reverse quoted in the literature is almost always in form of **reversed order of Signals received from superluminal sources.** This evidence is questionable at best because we have already seen in our tardyonic system that according to Relativity of Simultaneity, **the order of Signals received from Space-connected** *(or Space-separated)* **Sources does not always agree with their temporal order of origin.** More specifically, **two SPACE-CONNECTED Events at a Distance can appear** *(to Observers residing in different Inertial Systems)* **in THREE DIFFERENT WAYS, being either (1) simultaneous, (2) in one Time Order or (3) in Reverse Time Order.**

What's more: Relativity of Simultaneity links the two Monoverses *(separated by Light Barrier)* in a way already known to us because ***Space-connected Events on the** (Space-connected) **ITLs of tardyonic Travelers follows a pattern that is IDENTICAL to that of Space-connected Events on the** (Space-connected) **Worldlines of tachyonic Travelers. Therefore, in order to negotiate between the two Monoverses, all we need to do is to just SWITCH the KEY LABELS: ITLs and WLs !!!*** And there we have found an intuition-friendly connection between these **supposedly different but actually identical Spacetime geographies.** This sweeping generaliz-ation takes us back to familiar territory where our Special Relativity tool kit is all we need to confidently find our way around the Light Barrier.

Relativity of Simultaneity revisited

Relativity of Simultaneity is the key to Time Order, so let's review it again. In the next illustration, two Space-connected Events, *P* and *Q* are on *<A>*'s *ITL* and simultaneous to *<A>*. But the Map shows that to *<C>*, *P* comes before *Q*. To *<B>*, however, it is *Q* coming before *P*:

F9.1

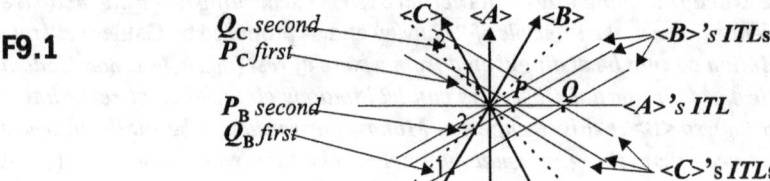

The differences in the timing (Time Order) of Events shown above are directly translatable to differences in Time Order of Events on the tachyonic WL but observed across the Light Barrier. The Events, *P* and *Q*, instead of being on the *ITL* of a Tardyonic Traveler are on the *WL* of a Tachyonic Traveler*!!!*

"Sneaking" across the Light Barrier

If superluminal *WL* is placed on the other side of the Light Barrier, so where on the *ST* Map would its *ITL* be? To obtain the answer logically, let's start with the Time-connected *WL* of a tardyonic Observer, *<B>* first stationary relative to *<A>*. Here her *x* and *t* Coordinates coincide with *<A>'s X* and *T*. At increasingly higher Velocities, *<B>*'s *WL* and *ITL* come closer toward each other and to the Lightline between them. Once the speed of Light is reached, *<B>*'s *WL* and *ITL* become superimposed not only to each other but also to the Lightline. In this position, all Events on *<B>*'s combined *WL* and *ITL* are **simultaneous** to *<B>* but are viewed as **sequential** by *<A>* in whose view *<B>* is now moving at the Velocity of Light. Note again that to *<A>*, all *<B>*'s Events are mapped separately and are definitely not simultaneous.

F9.2

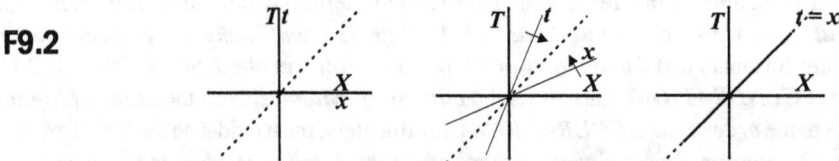

As we next move *<B>*'s *WL **across the Lightline*** into Space-connected domain *(Space-connected for <A>)*, her *ITL* also moves in opposite direction and across the Lightline into the Time-connected domain *(as viewed by <A>)*. In this way, *<B>*'s *WL* becomes Space-connected and her *ITL* becomes Time-connected. ***In this position, two Events on her WL can be considered either in the same or in Reverse Time Order by different tardyonic Observers.*** Finally, when *<B>* goes so fast that she transits two locations on *<A>*'s *ITL* at the same Moment, simultaneously from *<A>*'s perspective, ***the Velocity reached is "ABSOLUTE."*** At that Moment, *<B>*'s *WL* and *ITL* are positioned at 90 degrees from each other on the Map at locations exactly opposite from those of *<A>*, a peculiar state indeed: *<B>*'s *WL* coincides with *<A>*'s *ITL* and *<B>*'s *ITL* coincides with *<A>*'s *WL* :

F9.3

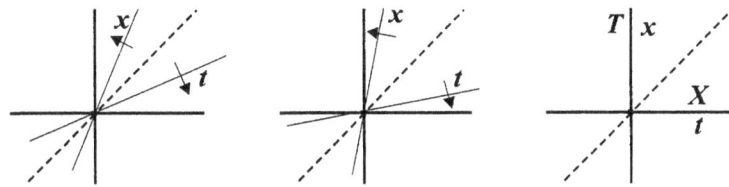

The next question is this: Do Events mapped on the Space-connected *WL* proceed from left to right on the Map or should Time Order be pointing from right to left instead *?* These two possibilities for the Time Order are illustrated next but one of them will be selected empirically as shown later in this chapter:

F9.4

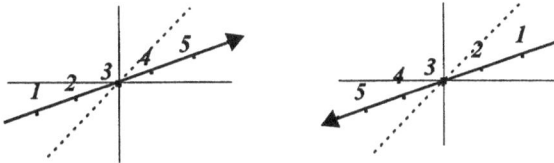

The deceiving or confusing order of Signal arrivals

Because the Reversed Order in which Signals arrive from *<B>* to *<A>* has been featured as evidence of Time Reversal, it is useful to take a look at the different ways these Signals arrive at the Observer. It is easy to show that the order of Signal arrival is a totally unreliable method for determining the Time Order of their origin. Only a small number of examples will suffice to make the point.

(1) Three Signals generated simultaneously are received sequentially:

F9.5

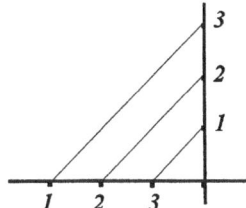

(2) Three Signals generated sequentially are received simultaneously:

F9.6

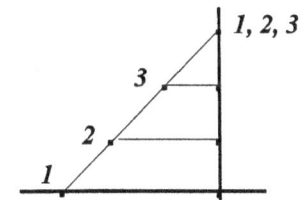

(3) Signals generated by two tachyonic Travelers moving in opposite directions along the same *WL* are received by a tardyonic Observer *in different Time Orders before and after the Passing Event:*

F9.7

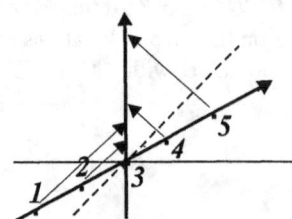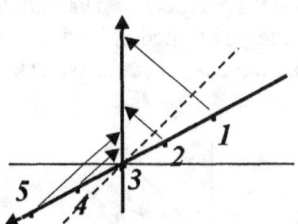

There has to be a better way to judge Time Order than the Signal arrival and, fortunately, there is. Read on.

A consistent method for establishing Time Order

The tool for determining the Time Order of our tachyonic Traveler, *<B>* is something already known to us. It is nothing other than our familiar Radar Probe known to us from our tardyonic Spacetime exercises. Let's apply this valuable expertise to Motion across the Light Barrier.

After *<B>* has passed *<A>*, *<A>*'s Radar Probe can no longer reach *<B>* whose Velocity exceeds that of Light. In the next illustration, Fig. 9.8(A), *(<A>)* sends out two Radar Signals blindly before Passing. There are two Reflections from *<B>* before Passing and received *in Reverse Order* and two after Passing received *in the Same Order*. But if these Reflections are connected to *<A>*'s *WL* by corresponding *ITL*s, then these come to *<A>* in the same Order and in complete agreement with his own Time Order. For that reason, the Time Order is always correctly displayed by *the ITL*s which give us a clear visual result that is not only simpler to read but is also the same both before and after Passing. With the process so simplified, we can easily read the Time Order directly off the Map *!!!*

F9.8

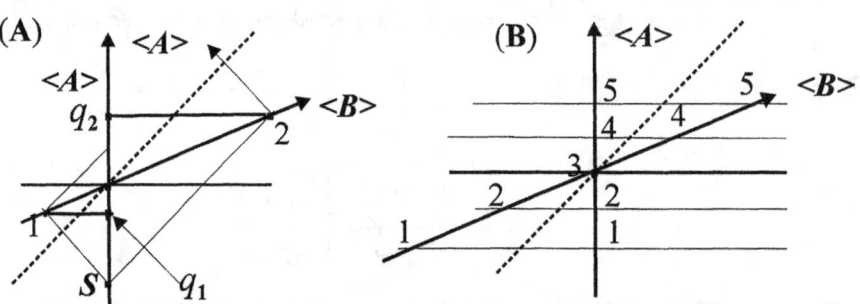

It is worth repeating that *the ORDER of SIGNALS received from <B> does NOT reliably determine the ORDER of their occurrence on <B>'s WL* and that *only ITLs can tell us the actual Time Order on <B>'s WL.*

At this time, we still have not yet established criteria by which we and *<A>* can "read" Time Order right off the Map. So, let us start with a set of observations about which we can be reasonably certain: if Motion is superluminal to one tardyonic Observer, it must be so to all tardyonic Observers. That's because Lightlines are nailed down firmly in the same place in the Home Frame of all Inertial Observers and are the only absolute traffic signs. What we need is an operational method to determine Time Order empirically. This may be *technically impossible* but, fortunately, *is simple in concept and easily done by topological examination of ST Maps.*

Establishing Time Order empirically !

Let's set up two Space Stations served by $<C>$ and $<D>$ and arrange $<B>$ to pass both of them at close range in a series of increasingly faster runs staged from a more remote Station not shown on the Map. *All runs start from the same direction, from left to right on the Map.* Now the details:

(1) Proximity Detectors are placed at Stations $<C>$ and $<D>$ which remain **(a)** equidistant from each other and **(b)** equidistant from $<A>$ who does not need to be positioned in the same Lineland.

(2) $<B>$ is engaged in 5 experimental runs *(done on paper, of course)* starting out at tardyonic Velocity and in each run she first passes $<C>$ then $<D>$ without hitting them.

(3) The Proximity Detectors with $<C>$ and $<D>$ are triggered by $<B>$ passing them and the resulting Signals are sent to $<A>$ and his equipment positioned at equal Distance from $<C>$ and $<D>$. Now the Time Order question is reduced to: *which Station was passed first and which was passed later.*

The Passing Events at $<C>$ are designated as O-Events and are automatically sent to $<A>$'s equipment that registers these as R_0-Events. Each passing of $<D>$ *(by $<B>$)* are also transmitted to $<A>$ who registers these as $R_{1,2,3,4, \text{ and } 5}$-Events.

The Velocities assigned for $<B>$ are: **(1)** *tardyonic,* resulting in R_0 and R_1-Events, **(2)** *luminal (Light-Velocity),* resulting in R_0 and R_2-Events, **(3)** *tachyonic,* resulting in R_0 and R_3-Events, **(4)** *tachyonic-Absolute,* resulting in R_0 and R_4-Events and, finally, **(5)** *faster-than-Absolute* resulting in R_0 and R_5-Events:

F9.9

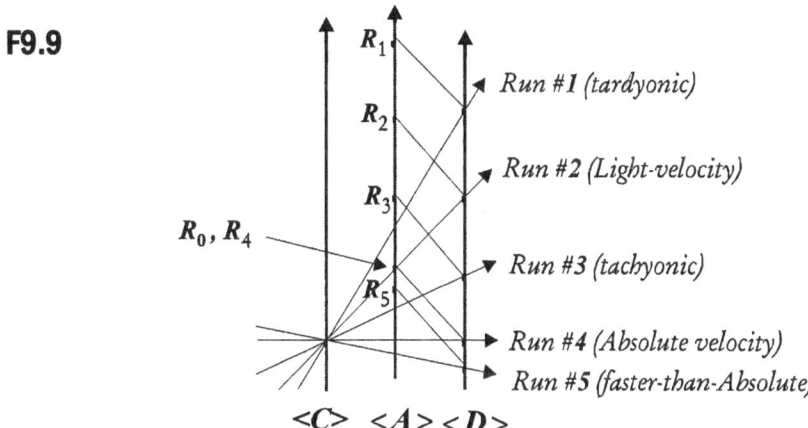

The results mapped above are self-explanatory. Run #4, however, deserves special comment. You can see that R_0 and R_4 occur *(for $<A>$)* simultaneously meaning that *(according to $<A>$),* *$<C>$ and $<D>$ were passed simultaneously (or at the same Moment).* This kind of Motion *(two places occupied at the same Moment)* is considered, as we have already noted, to be at *ABSOLUTE VELOCITY relative to $<A>$*. Run #5 is even faster-than-Absolute! In this case, R_5-Event happens *BEFORE* R_0-Event, a sure sign of *Time Reversal (as observed by $<A>$).*

By this empirical method, *we have established not only $<B>$'s Time Order but also the operational definition of Time Order.*

Never mind that the Velocities such as Absolute and faster are unheard of. All tachyonic Velocities belong to this unheard-of group. In this chapter, all kinds of wild possibilities are allowed. We are free to make up whatever we please as long as we keep everything logical and the registry of Events operationally consistent with Special Relativity.

Now, the preliminary results:

(1) *Time Order in any Frame of Reference is established only by the Order of Signal Generation* in the moving Home Frame and the determination of Time Order in Other Frames of Reference requires the ability *to map the order of Signal generation by ITLs of the Stationary Observer's Home Frame of Refernce.*

(2) The order of Signal arrival does not establish Time Order in the System where the Signals are generated. Moreover, the Order of Arrival is different before and after the Passing Event. The *ITLs* drawn through the tachyonic *WL* on the *ST* Map are the only reliable method of comparing Time Orders in the two Systems.

(3) When *<B>*'s Motion is proceeding from L-to-R on *<A>*'s *(and our)* Map and her *WL* is on the upward slope according to *<A>*'s upward sequence of *ITLs*, then the same Time Order is observed according to *<A>* if, and only if, *<B>*'s Motion is actually proceeding from L-to-R in her Home Frame and mapped by *<A>*.

(4) If all Events on *<B>*'s *WL* are simultaneous to *<A>*, then *<B>*'s Velocity observed by *<A>*, is *ABSOLUTE* and *<B>*'s *Time Order is indeterminate*, teetering between forward *(same)* and reverse *(backward)* Time Orders. So, any tachyonic Velocity slower than Absolute does not make *<B>*'s Time go backwards as viewed by *<A>* and by his friends, *<C>* and *<D>* *(all three being in the same Inertial System).*

(5) When *<B>*'s Motion is proceeding from L-to-R on the Map and is placed on the downward slope L-to-R, then *<B>*'s Velcocity is faster-than-Absolute and her Time Order is mapped on *<A>*'s *WL* as proceeding in Reverse Time Order but again if, and only if, *<B>*'s Motion is proceeding from L-to-R on the Map. Faster-than-Absolute Velocities produce Time Reversal but again if L-to-R Motion is attributed to *<B>* by *<A>*. We cannot yet tell how other Observers in different tardyonic Systems regard *<B>*'s Time Order and how *<B>*, in return, would view their Time directions but we'll find a definite connection between Time Order and Relativity of Simultaneity. How? Just read on.

(6) Time Order, like Absolute Velocity, is *RELATIVE* just like Simultaneity is relative with three possibilities: **(a)** The Time Order is indeterminate with Motion at Absolute Velocity, **(b)** The Time Order is the same when Motion is less than Absolute in Velocity and **(c)** The Time Order is reverse with Motion faster-than-Absolute *BUT (a big but):*

(7) *L-to-R down-slope* Motion on the Map can also be viewed by *<A>* as *R-to-L up-slope* Motion within the same Time Order but now moving in a Space direction opposite to what we saw in the L-to-R *(up-slope)* Motion, a directional turnaround in Space instead of in Time, thus *eliminating the Reverse Time concept altogether* and *converting all Reverse Time directions into Reverse Space directions.*

(8) Our empirical assignment of Time Order to tachyonic *<B>* may disagree with those who insist that once anything cgoes faster than Light, its Time Order *(as <A> views it)* would be immediately directed in Reverse, going backwards. But with *Space-direction Reversal as an alternative to Time (-direction) Reversal*, the topic of tachyonic Velocities threatens to "self-destruct" in confusion. The empiric approach adopted here is eminently simpler and is capable of pointing out most of the peculiarities of tachyonic behavior. Moreover, we need to avoid the choice between Space- and Time-direction Reversals. Picking one or the other does not change the general character of Time Order and at this stage of exploration, we can remain with the train of thought we have already adopted.

A minimal taste of math

In our "normal," tardyonic, slow-speed world, *Coordinate measurements of Space and Time Segments gives us the Invariant* $X_C < T_C$ and $X_C{}^2 + (iT_C)^2 = I^2$ which when squared always comes out with a minus sign in front of it, thus < 0. A square root of I^2, therefore, comes out as an imaginary number. Now with **tachyonic** $<B>$ whooshing past, $<A>$ finds that X_C is greater than T_C so that $X^2 + (iT)^2 = +I^2$. After $<A>$ takes a square root of this positive number, he gets a positive number for the Invariant, the Spacetime "Distance" from O to **Q**. This OQ in $<B>$'s Home Frame is both her Distance traveled which equals zero *(because she considered herself stationary)* and for her Proper Time, t_P she uses an imaginary number. To make Minkowski's Spacetime equation work across the Light Barrier, $<A>$ must assign to $<B>$'s Space an imaginary number value and to her Time a real number value. The real and imaginary values must, therefore, become switched around whenever observations proceed across the Light Barrier: Space on the other side of the Light Barrier is now "imaginary" and Time "real" in order to provide a real-number value for the Invariant: $(ix_P)^2 + t_P^2 = +I^2$ which, when a square root is extracted from it, gives us a positive number. Is it clear***???***

Restated: $X_C^2 - T_C^2 = +t_P > 0$, as obtained *(calculated)* by $<A>$ Finally, let $<B>$ calculate $<A>$'s T_P *(roles reversed)*: $x_C^2 + (it_C)^2 = +T_P^2$, X being **0**.

What you have seen in this math side-trip is this: **Superluminal Time, t** *(here Proper Time of superluminal $<B>$)* **calculated by** $<A>$ comes out a **"normal" or REAL number and Superluminal Space becomes imaginary.**

If the above "extremely simple" math was a puzzling, *you'll be relieved and delighted by the simplicity of the (no-math !!!) ST MAP METHOD.* That's why we'll be using it as our *topological tool* in the rest of this chapter. For the purpose of discussion, however, let us assume that the direction of Motion *(of moving participants such as $<B>$)* on our Map is arranged from left to right and we have seen that stationary $<A>$ with his friends, $<C>$ & $<D>$, have a workable, empirical method to determine all Time directions from their unique perspective. Moreover, $<A>$ & $<B>$ do not have to agree about their individual conclusions because Observer Dependence (*or Observer-Specificity*) permits harmonious disagreement without getting into fist fights*!* Don't try it in real life, however*!*

The significance of +t of <B> shown on the Map

In the "Minimal Dose Math" section we saw that *Distances of tardyonic Travelers behave very much like Time of tachyonic Travelers* as *both are located in the Space-connected domain OUTSIDE the converging-diverging Lightlines. (see Chapter III, Mapping Spacetime).* Similarly, tardyonic *WLs (Time)* and tachyonic *ITLs (Distances)* are mapped *INSIDE* the converging-diverging Lightlines. So *the real and imaginary character of numbers are actually determined by their (Time-and-Space) "connectedness"*:

F9.10

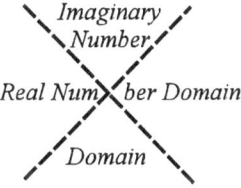

Imaginary Number
Real Number Domain
Domain

Let <B> determine <A>'s Time Order

Until now, we have seen $<A>$ observing and deciding what is $<B>$'s Time Order. Let's see how $<B>$ in turn determines $<A>$'s Time Order. Is it the same as her own*?* As the Order of received Signals is unreliable, we must do it by placing $<B>$'s *ITLs* across $<A>$'s *WL*. But now we see that $<B>$ finds Reversed Order of *ALL* of $<A>$'s Events, #1 - 5 on the next illustration:

F9.11

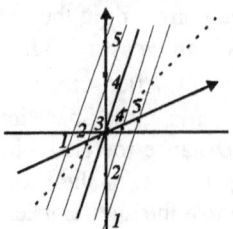

Initial conclusions recapitulated

In <B>'s first tachyonic run, ***all <B>'s Events were observed by <A> to be in the same Time Order but <A>'s Events were observed by <B> to be always in Reverse Time Order.*** So <A> and <B> are *always c onsistent b ut i n d isagreement about e ach o ther's T ime O rder* (*Observer-Dependency again following us also across the Time Barrier*).

By the way, <A>'s Radar Probe has serious limitations in practice because Radar Signals go much too slow compared with <B>'s own Velocity *(relative to <A>)*. But in <B>'s Home Frame, Signals reflected off <A> can return to <B> normally. Tachyonic Monoverse should appear entirely normal to <B> and we'll see soon enough that the *ST* Map will allow us a clear orientation across the Light Barrier.

Relativity always at work

According to Fig. 9.8, <A> observed <B>'s Time direction to be the same. But according to Fig. 9.11, <B> viewed <A>'s Time Order as being in Reverse, clearly ***a lack of mutual agreement !!!*** If, however, we place ***another tardyonic Observer, <C>*** on a path at a Velocity high enough so that his *WL* is mapped between <B>'s *ITL* and her *WL* *(between <B>'s ITL and the adjacent Lightline in between)* then according to <C>'s *ITL*s, it is <B>'s Time Arrow that is now pointed in ***Reverse Direction.***

F9.12

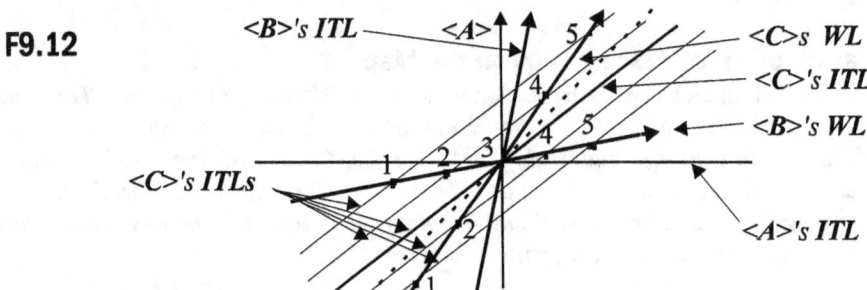

You can see next that while <B>'s Time Order was seen by <C>'s to be in Reverse, <C>'s Time Order is seen by <B> as the same.

F9.13

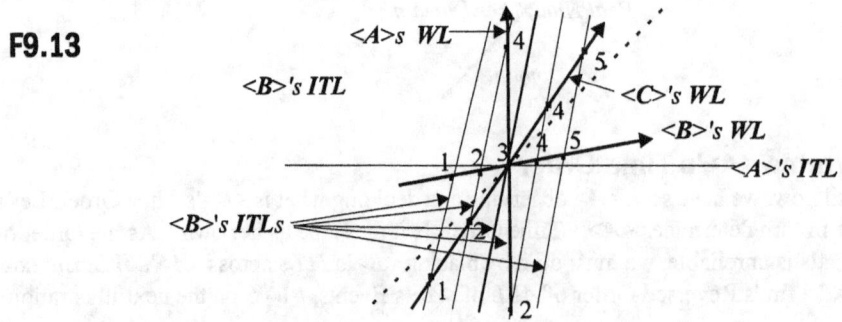

It is instructive to shift *<C>* into a stationary position on the Map *(in Fig. 9.14)* to view everything from his perspective. Note how *<C>*'s and *<B>*'s Coordinates become rearranged: *<B>*'s *WL* takes a plunge downward so that her Time directional Arrow points obliquely down as evidence that *<B>*'s Velocity according to *<C>* is *now FASTER than ABSOLUTE and in Reverse Time Order (!!!)* while to *<A>*, *<B>*'s Velocity was and still is less than Absolute. Time Order behaves exactly the same way. In other words, *Absolute Velocity itself is not always Absolute. It is "RELATIVE," the way Simultaneity is "relative."* All depends on the observed state of Motion relative to the Observer doing the observing. In addition, *whatever topological maneuver we might perform, Time Order as to two Observers always remains relative to these Observers* without regard as to whose Home Frame is made stationary on the *ST* Map. Very-very interesting*!!!* So, according to *<B>*, *<A>*'s Time Order is reversed for *<B>* but *<C>*'s Time Order is not.

F9.14

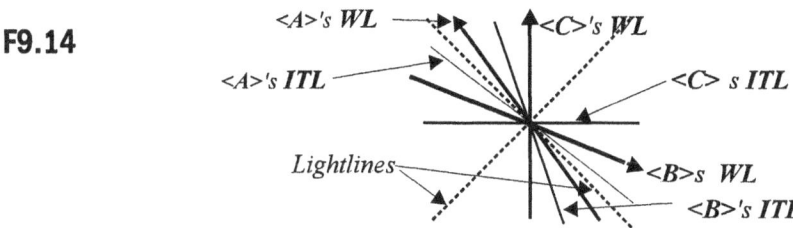

Let us look at the Coordinates of still another tardyonic Traveler, *<D>* placed so that her Path of Motion is moving in the 1-D Space and is mapped from right to left in Space Direction *(not Time Direction)* that *<A>* would consider to be *opposite* (in Space) to that of *<C>*:

F9.15

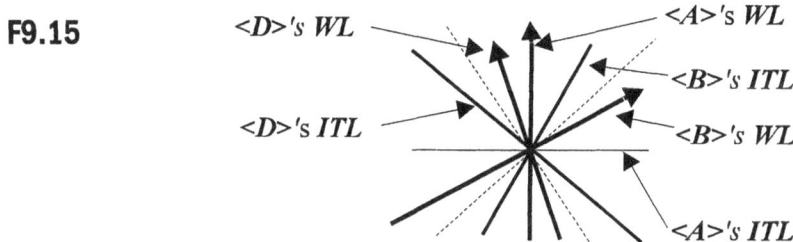

Again, let us see how everything would look from *<D>*'s perspective. You should see that *<B>*'s Time Direction now looks the same to *<D>* whose Time Direction is seen by *<B>* to be in Reverse:

F9.16

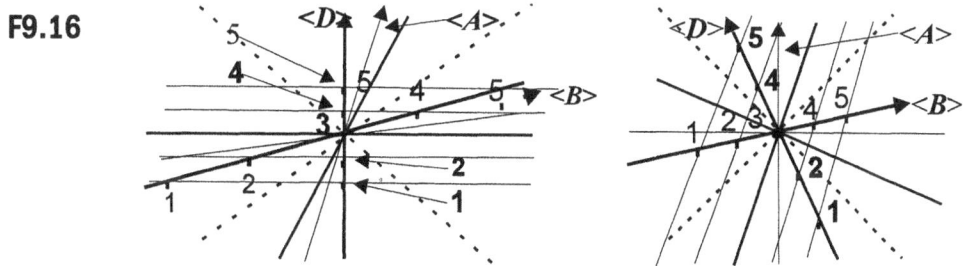

Let us next superimpose the next Observer's (that is *<E>*'s) *WL* exactly on top of *<B>*'s *ITL*. Can we do it? Well, let's just go ahead and do it:

F9.17

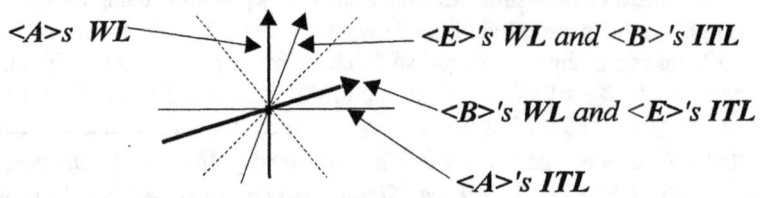

This third possibility is also shown from *<E>*'s point of view on the next illustration. We should see that their (*<E>*'s *and <B>*'s) Coordinates have now become switched so that *<E>*'s *WL* coincides with *<B>*'s *ITL* and vice versa: $X_E = t_B$ and $T_E = x_B$. *<B>*'s velocity is now seen by *<E>* as **Absolute** *(and vice versa):*

F9.18

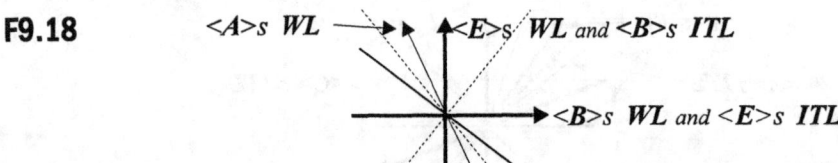

Three Time directions: two Time-connected and one Space-connected

The preceding exercises may not have made everything clear but some kind of a pattern did seem to emerge. Let's summarize it visually in the Map below. The Time Order relationships between the different Tardyonic Observers *(<A>, <C>, <D>, <E>)* should now be better understood:

F9.19

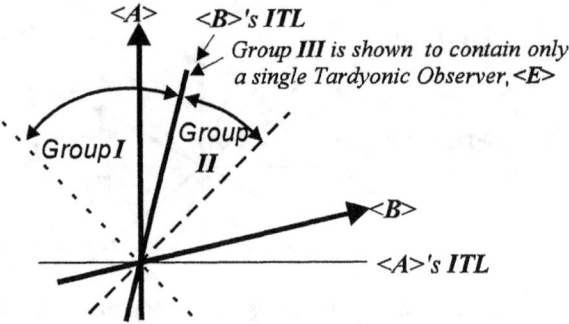

(1) Observers in **Group I** *(<A> and <D> with their WLs on the left side of <B>'s ITL)* view *<B>* as being in the same Time Order while *<B>* views *<A>* and *<D>* as going faster than Absolute Velocity and being in Reverse Time Order.

(2) Observers in **Group II** *(only <C> with his WL on the right side of <B>'s ITL)* view *<B>* as going **faster than Absolute Velocity and in Reverse Time Order** while *<B>* views them to be slower than Absolute and in the same Time Order.

(3) We can position only one Time-connected Observer, *<E>* to represent **Group III.** His *WL* falls on *<B>*'s *ITL* so he and *<B>* view each other as **moving at Absolute Velocity** with their Time Orders mutually indeterminate. The entire *<E>*'s Time duration along *<B>*'s *WL* also lasted no Time at all. On *<E>*'s *WL*, all *<B>*'s Events also occur simultaneously, in less than the proverbial blink of the eye *(see comment #8 toward the end of this chapter)*. Both *(<E> and <B>)* are mutually without any Time direction *(and duration)* to each other. All these strange statements about *<E>* and *<A>* as well as about *<B>* and *<E>* are true because their Velocities must be mutually Absolute.

Now hear ye, hear ye *!!!*

Now, the time has come to pronounce our first verdict about Time Reversal, so *hold your breath:*

(1) *A Spacetime Traveler whose Worldline is Space-connected is going faster than Light to all tardyonic/Time-connected Travelers.*

(2) *Those Time-connected Travelers who find Space-connected Traveler's Worldline located BEYOND their own Isotemporal Lines* (between that and the next Lightline) *find the Space-connected Traveler not only going backward in Time but also FASTER than ABSOLUTE VELOCITY.*

(3) *The other Time-connected Travelers who find a Space-connected WL between their ITL and the Lightline find the Space-connected Traveler merely in need of a speeding ticket for going faster than Light, but no penalty for going backward* (in Time, that is).

(4) *A tardyonic Traveler whose WL falls upon the ITL of a tachyonic Traveler finds their X and T Coordinates mutually reversed and both view each other as possessing Absolute Velocity.*

Thus, GOING IN THE "WRONG" DIRECTION IN TIME ALWAYS INVOLVES GOING FASTER THAN ABSOLUTE VELOCITY.

(5) But *Absolute Velocity itself is RELATIVE*, that is Absolute, less than Absolute or more than Absolute depending on the relative Velocities of the two Observers.

(6) By the way, **(a)** *if we assume that all tachyonic Observers are on Reverse Time Order the moment they exceed the Velocity of Light, the set of rules described above need simply switched.* But then, the initial *Time Reversal is changed back to "normal" the moment Absolute Velocity is exceeded (!!!).* Other considerations remain unchanged. And **(b)** if, instead of faster-than-Absolute Motion, we consider it as happening in Reverse direction in Space, we abolish both faster-than-Absolute Motion and Reverse Time Order altogether *!!!*

Now *Quick, resume breathing* before you suffocate*!* But Time Order still seems a little mysterious, so we better go over this matter again but in a different way so that we can recognize more readily which *TACHYONIC* Traveler in what circumstance is in Reverse Time Order in relation to which *TARDYONIC* Traveler and vice versa.

Let's do it one more time

Let us place the *WL* of Observer, <A> in the usual, stationary position and plot a sample of *WL*s from <F> to <N> in positions representative of all categories of Velocities but now we'll mark them with arrowheads to flag their Time directions.

F9.20

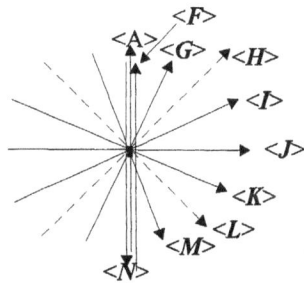

The nine different Spacetime Travelers shown above can be classified according to their "connectedness" as to <A>:

(1) Time-connected: <F>, <G>, <M>, <N>

(2) Light-connected: <H>, <L>

(3) Space-connected: <I>, <J>, <K>

Let us next see how these nine Time directions at these different Velocities compare with that of *<A>*:

<F>'s Time direction same. After all, *<F>* is not moving relative to *<A>*.

<G>'s Time direction same. Both *<A>* and *<G>* are Time-connected.

<H>'s Time direction here is a puzzle. What is it? A special case. Question: What is the Time direction of someone whose clock is not ticking and whose Time is not progressing at all? Answer: The Time is stalled and it has *no direction*. To think of it, there should not be an arrowhead attached to *<H>*'s *WL*. Maybe *<H>*, being on Lightline, is located outside Time or perhaps outside the four Spacetime Coordinates on the stagnating border between Tardyonic and Tachyonic Monoverses intersecting enough with our 4-D Monoverse to become (*with some difficulty*) observable to us.

<I>'s Time direction is the same as with our first tachyonic case, *<B>* in this chapter: So *<I>*'s Time Order is the same to *<A>* even though *<I>* views *<A>* as going backward in Time, a complete lack of mutual agreement! This discrepancy was something entirely new and unexpected when we first encountered it, so let's visualize it again, first as viewed by *<A>*, then by *<I>* on a roles-reversed Map and, finally, by means of *<I>*'s *ITL*s:

F9.21

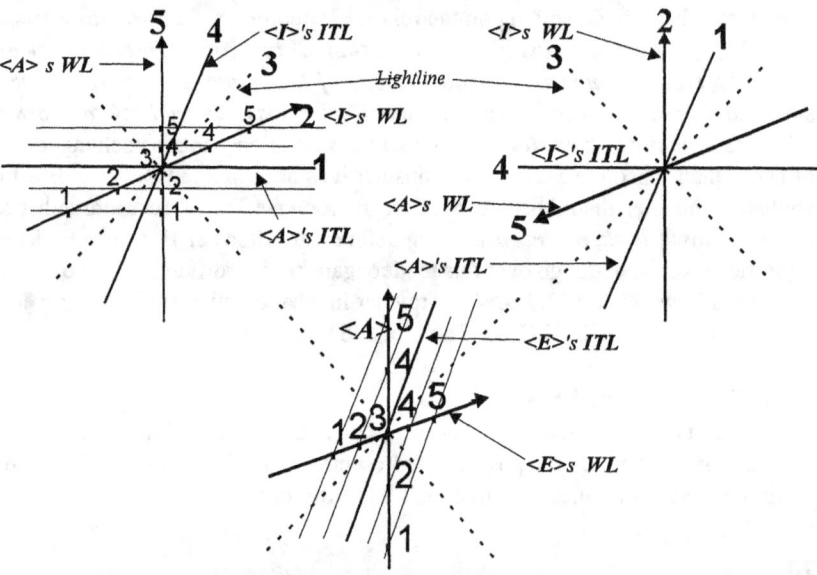

An additional comment is needed about how to compare any roles-reversed *ST* Maps In Fig. 9.21, the first two Maps are drawn from *<A>*'s and *<I>*'s two different points of view. There are five key lines on both upper Maps counting counter-clockwise: **(1)** *<A>*'s *ITL*, **(2)** -*<I>*'s *WL*, **(3)** Lightline, **(4)** *<I>*'s *ITL*, and **(5)** *<A>*'s *WL*.

Now, if we rotate the first Map counter-clockwise until *<I>*'s *WL* is placed into vertical (*stationary*) position and then push *<I>*'s *ITL* into horizontal position (*at right angle to WL),* we also *PUSH <A>*'s *WL* to a position below the (*now*) horizontally placed *<I>*'s *ITL (!!!)* and into faster-than-Absolute Velocity domain. Also, *<I>*'s *WL* is now in the imaginary number domain (*in a roles-reversed setting; also see p. IX-10: "The Significance of +t of <B>").*

<J>'s Time direction, what is it? Well, this is another special case we have seen before. It is hard to say what to make of it. As viewed by *<A>*, *<J>*'s Time, regardless of its

direction, registers *NO TIME* on <*A*>'s clock. The entire *WL* or Time Axis of <*J*> is super-imposed on <*A*>'s *ITL*. What this means is that everything <*J*> ever does *(sequentially)* is, of all things, *SIMULTANEOUS* to <*A*> lasting Zero Time *!!!* The Time Order is *INDETERMI-NATE (see "Group" III, Figure 9.19 on page IX-12)* and <*J*> *views* <*A*> *the same way,* so that:

$$\infty t_{\text{J}} = 0 T_{\text{A}}$$

$$\infty T_{\text{A}} = 0 t_{\text{J}}$$

Note *AGAIN* that *ABSOLUTE VELOCITY is NOT ABSOLUTE to EVERY Observer!* As Velocity is "relative," *Absolute Velocity is also "ABSOLUTELY RELATIVE,"* it is *absolute to SOME Observers but not to all. A Velocity is either LESS THAN ABSOLUTE, ABSOLUTE or GREATER THAN ABSOLUTE to DIFFERENT OBSERVERS in the DIFFERENT TARDYONIC GROUPS of OBSERVERS.* And differences of opinion based on observations are normal*!* Observer-Dependancy *(Specificity)* again*!* Is it Clear*???* By the way, did you ever see clearly through fog before it lifted*?*

<*K*>'s Time direction is Reverse as viewed by <*A*> because <*K*>'s Velocity is greater than Absolute to <*A*>. *BUT*, please, consider how <*A*> and <*I*> viewed each other. *(Fig. 9.21, p. IX-14)*. Don't you see a similarity between <*I*> and <*K*> ? Except that here <*A*> views <*K*> as being in Reverse Time Order and <*K*> views <*A*> as having the same Time Order:

F9.22

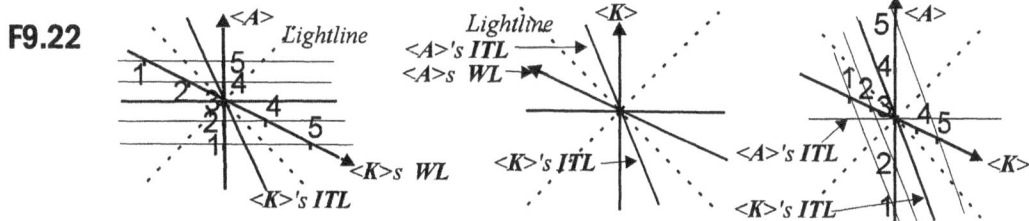

<*L*> The Path is on Lightline, situation same as with <*H*>.

<*M*> - Time order here is Reverse. The interesting thing here is that <*M*> is entirely Time-connected, just like <*A*>. This guarantees that Time directions are always opposite and in mutual agreement. Recall that with both <*I*> and <*K*>, there is mutual disagreement with <*A*>. The reversed-role positions of <*A*> and <*M*> need to be shown:

F9.23

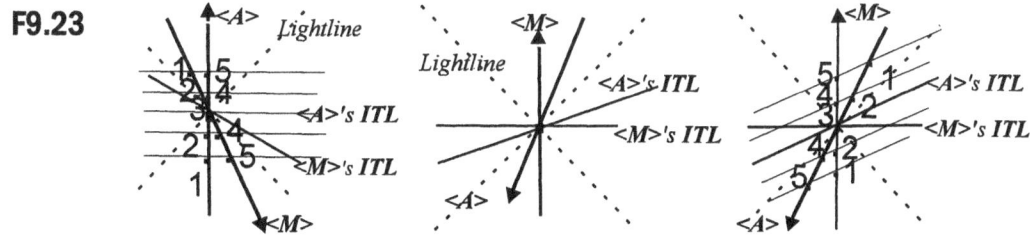

<*N*> - The Reverse Time direction is agreed mutually as expected, just as in the case of Observer, <*M*>. That's because <*N*> and <*A*> share a common Time-connectedness, see Figure 9.20 again for orientation.

The peculiar thing here is the total absence of relative Motion between them that otherwise would quickly separate them in Space*!!!* Both are stationary to each other and they could even be sitting side-by-side on a park bench having a picnic together. The trouble is that while <A> is putting food into his mouth, he would see <N> taking it out of his. And <N> would see <A> doing exactly the same thing. If both happen to speak the same language, they would not understand each other without taping each other's words first then playing the tape backwards. Each would see the other walking backward. If one of them would drop a glass and break it, the other would see the glass fragments on the pavement suddenly reassemble and the now intact glass would inexplicably fly up into the outstretched hand *(that dropped it)*. While <A> is getting older, <N> is getting younger but this process would be too slow to be obvious. A number of odd and funny things taking place in this setting would make the day a unique situation comedy.

We did not consider the possiblility that a second Traveler, <J_2> heading in a direction of Space *(according to <A>)*, opposite to that of the first <J> with whom he would make up another odd couple, <J_1> and <J_2>, very much like <A> and <N>. But both of them would also be indeterminate to <A> as to Time Order. On second thought, each of the other Observers from <G> through <M> could have a stationary companion going in the exactly opposite direction in Time similar to <A> and <N>.

After reviewing these representative examples (*<A> paired off with <F> to <N>*), we should be able to set down additional rules about Time Arrows in Spacetime. In view of the importance of Time Reversal as a reward for merely going faster than Light according to so many books and magazine articles, the extra attention given to Time Reversal in this chapter was certainly justified. It looks like the additional rules do not add up as simply as expected but let's do it anyway. So, here they are:

The final traffic rules for Time

(1) Not all Tardyonic Travelers share the same Time Order with tachyonic Travelers: <F> and <G> agree with <A> but <M> and <N> disagree with <A>.

(2) All Tachyonic Travelers shown on our Map share the same Time Order and those we did not put on our Map *(mirror images to <I> and <K> pointed in opposite direction in Space-connected domain)* would not have agreed with <I> and <K> about Time Order either.

(3) A tardyonic Time-connected and a tachyonic Space-connected Traveler never agree about each other's Time Order. If one considers the other as being in Reverse Time Order then the other would consider the first one as being in the same Time Order and vice versa.

(4) Among two Observers both oppositely Time-connected or Space-connected, there is mutual disagreement about their specific Time directions but they would be in mutual agreement that the "other one" is in Reverse Time Order by moving faster than Absolute Velocity.

A disagreement between tardyonic and tachyonic Observers about each other's Time Order in certain situations and the unreliability of the order of Signal arrival as to Time Order should have been foreseen. What is involved here is this: ***ABSOLUTE VELOCITY and SIMULTANEITY are two sides of the same coin and both are "relative" for the same reasons.***

(5) Light-connected *ST* Travelers have *NO Time Order* at all as their clocks are stalled, not running.

(6) In case of Absolute Velocities, the observability of each other is nil as no observations can be concluded in zero time. So, they are mutually not in a position to make or compare observations.

(7) *Reverse Time direction is always associated with observed Velocity that is greater than Absolute.* But: *Relativity also governs Absolute Velocity!!!* What is Absolute Velocity to one Observer may be greater, the same or less than Absolute Velocity to another Observer.

There is a final way to visually summarize Time relationships between <A> and <B> at different tardyonic or tachyonic Velocities and compare their differering views of each other as to when the observed Time Order becomes reversed. This is shown in the next illustration. Please, note that: **(1)** to *<A>*, *<B>*'s Time Order is always the *same* as his own when *<B>*'s **WL** is located in Sections I and II *(above his horizontal ITL)* and always in *reverse* when *<B>*'s **WL** is located in sections III and IV *(below his horizontal ITL);* but **(2)** *<B>*, located *(according to <A>)* in Sections II and IV, considers *<A>* to be in Reverse Time Order; and **(3)** when *<B>* *(again according to <A>)* is in Section III, she considers *<A>* to be in the same Time Order; and, finally, **(4)** in Sections II and III, *<A>* and *<B >* always mutually disagree.

F9.24

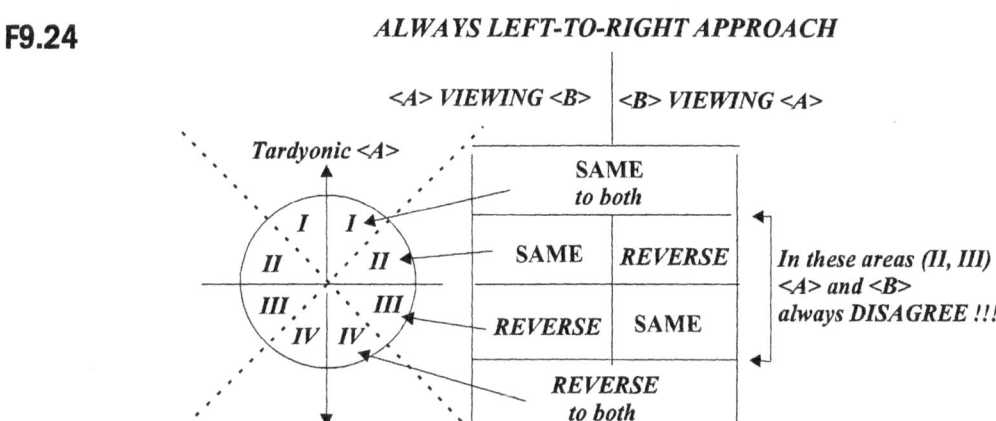

The only alternative view

In case *<A >* considers *<B >* to be in Reverse Time Order the Moment she crashes the Time Barrier *(a situation we did not consider)* then, in Sections II and III, their views of each other simply become switched so that in Section II, *<A >* views *<B >* to be in Reverse Order and in III in the same Order. *<B>*'s views about *<A >* also become switched: in II, *<B>* views *<A>* to be in the same Time Order and in III, *<B>* views *<A>* to be in Reverse Order.

Absolute Velocity = Absolutely fascinating

Of all the tachyonic Velocities, Absolute Velocity presents unusual possibilities. It may be interesting to add a few more comments about topics that could not be discussed at any length. These cross over into areas of physics we have no business entering. However, this topic itself is intriguing enough and thought provoking. As we do not have to get anything right in this chapter, we have perfect freedom to say anything that comes to mind:

(1) At Absolute Velocity, *<B>* can be in two or more *<A>*'s places at once, at the same split attosecond or actually no Time at all. But being in more than one place at the same Time happens only in *OUR* Time Frame. In *<B>*'s own Monoverse, these two or more places are transited sequentially, one at a time. The situation is quite the opposite from Light occupying all places in no Time at *all in its own Time Frame* while for us it goes through them one after the other. Light's clock, if it has one, is not ticking, not moving at all. Tachyonic Time at Absolute Velocity, in contrast, proceeds rather normally in its own Home Frame.

(2) The clock of a Tachyonic Traveler moving at Absolute Velocity is tracing out a Worldline on our Space Axis and zips through our entire tardyonic universe in zero Time: $t = X$ while $x = T$. All this means that *<B>* can spend a great deal of Time at Absolute Velocity which to her would seem like being stationary, not moving at all while her own Time proceeds totally at normal rate. The Distance traveled by *<B>* on our Space Axis can be short or long, any length at all, depending entirely on *<B>*'s duration of existence. What she would register on our Time Axis depends entirely upon the physical size of her Spaceship divided by the Velocity of Light.

(3) In physics, an absolutely fast tachyonic particle can be in many of **OUR** places at the same *(our)* Time. It can spend a very long Time traveling in its own Time Frame while our own clock does not register hardly any Time at all. In the atom, all particles may be in *(orbiting)* Motion at nearly the speed of Light or faster, even at Absolute Velocity. If one of these particles has the occasional talent going tachyonically backward in Time, it can keep going back and forth in Time without net advance on our clocks or it can advance ratchet-like instead with irreversible net forward motion in steps that are the smallest possible Time Segments *(Time Quanta)* too brief for us to measure. So in our view, Time still flows smoothly like water in a mythical river.

(4) A tachyon, passing through our Monoverse can make itself visible by emitting Signals all along its trail in a Wilson's cloud chamber *(or its modern equivalent)*. But for us, it would seem to be two identical particles originating at the *(nearby)* Passing Moment and then seemingly going in two opposite directions. The overall result is similar to two photons moving apart from each other from the site of their apparent origin powered by the destruction of a particle by its antiparticle. Or there may be an apparent creation of two opposite particles like an electron-positron pair from energy liberated by some other interaction:

F8.25

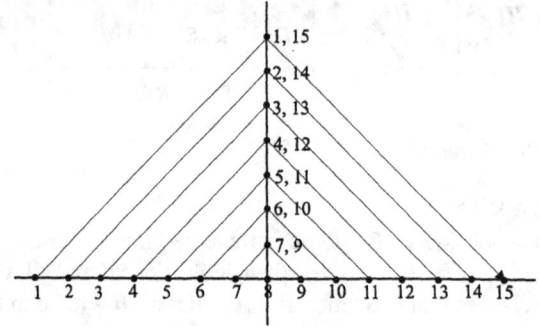

A tachyon at less than Absolute Velocity can also create an illusion of two unidentical particles flying apart at unequal velocities:

F8.26

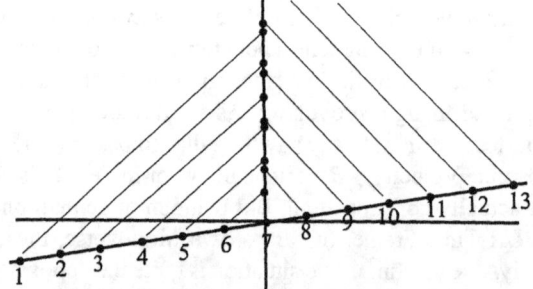

(5) In the Tardyonic Monoverse, the three Space Coordinates enter Minkowski's Geometry equation as real numbers and the Time Segment is given as an imaginary number:

$$I^2 = X^2 + Y^2 + Z^2 + (iT)^2$$

The tardyonic Time Coordinate moves into Space-connected domain as <**B**> becomes tachyonic and her **WL** also crosses over to the other side of the Lightline. In this case, Time value in our calculations becomes a real number, Space values become imaginary numbers and Minkowski would then write his Invariant equation as follows:

$$I^2 = (iX)^2 + (iY)^2 + (iZ)^2 + T^2$$

which Hamilton, the Irish mathematician, would write differently in his own *(super*-complex, *quaternian)* number system:

$$I^2 = (iX)^2 + (jY)^2 + (kZ)^2 + T^2$$

The above-mentioned abstract number system plays out in a Complex Space with three imaginary Coordinates and one real one, a system sometimes utilized explicitly in some areas of subatomic physics, lending credence to the hint that the quantum world may well be populated by ghostly denizens going at least some of the Time or in some places faster than Light, maybe even backward in Time.

(6) Backward-in-Time progression implies Velocities even faster than Absolute, an extravagance only privileged particles can boast.

(7) In the subatomic world, a faster-than-Light Velocity may be the key to preferred travel routes. Maybe there are renegade electrons after all. Maybe, nonlocality does hold true and maybe there was a superluminal expansion after all at a "Time" when the cosmic inventory consisted of all photons or other non-matter and when gravitation did not yet have a firm grip on everything existing at the Moment. Otherwise, our own Monoverse would have ended up as just another transient black hole in some lonely heavens.

(8) If Absolute Velocity in one Frame does not involve passage of Time in some other Frame, isn't it possible that a universe could be created on "borrowed" energy, allowed to last billions or trillions of its own years before becoming extinct? But in the Frame of Reference from which it escaped on a Time Loop, it lived out its virtual existence in the minutest amount of Time consistent with Heisenberg's Uncertainty Principle *(here not defined but look it up and ponder).* A *(near-)*infinity of Time in one Frame of Existence can measure *(almost)* no Time at all in another.

A wake-up call to Reality

Now a concluding remark to shock you out of this Dream World of Imagination. Everything in this chapter was nothing but *pure fabrication*, impossible in the real world and *contrary to all known natural laws.* The presentation was sprinkled liberally with *relativistic buzzwords* to make it impressive, accompanied by visual persuasion to make it convincing and in some places was dressed up in a skimpy *mathematical attire* to appear attractive, all in all something like *a big April Fool's joke.* Don't believe a word of it! Now, go to the next chapter and *re-enter Reality.*

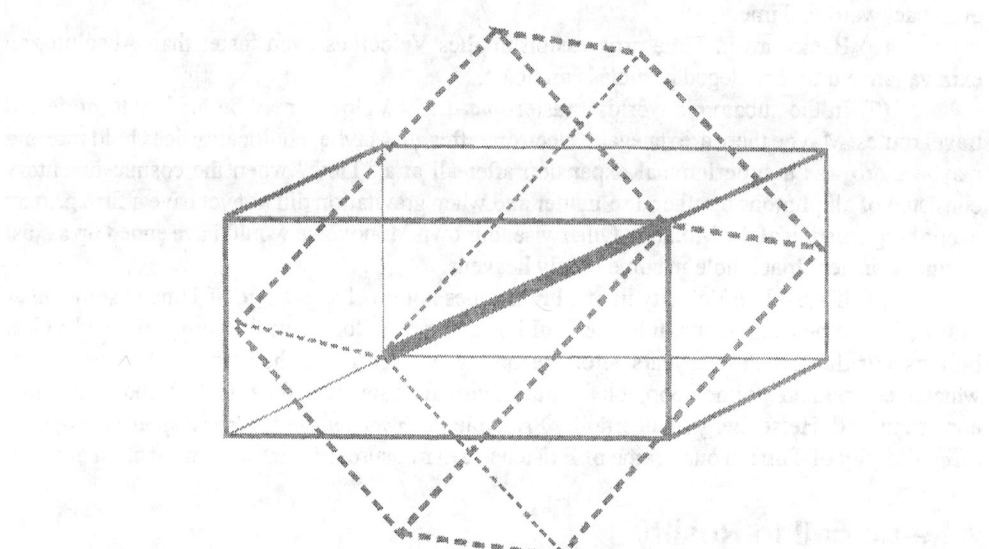

X - THE M-M EXPERIMENT

A common-sense model of the Michelson-Morley Experiment

In the first half of Doppler Acoustics, the moving train experiment carried both the *Source (of the Signal)* and also the *Receiver (Observer)* of Sound Signals, the Receiver standing proxy for us. In the virtual experiments, both participants were made to move through the air at a constant Velocity. It was pointed out that the *Air Wind* was responsible for changing the *True Velocity* of the signal through that medium, the air, into an observable *"Effective" Velocity* measured across a fixed Space Segment *(Distance)* that also was moving through the same medium, the air. The measured Sound Velocity was found to be either the *sum or difference* of the two Velocities: the Sound Velocity and the train Velocity. The virtual setup of the experiment was then used *to challenge us to imagine that the air around the train was rendered undetectable.* The *analogy with the Michaelson-Morley Experiment (MME)* was duly pointed out. Now we have finally arrived at the historic experiment and our task here is to find out what happened to the *ETHER WIND (and the "Effective" Velocity of Light).*

As we have already done with other effects of Special Relativity, the Michelson-Morley Experiment (*MME*) carried out in the 1890'ies can also be better understood if we first examine a simpler model made up of elements of familiar experience. Rather than using fast airplanes or Space vehicles, a suitable model here is something as ordinary as a *BOAT RACE!*

The idea behind it is the rather simple though not so obvious fact that *it takes less Time for a boat to travel across the stream and back than it takes (for the same boat) to travel an equal Distance upstream and back.* Once we develop the mathematics of this arrangement, the broader implications of the *MME* are easier to grasp.

To set up the boat race, we need **(1)** two identical *boats* to race at the same time or the same boat running the two different courses consecutively, **(2)** roped-off *paths* of equal length set at 90 degree angles to each other, one oriented across the stream and the other up-and-down the stream, and finally **(3)** a *river* sufficiently wide and the current swift enough to bring out differences in the duration of the boat races:

F10.1

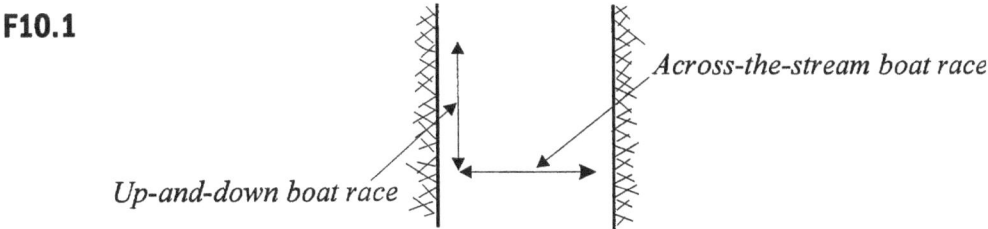

Across-the-stream boat race

Up-and-down boat race

Let us place a *BOAT* on the river, assign it a *cruising Velocity of 5 miles per hour* and let the *CURRENT Velocity be 3 miles per hour.* These figures are chosen so that practically all calculations could be done in whole numbers. Note that the Velocity of the boat is its Velocity through water, the medium through which it travels and which is causing a *WATER WIND* of 5 mi/hr to "blow" past the boat. Likewise, the Velocity of the current is its Velocity relative to the riverbank to which our boat-racing setup is firmly oriented and where the station for the official judges of the race is located.

The primary purpose of the race is to compare the total Time spent traveling round-trip across the current *(Transverse Motion)* with the Time traveled up-and-down the current *(Parallel Motion).* We'll tackle the simpler case of Parallel Motion first.

Parallel Motion

With true Velocity in water fixed at 5 miles per hour struggling *AGAINST the current*, the boat can achieve an *EFFECTIVE VELOCITY* that is less than 5 mi/hr. Let us see: in 1 hour the boat travels 5 miles in water but the current brings it back 3 miles. The resultant *Effective Velocity is a mere 5 - 3 m/hr = 2 mi/hr.*

F 10.2

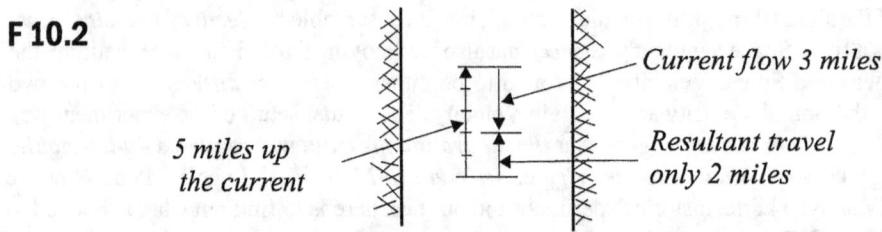

Current flow 3 miles

5 miles up
the current

Resultant travel
only 2 miles

Going *DOWNSTREAM* is easier. In one hour, the boat travels 5 miles in water and the current takes it another 3 miles further, giving the boat an *Effective Velocity of 5 + 3 = 8 mi/hr:*

F10.3

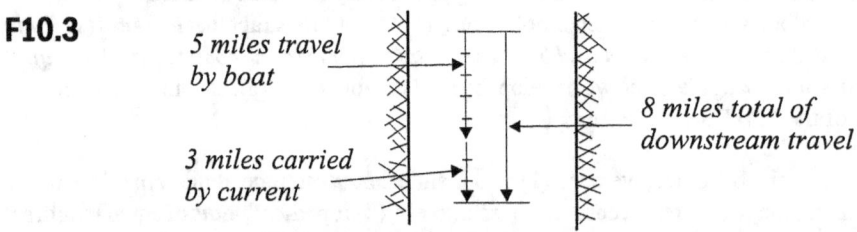

5 miles travel
by boat

8 miles total of
downstream travel

3 miles carried
by current

It should be obvious that the boat goes faster downstream than upstream. Now comes the important observation: *The Effective Velocity actually achieved is either the SUM (downstream) or the DIFFERENCE (upstream) of the two component Velocities.*

Transverse Motion

Let us ask: what happens to the boat if it is aimed directly across the river at 90 degrees to the riverbank? In 1 hour it travels 5 miles across the water but in 1 hour the current moves it also downstream 3 miles and off its course:

F10.4

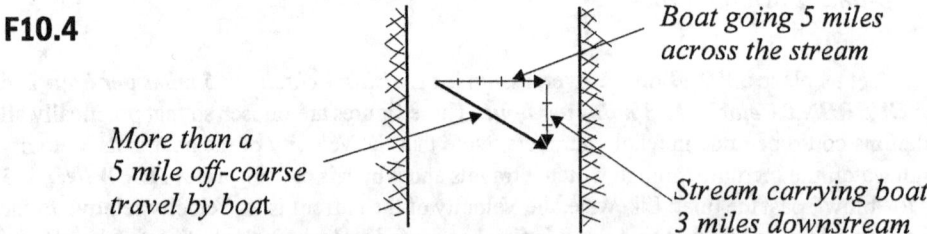

Boat going 5 miles
across the stream

More than a
5 mile off-course
travel by boat

Stream carrying boat
3 miles downstream

To travel directly across the river, the boat has to be *pointed sligtly upstream* in order to compensate for the downstream drift. Note that the resultant length of travel relative to the river bank is *not* the same as when the boat is aimed directly across the river at 90 degrees to the water current.

F10. 5

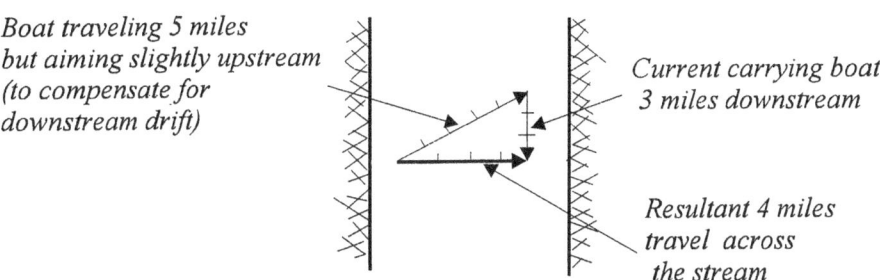

Boat traveling 5 miles but aiming slightly upstream (to compensate for downstream drift)

Current carrying boat 3 miles downstream

Resultant 4 miles travel across the stream

Now the boat has traveled 5 miles in water and the current has taken it downstream 3 miles. The actual path traveled straight across the stream, calculated with the help of the Pythagoras Theorem, is:

F10.6

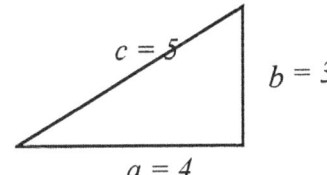

$c = 5$ $b = 3$ $a = 4$

So the round-trip in both directions consists of two legs, each 4 miles long and properly marked off by buoys anchored to the river bottom.

Let us restate the basic, <u>unmodified velocities</u>:
(1) Velocity of the boat in water: 5 mi/hr
(2) Velocity of water current: 3 mi/hr

The "<u>actual</u>" or **Effective Velocities** of the boat in those different runs *(combining the basic velocity of the boat in water with the velocity of the water current)* come out to be:

(1) 4 miles Upstream at 2 mi/hr
(2) 4 miles Downstream at 8 mi/hr
(3) 4 miles Across the Stream in both directions at 4 mi/hr

Round-trip times

Now we have all the information needed to calculate the amount of Timespent racing the boat in both parallel and transverse round trips with *the length of each leg of the round-trips kept at 4 miles.*

PARALLEL MOTION
(1) 4 mi upstream at 2 mi/hr: 2 hours
(2) 4 mi downstream at 8 mi/hr: ½ hour

TRANSVERSE MOTION
(1) 4 mi across the current at 4 mi/hr: 1 hour
(2) 4 mi back at same velocity: 1 hour

TOTAL ROUND-TRIP TRAVEL TIMES
(1) Parallel round-trip: $T_P = 2\frac{1}{2}$ hours
(2) Transverse round-trip: $T_T = 2$ hours

You can now see with your own eyes that there is a noticeable difference in the round-trip times between parallel and transverse directions. Therefore, the less Time-consuming transverse direction has definite advantage in all races:

$$T_P > T_T$$

while the same kind of racing in the absence of water current effect *(Water Wind!)* would give us:

$$T_P = T_T$$

Converting boat trip results to MME

Let us next describe this boat race experience in those mathematical terms that can be entered directly into equations applicable to the *MME* with Light *(instead of boats)* propagating in *ether* instead of *water*:

c = Velocity of boats relative to water *(analogous to the Velocity of Light through ether)*

v = Velocity of the water current or Water Wind in the river *(analogous to Velocity of the Earth through ether producing a flow of ether wind blowing through the test setup)*.

D = Distance traveled by the boats in one leg of the round trip both in parallel and transverse travel through the Water Wind *(same as Light traveling through Ether Wind)*:

T_P = Time spent in parallel round-trip travel.

T_T = Time spent in transverse round-trip travel.

$c - v$ = upstream velocity in parallel travel

$c + v$ = downstream velocity in parallel travel

$\sqrt{c^2 - v^2}$ = across the stream, or transverse, Velocity *(same in both directions)*

Now we can derive the *travel durations* for the entire round-trip race knowing that T equals the Distance traveled divided by the Effective Velocity:

$$T_P = \frac{D}{c-v} + \frac{D}{c+v} = \frac{D(c+v) + D(c-v)}{(c-v)(c+v)} = \frac{Dc + Dv + Dc - Dv}{c^2 - v^2} = \frac{2Dc}{c^2 - v^2}$$

$$T_T = \frac{D}{\sqrt{c^2 - v^2}} + \frac{D}{\sqrt{c^2 - v^2}} = \frac{2D}{\sqrt{c^2 - v^2}}$$

Again, comparing round-trip Times applicable to MME

Comparing the parallel round-trip Time with the transverse roundtrip Time could be done in different ways. First, we could subtract the smaller from the larger to obtain the *difference* in travel Times. However, it is more useful to express the comparison in form of a *ratio*. A ratio remains the same for all parts of the two different travel directions and does not depend on any given length of the run. Thus, *the ratio of parallel to transverse race Time is the most useful basis for comparing measurements*:

$$\frac{T_P}{T_T} = \frac{\dfrac{2Dc}{c^2-v^2}}{\dfrac{2D}{\sqrt{c^2-v^2}}} = \frac{c\sqrt{c^2-v^2}}{c^2-v^2} = \frac{c\sqrt{c^2-v^2}}{(\sqrt{c^2-v^2})(\sqrt{c^2-v^2})} = \frac{c}{\sqrt{c^2-v^2}}$$

The above result can be tidied up by dividing both the Numerator and the Denominator by the velocity of Light, c:

$$\frac{\dfrac{c}{c}}{\dfrac{[\sqrt{c^2-v^2}]}{c}} = \frac{1}{\sqrt{1-\dfrac{v^2}{c^2}}}$$

If we express v as a fraction of c, the final equation becomes more familiar:

$$\frac{T_P}{T_T} = \frac{1}{\sqrt{1-v^2}}$$

Of course, the transverse Time divided by parallel Time is the inverse of the above result:

$$\frac{T_T}{T_P} = \sqrt{1-v^2}$$

Final result

From this we can state that:

$$T_T = T_P\sqrt{1-v^2} \quad \text{and that} \quad T_P = \frac{T_T}{\sqrt{1-v^2}}$$

Recall that $T_P > T_T$. This relationship holds for all values of v *(c being a constant!)* as long as $v > 0$. In our boat race experiment, the magnitudes of c and v were clearly unequal. If, however, the speed of the current, v is only a small fraction of c, say 1/10,000-th of the boat Velocity (c), it would take a rather accurate instrument setup to register the small but still a significant round-trip Time difference *(expressed as ratio)*.

Another MME analogy

To explore just this kind of situation before switching from racing boats on water to racing photons through ether, let us take a look at a modification of the above-described double boat race. Suppose that the entire setup was located on a very large and quiet river but the water was somehow rendered completely undetectable. With the above boat race setup, it should be easy, in principle, to determine if the boat racing rig was motionless on the water or was traveling quietly through the unobservable waters. This was exactly the problem facing Michaelson and Morley when they took up designing their experiment to detect the existence of "ether" that everyone was certain it was there even if no-one had been able to find it.

What was MME to prove

The question posed in astronomical terms was this: Is it possible to detect Earth's movement through the hypothetical and so-far undetectable ether? The movement of the Earth around the Sun was believed to create an **ETHER WIND** quite analogous to the **Air Wind** blowing around the moving train *(in Chapter V)* and also to the **"Water Wind"** flowing through the boat race setup.

During the 19th century, it was commonly accepted that Space was filled with stationary ether serving as a medium through which Light waves could propagate just like water was needed for propagating ocean waves. It had to be extremely light so that all celestial bodies could pass through without encountering the least bit of mechanical resistance. At the same time, it also had to be exceedingly hard and transparent to transmit Light at such difficult-to-believe, extraordinary Velocity. These properties added up to something extremely fantastic and unbelievably difficult to imagine. But there it was, a substance believed necessary for the propagation of Light. Michelson and Morley took up the challenge to prove that it really existed.

The MME instrumentation

The instrumentation used by *M & M* was essentially an improved Fizeau Spectrometer used by Michaelson earlier to measure the wavelength of Light to a degree of accuracy never before achieved. Because of its historical importance, let us take a close look at the setup.

The first component of the apparatus was a Light Source. From there, a beam of Light was sent to a half-mirrored glass that reflected half of the light at 90 degrees to the side, allowing the other half to go straight through *(ignoring the slight parallel deflections due to double refraction by the thickness of the glass and other optical problems needing adjustment)*. At this point, we have obtained *two Light beams directed at 90 degrees from each other* and analogous to the boats racing across and parallel to the stream:

F 10.7

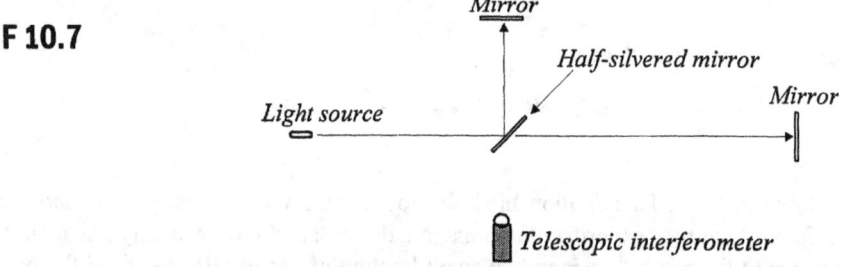

The two fully silvered mirrors would reflect the two perpendicular Light beams back to the same half-silvered mirror which, again, would allow half of each returning beam to go straight through, the other half being reflected at right angles:

F10.8

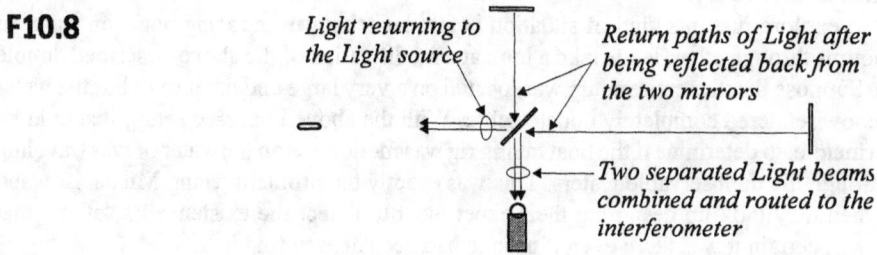

Thus one half of each secondary Light beam would go back toward the Light source, *the other half of each beam would be directed into a telescopic interferometer* (*Fizeau's spectrometer adapted to analyze the phase coincidence of the two incoming, reunited Light beams*) to be analyzed for the interference *(of Light)* on which the detection of the Ether Wind depended.

If the two separate Light beams would enter the interferometer in the same phase, a look through the eyepiece of the interferometer would reveal a *brightly lit field*. If, on the other hand, they were 180 degrees out of phase, there would be total cancellation of the waves and a *dark field* would be seen instead. For partial phase coincidence, an intermediate bright-dark field would be seen. The two paths did not need to be the same in length. One of the mirrors was equipped with an accurate micrometer screw allowing that path to be lengthened or shortened to adjust the brightness or darkness of the telescopic field so that the amount of adjustment from one bright *(or dark)* field to the next could be accurately measured.

By means of the micrometer screw adjustment, moving one of the reflecting mirrors from one bright field to the next *(or from one dark field to the next)*, Michelson was previously able to measure the wave length of Light with unprecedented accuracy. The same setup was used in the *MME* and, additionally, both Light paths were lengthened by multiple back-and-forth reflections between two pairs of mirrors before combining both beams in the interferometer. To keep everything firmly fixed, the apparatus was built *on a heavy stone slab*. To isolate it from incidental vibrations caused by traffic, nearby machinery and people walking in the lab, the stone slab was *floated on a bath of mercury*. The slab could be *rotated* to point the perpendicular Light paths in any direction desired but always remaining parallel to the Earth's surface. Additionally, *the experiment was repeated at different hours of the day and night* to take advantage of the different positions of the Earth in its rotation around its axis and also in *different seasons* while the setup *(on the Earth)* was located at different positions in its orbit around the Sun.

Considering MME results

Recall that the boat in the parallel race required more time to complete its round trip. The expected additional Time needed for Light to complete its parallel journey was estimated to be well within the capability of the interferometer to register a change in the brightness/darkness in the viewer. But the equipment failed to register any change at all in the telescopic field. M & M tried it repeatedly and in various positions of the Earth in different hours of the day and in different seasons of the year. There was *NO CHANGE* in the viewer at any time. Nothing. The reluctant conclusion: *there was NO ETHER WIND!* Note that *the experiment did not directly conclude anything about the existence of the ether!* Improving the experimental design and instrumentation did not help. The experimenters were bold and *reported their "failure"* thereby dumbfounding themselves and everyone else.

Now the theorists got busy inventing explanations in an effort to save the existence of ether. Different conclusions were proposed:

(1) Can't understand it! Something must have gone wrong with the experiment.

(2) The Ether Wind blowing through the equipment flattened it in the direction of motion and in proportion to the force of its penetration related to Velocity, more so in the parallel than in the transverse direction thereby canceling out the difference in the expected round-trip Times: *the Lorentz (also known as Lorentz-Fitzgerald) Contraction.* If Observers in moving Systems were unable to notice any of this foreshortening it was only because their *measuring sticks also "contracted"* exactly in the same direction and to the same proportion.

(3) The ether remained wrapped around all material bodies like a padding and did not allow the Ether Wind to blow through the test equipment *the Ether Drag.*

(4) The *ether* *(was there but)* **had no effect on the transmission of Light.**

(5) *Why suppose that the ether existed* in the first place. It was not needed for any other purpose and, evidently, not for transmission of Light.

It was hard for scientists to give up on ether. Maxwell's theory of electromagnetism was somewhat mistakenly believed to depend on it, so it had to exist. The ether wrap (*or drag*) was still a possibility and it took quite a few more experiments with repeatedly updated equipment well into the 1930-ies to finally put the ether question to *rest (in peace!).*

The fallout of the negative *MME* was positive, however. It got rid of the troublesome concept of the ether and hastened the hesitant acceptance of the still incomplete theory of Special Relativity already being worked out little by little by Lorentz who managed to carry it through Lorentz Transformation stage, leaving it for Einstein merely to restate the entire theory in a completely general form with a few extra details added.

XI - ACCELERATION FROM 2-D MOTION

The trouble with Lineland

Except for a little cheating at the *O*-Event to avoid collisions and to synchronize clocks, our virtual experiments were always staged in a Space of just one dimension. And true to promises, elementary algebra was all we needed. Enjoying just a little more cosmic elbow room was nice but we totally overlooked how dangerous a place it was? As pointed out in the Preface, *a 1-D Lineland is like a one-lane highway featuring two-way traffic*. This 1-D Space is a very strange place, indeed. It is where the freedom of choice as to location is confined along a straight, thin line and anything that moves is also limited in the choice of direction and inevitably finds itself on a collision course with everything in its forward path. An Observer making a study of the natural events in this restricted setting is always in grave danger of being hit, smashed into, run over or through by the objects of his attention. Only by limiting our dangerous experiments to the imaginary world of Virtual Reality have we not only saved our own lives but also those of all the other participants.

2-D and 3-D solutions

In the real world of 3-D Space, and this holds true also for 2-D Space *(or surface)*, catastrophic events of this kind are usually averted either by stepping aside at the right Moment or by arranging everything in advance so that all moving objects pass at a safe Distance. So the cosmic elbow room does come handy but, as we'll see next, at a price. Getting into more than one dimension of Space puts us into *a different mathematical territory*, one without our familiar landmarks. Let's enter and look.

Consequences of 2-D setting

As long as Motion proceeds along a straight line and the Observer remains at some Distance from its path, all observations would be conducted from outside of the one-dimensional world of Motion we have studied so far. An outside location like this need not remain geometrically isolated but can be linked with any one-dimensional path of Motion by means of a two-dimensional surface placed so that it trans-sects and contains them both. We can, therefore, generalize that *if a moving body misses the Observer, the Space in which the Motion is observed is NOT one-dimensional but (at least) TWO-DIMENSIONAL.*

In our experience, this two-dimensionality of observed Motion is the rule rather than the exception as most moving objects fail to run into us. However, what seems an advantage in one sense merely serves to complicate the application of Special Relativity to those moving objects we see in our real world. This is so because, strictly speaking, *the mathematics of Special Relativity we have so far learned was applied only to Uniform Motion in just ONE-DIMENSION of SPACE.* The single most important consequence of the two-dimensionality of observed Motion is this: *Motion observed at any Distance from its path NO LONGER appears UNIFORM in VELOCITY. It becomes transformed into Motion with relative, linear velocity that ALWAYS keeps CHANGING* as observed from our separate vantage point.

This surprising new development needs illustrated by an example from everyday life. Suppose you are standing in the middle of a narrow road. A car still some distance away is coming towards you. As long as you stay where you are, the observed Velocity of the car is not only the same as its Ground Velocity but it also remains unchanged right up to the Moment it crashes into you. If, however, you prudently place yourself beside the road, off the traveled part, what you see is something entirely different. *Even as the car gets closer and closer, it cannot get any closer than your Distance from the road. The approaching Relative Velocity keeps slowing down, even as its Ground Velocity remains unchanged until, at the exact Moment of Passing, it momentarily slows to a complete stop. Immediately thereafter, its Distance from you starts to increase again and the apparent Velocity* (relative to you) *starts to pick up as the car is now going away from you. Finally, at a considerable Distance, its Velocity relative to you is again almost the same as its Ground Velocity.*

Velocity observed in 2-D

This roadside scenario must be quite familiar to everyone. But it takes more than casual reflection to realize that in addition to a **Constant Velocity relative to the ground**, the car also has a **Variable Relative Velocity relative to the roadside Observer**. These two Velocities are definitely **NOT** the same. In our commonsense judgement, such hairsplitting distinctions are not important because our past experience, in and out of all kinds of vehicles, has made us pay attention only to Ground Velocities.

An interesting sideline of the passing car experience (*which we'll be taking up in the last chapter*) is this: when the car gets close enough, the Observer has to turn his eyes to follow its path. This feature of passing vehicles is again so mundane that it is usually overlooked. But it serves to point out that there is another kind of Velocity that passing Motion can possess: **ANGULAR VELOCITY!** This Angular Velocity is greatest where the Relative **RADIAL VELOCITY** is lowest, that is, at the Passing Moment. We may, therefore, conclude that **what the car loses or gains in Radial Motion, it gains or loses in Angular Motion.** In this way, **the total quantity of observable Motion always remains unchanged and is equal to GROUND VELOCITY:**

Note that the Ground Velocity is almost equal to Radial Velocity if viewed from a great Distance and if the small component of Angular Motion is ignored. Radial Velocity, as noted, is zero at the exact Moment of Passing and the existing Motion is momentarily Angular Motion. A pure Angular Motion is Circular Motion without any Radial component directed toward or away from the center of the circle. This kind of Velocity constitutes a special category and does not belong to our study.

Mapping 2-D Motion

It is, again, instructive to present visually what has been said in words, first by **drawing two roads on the GROUND MAP** then **plotting the same experiment on a SPACETIME MAP**. The two roads allow us to show the two alternative positions of the Observer, first in the middle of the road then also beside the road. The two observing positions are placed differently as to the Path #1 and #2. On the Spacetime Map, the (*Stationary*) Observer's Worldline and the Worldlines of the two cars are plotted. The Path #1 is not shown as a straight line going to the other side of the Stationary Observer but is kept on the same side after being "reflected" off without loss of Velocity. The Path #2 represents 2-D Motion.

F11.1

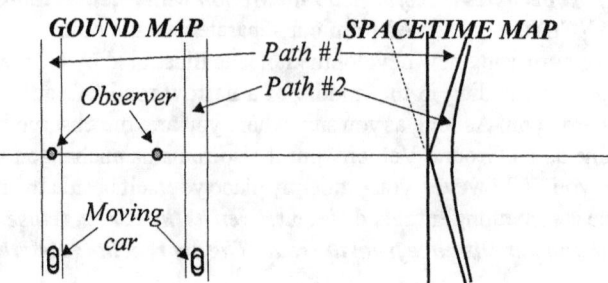

As the Ground Velocity of the car remains constant, we can consider the Ground Distance traveled proportional to the Time spent traveling. Thus, a line parallel to Path #1 but drawn through the Observer's location can serve as the Observer's Time Axis and *the Ground Distance traveled by the car but measured in succession from the position of the Stationary Observer is the relative Distance actually traveled.*

Plotting the Paths of the two Space vehicles at a much faster speed, we obtain a Map with features closer to our previous experience in 1-D Space. On the next *ST* Map, we see the vertical Time Axis of the stationary Observer, *<A>* and a series of horizontal Isotemporal Lines crossing the different Moments on his Time Axis. Note that with the Path#2, the Passing Distance, *R* is shown. The Distances *(from the Observer)* of the two cars before, at and after passing are shown, first on the Ground Map then on the *ST* Map. Note that *the PASSING DISTANCE, R is zero for Path #1* but with *the road-side scenario of Path #2, R > 0. The Ground and ST Maps here are true-to-scale (moving <B>'s observations not mapped):*

F11.2

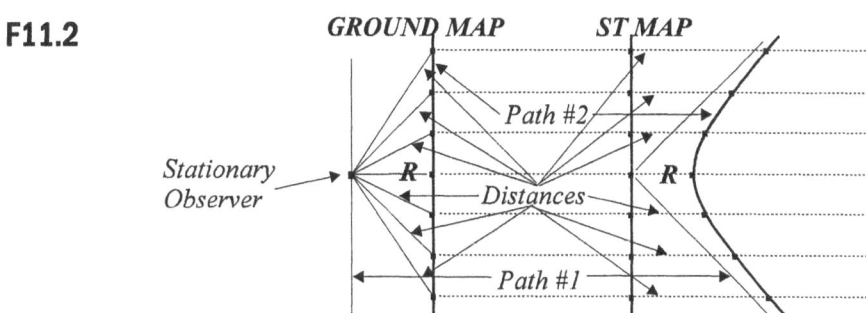

With a ruler and compass, we can measure the changing Distances *(X)* of the moving vehicle from the stationary road-side Observer at different Moments. These Distances can be plotted on horizontal *ITL*s crossing the vertical *WL* and forming a *HYPERBOLIC CURVE*. In Fig. 11.3, the upper half of the hyperbolic curve is shown. The *X* at its smallest magnitude is the *Passing Distance*, *R*:

F11.3

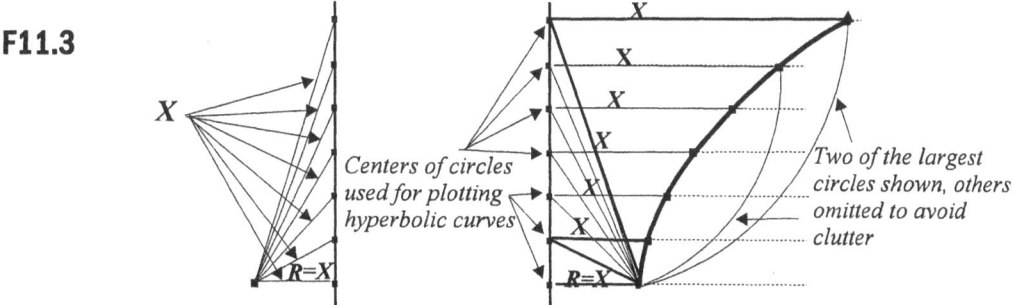

At the exact Passing Moment, both participants are *momentarily motionless to each other* and, therefore, momentarily also in the same Inertial System. This Passing Moment can be used, as we have always done, to synchronize the clocks of the two participants. We'll ignore the Path #1 and the Observer in one-dimensional Space from now on and deal only with the 2-D Motion in the rest of this chapter with special attention paid to *<A>* observing a moving *<B>*'s vehicle on Path #2 in two-dimensional Virtual Experiments.

2-D Motion => continuous Acceleration

By including on the Map all the <*B*>'s Distances before and after the Passing Moment, we'll obtain a hyperbolic curve with its outgoing lines diverging, gradually approaching the straight lines of Path #1. The very important point is this: *the smoothly continuous change of slope means a smoothly continuous change in Relative Velocity*. This change in Velocity is *ACCELERATION*, occurring over the entire Path #2 and is always *directed away from the Observer*. In approaching mode, it produces deceleration *(a continuous reduction in Relative Radial Velocity)* then, after passing, acceleration *(a continuous increase in relative Radial Velocity)*.

Visualizing 2-D Motion

What is shown on the *ST* Map can be illustrated also with a *stack of 2-D surfaces*, one placed upon the other and each marked as to where the Observer and the two vehicles *(shown in Fig. 11.1)* are momentarily located. To reduce clutter, only the bottom, the top and the middle surface are shown. Of those, the middle surface is the most informative. It shows the Passing Moment when Path #1 occupies the same spot with the Observer while Path #2 stays at a safe Distance. Time is shown on the vertical Axis with the Observer's *WL* going straight up, the *WL*s of the two cars *(in uniform notion)* slanted but parallel. The Velocities involved are exaggerated to bring out and clearly illustrate the difference between 1-D and 2-D Motion:

F11.4

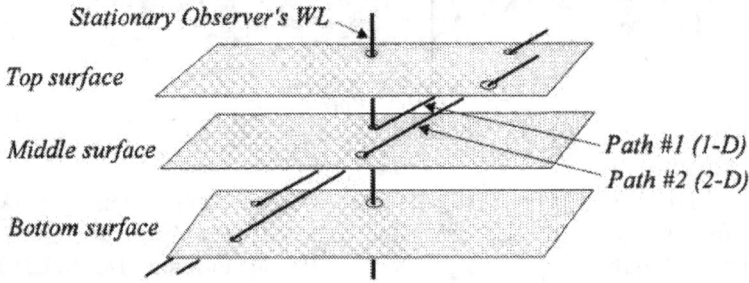

Stationary Observer's WL

Top surface

Middle surface — Path #1 (1-D) / Path #2 (2-D)

Bottom surface

The big question

It is here that an increasingly nagging question intrudes: *Do we really have a constant, unchanging, continuous acceleration, a rate of Velocity change that remains exactly the same no matter how long it continues?*

A big turnaround !!!

Answering the question is easier after it is turned around and changed into a different question, a rhetorical one: If equal increments of Velocity are added to existing Velocity, how do we add them up? Aha, let us refer to what we learned in Chapter VIII, *"Adding Velocities."* *Remember that Velocities do not really add arithmetically, they add RELATIVISTICALLY! What we obtain is this: THE SAME INCREMENTS of ACCELERATION do NOT keep adding THE SAME AMOUNT of OBSERVABLE VELOCITY to the existing Motion.* If it could, a moving object would eventually surpass the Velocity of Light itself! And we have already found out that we cannot break the cosmic speed record. In general, we can only say that whatever equal increments we keep adding to an existing Velocity in a 1-D Space, we can never exceed the Velocity of Light! Remember: $w = (v+u)/(1+vu)$. In the specific instance at hand, continued acceleration cannot make Relative Velocity exceed the *(original)* Ground Velocity.

All kinds of uniform, linear Motion in 2-D Space and all uniformly accelerated Motion in 1-D Space are, therefore, described *(as already said)* by the hyperbolic curve.

If all the above wasn't enough of a surprise, then hear this: *Even the relative (radial) Motion of Light itself in a two-dimensional setting would be HYPERBOLIC, undergoing uniform DECELERATION relative to the Observer positioned away from its path, coming to a complete, momentary STOP at the Passing Moment and then continuing to accelerate uniformly while beginning to move away from the Observer, eventually reaching NEAR-LIGHT VELOCITY and all this is happening relative to the Observer standing away from its path.* This, indeed, is possible because *Radial Velocity of Light in a 2-D Path and not hitting the Observer NEVER equals its own Ground Velocity.* Only the Velocity of Light relative to an Observer *standing in its very path (in 1-D Space)* is 1.0 and *it is THAT Velocity that cannot ever be reduced.* The *(radial)* Velocity of Light relative to an Observer standing *AWAY* from its path in a 2-D setting is *ALWAYS LESS THAN 1.0* by a certain amount greater than zero and can, therefore, *NEVER be exactly 1.0. It is THIS Relative Linear (Radial) Velocity of Light that can be reduced to ZERO, but only MOMENTARILY so, only at the Moment of Passing !!!*

A convergence of math

At this point in our discussion, and pay close attention to this: *the Spacetime mathematics of two-dimensional, constant-velocity Motion and one-dimensional, constantly accelerating Motion converge, become IDENTICAL!!!* We will, however, continue with the 2-D Motion we started out with for the sake of continuity of concept, ease of geometrical argument and simplified mathematics. We also need to realize that all conclusions drawn will be fully applicable to one-dimensional Motion with constant acceleration.

Please, digest the last three paragraphs very carefully and continue only after you recover from shock, seizures or other side effects of a *delayed AHA experience!* If all this gave you trouble, the rest could be fatal! It's no fun ending your life in a thick mental fog.

Velocity at any Moment during Acceleration

As the Relative Velocity of a passing object always keeps changing, we need to know how to calculate it for any location along the Path #2. Unfortunately, if we succeed doing it for one location, it will not be the same at some Distance or a Time later along its path. That's because it always keeps changing, never stays the same. This is why we previously avoided considering Motion close to the Observer whom we generously saved from dangerous collisions. Also the math near the Passing Moment would have been prohibitively difficult. But now we are stuck, have no choice, will have to do it at any cost or never get close to $E = mc^2$. A real tough problem.

To get our feet firmly to the ground, we must see what information we can get. For a survey, let's return to our Ground and *ST* Maps and show the parameters that can be measured. These are X and R obtained by standard Radar Probe and T by using a clock:

F11.5

GROUND MAP   *SPACETIME MAP*

X   WL *of passing vehicle*

Fixed position of Observer   X   T *or vT*   T   Observer's WL

R   R

Passing moment

Definitions:

X = Distance of the moving vehicle from the stationary Observer at any Moment, O being the anchoring Event of Passing.

T = Time elapsed from the Passing Moment, O to another Moment when the Distance of the moving vehicle was obtained. As usual, only the result of measurement, X *(without the Radar Signal lines)* is mapped.

R = the nearest Distance of the vehicle at the Moment of Passing.

Note that the Time measured by stationary Observer is not really the Time clocked by the Observer in the moving vehicle *(the clock riding with the moving Observer therein)*. That Time, *t,* measurable only inside the vehicle, cannot be measured by the Observer on the ground. We'll postpone calculating this Time Segment until after we are through with the Time (T) measured by the stationary Observer's clock and are eager to tackle the problem of Time clocked in the accelerating vehicle *(with the Observer therein)* in the next chapter.

More definitions:

v = Ground Velocity, a constant magnitude derived from X, T and R, the data available to the stationary Observer firsthand.

w = *(Radial)* Velocity relative to the Observer, always changing, the result of progressive increases *(or decreases in approaching mode)* in relative Velocities.

Crunching equations inventively

According to the Ground Map in the last illustration *(Fig. 11.5)*, the Ground Distance traveled by the vehicle equals vT *(at slow speeds, T very closely equals t)*. On the Ground Map, the T, X and R are the sides of a right-angle triangle. We can, therefore, conclude that even without knowing the Velocity, the value of X can be calculated:

$$X^2 = T^2 + R^2 \quad \text{and} \quad X = \sqrt{T^2 + R^2}$$

If we really want to be fussy about obtaining Ground Velocity from the Ground Distance, vT traveled by the vehicle at Velocity, v we'll have to change the above equation into the following:

$$X^2 = (vT)^2 + R^2, \qquad (vT)^2 = X^2 - R^2$$

Proceeding from this basic equation, *we can derive v, the unchanging* (constant) *Ground Velocity:*

$$v^2 = \frac{X^2 - R^2}{T^2} \qquad v = \sqrt{\frac{X^2 - R^2}{T^2}}$$

Actually, each one of the four parameters *(X, T, R, v)* is derivable from the other three. Two of these *(v and R)* are **CONSTANT** for any single *(virtual)* experiment leaving us with **only two true variables**, X and T which, however, always change together. Now, X is entirely dependent upon, and derivable from, T. So, there is really just **ONE TRUE VARIABLE, and this is T, the TIME clocked by the Observer on the ground!**

Using our initial equation, $X^2 = (vT)^2 + R^2$, we can easily derive X, one of our key parameters:

$$X = \sqrt{(vT)^2 + R^2}$$

Adding incremental Velocities continuously produces Radial Velocity that eventually approaches the Velocity of Light, so *the true Ground Velocity applicable to all the possible circumstances with continuous acceleration without limit must be the Velocity of Light!!!*

With Ground Velocity, $v = c = 1$, the equation for X becomes greatly simplified by the elimination of v:

$$X = \sqrt{T^2 + R^2}$$

Problems, problems, problems

We have not yet cleared all the important hurdles. Now, the relative Radial Velocity in the above setting is the increase in the vehicle Distance *(observed by the ground-based Observer)* divided by the Time it took to obtain that increase. To calculate the Relative Velocity, w between the Observer and the passing car, we best start at zero Time, the Moment of Passing. Let us designate two Distances along the Path #2. The first *(Distance)* is that at the Moment of Passing *(of the vehicle)* at O. The second can be measured at any other Moment selected arbitrarily such as T_1 at a short Distance, vT_1 after O:

F11.6

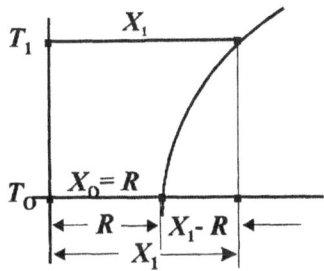

On the above Map, we have the moving vehicle *(with Observer, <B> inside)* at two Distances from the Observer, *<A>* on the ground: (1) X_0 which happens to equal R, and X_1 which contains the Distance increase from X_0 to X_1. The average of the continuously changing Velocity, w over the given Distance, $X_1 - R$, would then be the Distance traveled divided by the Time of travel, T_1:

$$w_1 = \frac{X_1 - R}{T_1}$$

But this is a *Continuously Accelerating Velocity, w*. The best we can do is to derive w_1 that is only the average Relative Velocity at halfway point between O and T_1, that is at $T_{1/2}$:

F11.7

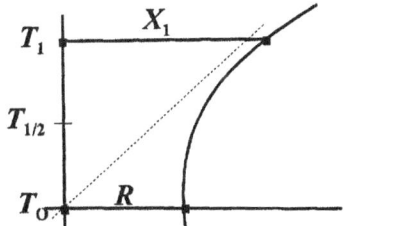

$$\frac{X_1 - R}{T_1} = w \quad \text{approximately at} \quad \frac{1}{2}T_1$$

Next, let us select another Moment, T_2 and repeat the process of obtaining the Velocity at $T_{1\frac{1}{2}}$ which is halfway between T_1 and T_2, the average Velocity between T_1 and T_2:

F11.8

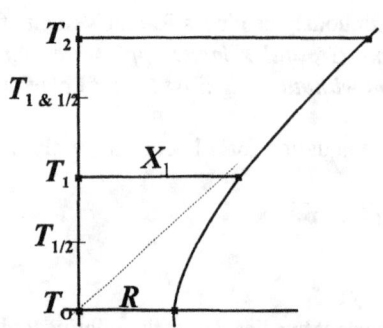

$$\frac{X_2 - X_1}{T_2 - T_1} = w \quad \text{at} \quad 1\frac{1}{2}T_1$$

With a good number-crunching calculator and plenty of patience, we could take much smaller increments between successive Moments on <A>'s *WL* in order to keep our margin of error as small as we wish and calculate the average Velocity relative to the Observer for any number of successive Time Segments. You can see, however, that this gets us bogged down in tedious calculations just to map <B>'s Velocities along her path of travel. Still, we would be getting only approximate values at best. Try as hard as we may, we just don't have a general solution for w at any point along the Path #2, a seemingly hopeless situation.

There has to be *a general solution for w.* So hold on and pay close attention to the argument that follows while marshaling all our powers of imagination for the job. We'll use an intuition-driven shortcut method, one that calls for some conjectural *(let us not say sloppy)* but ingenious math. We'll be forgiven if we come up with a beautifully correct solution.

A hint at the solution

To get started, recall that **(1)** *at the exact Moment of Passing, the Relative* *(Radial)* *Velocity equals a BIG, FAT ZERO: w = 0!* and that **(2)** *at a very great Distance, the RELAT-IVE (Accelerating) VELOCITY comes very-very close to the GROUND VELOCITY, w = v* . Next, let us stare very hard at the Ground and *ST* Maps *(Fig. 11.5)* in search for some idiotically simple combination of available parameters: v, T, X, R to help us out. More specifically, let us ask if there is, perhaps, a simple combination of these such as sum, difference, product, ratio, anything that can give us $w = 0$ at the Moment of Passing and $w = v$ at infinite *Distance?* Natural laws often come in quantitative relationships that can be stated simply and in whole numbers. Let's bet that this rule will hold for us here.

To simplify the problem even further, we need to restate our aim. Let us put on our thinking caps and find a *MULTIPLIER, M* for the unchanging Ground Velocity, v so that: *w (the Accelerating Velocity) = M (Multiplier)* times v *(the Ground Velocity)* **for all values of T.**

More specifically, we need a Multiplier that at moment, *O* would be zero and an infinite Time *(and at infinite Distance)* later would equal 1.0 *!!!*

What candidates do we have? At the moment of passing, $T = 0$. Here the ratio T/R as well as T/X *(out of all the remaining available parameters, X, T, and R)* would do and would not be influenced by our choice of measuring units. Using either of these as Multiplier, we get Mv only in two different ways and the product, Mv at the Passing Moment always comes up zero:

$$\frac{T}{R} v = 0, \quad \text{also} \quad \frac{T}{X} v = 0$$

This is so because at the Passing Moment, the ratio of T_0 to any other quantity would be equal to zero and we can use only T and either X or R in the Multiplier.

Final pick of the Multiplier

At moment, T_O either Multiplier will do but T an infinity of Time later is extremely close to being equal to X just as w is extremely close to v, so the ratio T/X, and *NOT T/R*, is almost equal to 1 and is exactly what we need to make $w = v$, so:

$$M = \frac{T}{X}$$

Let's accept the ratio T/X and not T/R as our Multiplier, correct at both extremes of Time and Distance. Again:

$$w = Mv. \quad \text{But as} \quad M = \frac{T}{X}, \quad \text{so} \quad w = \left(\frac{T}{X}\right)v.$$

We have finally found the badly needed but still provisional equation for the Relative Velocity correct not only at the Moment of Passing but, potentially, also *at any point along the Path #2* regardless of the amount of Time from T_0 or of the Distance from the Observer. Our search has come to a happy ending. Our Multiplier varies obediently between 0 and 1. These values set its limits. The Velocity, w, therefore, also varies between 0 and 1, exactly as ordered.

The basic equation can now be written in a much more general way making the velocity, *w a FUNCTION of Time (function for us being a new but not too complicated a math tool)*, also written as $w = f(T)$, thus:

$$w = \left(\frac{T}{X}\right)v$$

At Ground Velocities less than c, we are in a bind because we cannot measure v directly. But we know that $vT = \sqrt{X^2 - R^2}$, so that in a 2-D virtual passing experiment, w will eventually approach v but cannot ever exceed it. The Ground Velocity, therefore, *(as it is developed from $vT = \sqrt{X^2-R^2}$)* is derivable in principle from data available to the ground-based Observer:

$$v = \sqrt{\frac{X^2 - R^2}{T^2}}$$

Inching toward the solution

For any Velocity, v less than c we can use $w = v\ (T/X)$. But once we realize that *any Velocity in 1-D Space with constant acceleration, given enough Time, will end up almost equal to the Velocity of Light. Therefore, the Ground Velocity, v in the 1-D setting is almost equal to that of Light, v = c = 1.* Take all the time you need to digest all that has been stated above and only proceed once you have absorbed it well.

Proceeding from the initial equation above, with $v < c$ and our Multiplier, $M = \dfrac{T}{X}$

we get: $\quad Mv = w = \dfrac{T}{X}\sqrt{\dfrac{X^2-R^2}{T^2}} = \sqrt{\dfrac{T^2(X^2-R^2)}{X^2T^2}}$

Now, the T^2 in the Numerator and in the Denominator cancel out, giving us:

$$w = \sqrt{1 - \left(\frac{R}{X}\right)^2}$$

By replacing R^2 with its previously derived value, $X^2 - T^2$, we'll get:

$$w = \sqrt{1 - \frac{X^2 - T^2}{X^2}} = \sqrt{\frac{X^2 - (X^2 - T^2)}{X^2}} = \sqrt{\frac{X^2 - X^2 + T^2}{X^2}} = \sqrt{\frac{X^2}{T^2}} = \frac{X}{T}$$

Restated: $w = \frac{X}{T}$. That is the result if the Ground Velocity, $v = c$.

A simple way is to start with our initial equation for v but without the Multiplier. We can argue that at near-infinity value of X, the magnitude of R becomes so small compared to that of X that it becomes insignificant, very close to zero and can be eliminated from the equation. So, the value of v assumes a form that is very familiar to us, as you'll see:

$$v = \sqrt{\frac{X^2 - 0^2}{T^2}} = \sqrt{\frac{X^2}{T^2}} = \frac{X}{T} \text{ , which compares well with } Mv \text{ when } v = c:$$

$$w = \frac{T}{X} c = \frac{c}{X} T = \frac{1}{X} T = \frac{T}{X}$$

Now, this final result is really odd: *a Velocity that is Time divided by Distance!!! Just the opposite, or INVERSE, of our known procedure for calculating Velocities: $v = X/T$!!!* Have we gone mad or is this a spooky world? Calm down, be reassured ! It is nothing of the kind! In the chapter, "Mapping Spacetime," we always dealt with Motion in the Time-connected Realm of the Event, O *(the Passing Moment)*. Here, on the other hand, we have been dealing with Motion in the Space-connected Realm of the Event, O. The switch from Time-connected to Space-connected Realm changes our orientation. In the next chapter, you'll find clarification of this most puzzling switch. In the meantime, just be reassured: we are still in full command of our senses.

Proceeding now with the equation for w, we can first reduce the measured parameters to the least number of items, just T and the constant, unchanging R by replacing X *(that has to be measured inconveniently at the same Moment with T)* with its derived value, $X = \sqrt{T^2 + R^2}$ *(according to our previous calculations)* and also realizing that all Accelerating Velocities end up being almost equal to $c = 1$ at $T = \infty$, we can say that the stationary Observer's only independent variable is really T because the value of R always remains the same and is, therefore, a constant. Now we can finally write down the instantaneous, Accelerating Velocity, w *calculated only from T and R:*

$$w = \frac{1}{\sqrt{T^2 + R^2}} T \text{ which can also be written as: } w = \frac{T}{\sqrt{T^2 + R^2}}$$

What an enormous simplification *(but do you really think it is simpler?)!!!* At least, we have T as the only true variable with R being a constant. But now we'll be able to consider the 2-D kinematics in the setting of passing Light beams in place of slow-moving vehicles. Accelerations of any magnitude, if continued long enough, will end up with Velocities approaching that of Light anyway, so there is nothing different about two-dimensionality of a passing Light beam compared to constantly accelerating Motion in one-dimensional Space. Note, please, that relative Radial Velocity of a passing *(non-colliding)* Light beam never reaches the full value of c but always stays just a little below that. So, again, *if the Light beam does not hit the Observer, its relative*

(Radial) **Velocity is LESS THAN 1.0 !!!** Clear? Isn't it? Doesn't it *(the very unorthodox statement)* seem a little more friendly by now?

Next, the Acceleration

Our next task is to develop **the mathematical description of Acceleration.** We have already seen that with uniform Acceleration, the progressive reduction of Velocity increments is due to relativistic *(in contrast to arithmetic)* addition of Velocities. Now, Acceleration, **a** is quantitatively related to the parameters, **T** and **R**, only to those. With **R = 0**, the Motion is 1-D. With **R > 0**, the Motion is 2-D. The change in Relative Velocity (**w**) is greater for smaller **R** values and smaller for greater **R** values. **Acceleration is, therefore, inversely related to the magnitude of R** for any value of **T** and this gives us a preliminary, totally intuitive, conceptual solution:

$$a = \frac{1}{R}$$

Intuition is fine but we need to obtain acceleration by a more rational process. Now, in classical mechanics, $v = aT$ and $a = v/T$. The same relationship must exist between **a** and **w**. Let's see what we'll get if $w = aT$:

$$a = \frac{w}{T} = \frac{\frac{T}{X}}{T} = \frac{T}{XT} = \frac{1}{X}$$

from which it follows that:

$$\text{as} \quad X = \sqrt{T^2 + R^2}, \quad \text{so} \quad a = \frac{1}{\sqrt{T^2 + R^2}}$$

The actual Velocity, **w** at the Moment of Passing *(O-Event)* is zero and the Time, T_O is zero. But look, despite zero Velocity and zero Time at Event, **O,** acceleration is still fully operative and at a constant magnitude even in the momentary absence of actual Velocity. Therefore:

$$a = \frac{1}{\sqrt{0^2 + R^2}} = \frac{1}{\sqrt{R^2}} = \frac{1}{R} \quad \text{Restated:} \quad a = \frac{1}{R}$$

Hurrah! A celebration is due. Open the champagne bottle, poor some beer! Anyone for pretzels or peanuts? Perhaps caviar? We've got the long-sought-for equations for **(1)** constantly changing Relative Velocity at any instant and also for **(2)** Constant Acceleration at all times for both the 2-D and 1-D scenarios *(uniform Motion in 2-D setting and the constantly accelerating Motion in 1-D setting)*.

Plotting the complete accelerating Motion Map

It is best to pause a while and look at what we've got and find out what else we can dig for. While working with acceleration scenarios from **O**-Event up, we completely disregarded the decelerating part preceding the **O**-Event. We are lucky, however, not having to redo our math for the decelerating part because:

(1) The approaching Motion behaves exactly the same way as the receding Motion, both being mirror images of each other.

(2) Instantaneous Velocities can be calculated from the Time markers counted ***downward*** from the ***O***-Event.

(3) Acceleration is constant, the same at all times and, as always, directed away from the stationary Observer. In approaching mode, it results in a constantly diminishing Velocity, or deceleration.

(4) The approaching and receding parts of the entire Motion can be plotted on the *ST* Map as a complete hyperbolic curve, not just the upper half of it.

What about soft landings?

In 2-D setting, the most logical, basic Ground Velocity is $v = c$. But the Velocity relative to *<A>* *(or us)* even at infinite Distance was always just a little less than c, $v < c$, even in our usual 1-D setting. This is another piece of evidence that the same math is applicable to both settings.

The fact that the approaching relative Motion comes to a complete stop before beginning to move away brings up an interesting possibility. Can we, perhaps, ***take advantage of the momentary absence of relative Motion and bring the decelerating vehicle to some kind of a soft landing*** a manageable Distance ahead instead of immediately speeding away? Off-hand, anything not prohibited by laws of nature should be possible. It may need quite a lot of planning, however. But then, our mathematics has to be reworked as well as to accommodate the change in travel plans.

From purely navigational angle, the approaching spaceship will have to reduce its engine thrust near the *R*-Location and gradually slow down descent until touchdown. From slightly before *R*-Moment on, the value of Constant Acceleration *(or deceleration)* must be changed completely and our well-worked-out math has to be replaced just before the *O*-Moment by soft-landing math that is no longer part of our Relativity study. One thing is certain: Soft landings are possible and have already been performed many times over.

Simultaneity relationships

A routine but a more intriguing question is this: what on *<B>*'s *WL* is simultaneous to *<A>* and vice versa? Of course, *<A>* and *<B_a>* cannot ever agree on anything pertaining to Simultaneity. Relativity of Simultaneity never sleeps. As usual, Isotemporal Lines *(without the clutter of Radar Probe Signal lines)* should settle all questions. For orientation, let us plot the entire hyperbolic travel path, draw a few *<A>*'s *ITL*s and number the corresponding Event markers on *<A>*'s *WL*. This gives us a rather straightforward Map that is already familiar to us from Fig. 11.2:

F11.9

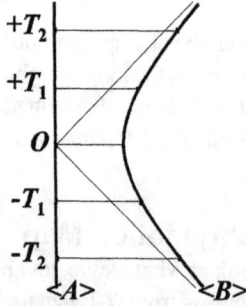

To show the isotemporal relationships from *<B_a>*'s point of view let's draw *TANGENT lines* to *<B_a>*'s curved *WL* at equal Time Segments on *<B>*'s *WL* then draw *<B_a>*'s *ITL*s as heavier lines through those tangent-touching locations connecting to *<A>*'s *WL*. To keep the potential clutter of lines to a minimum, we'll select only five Event markers on $<B_a>$'s *WL* and delete the hyperbolic curve *(but not the Event markers)*. You'll be astonished by the results:

F11.10

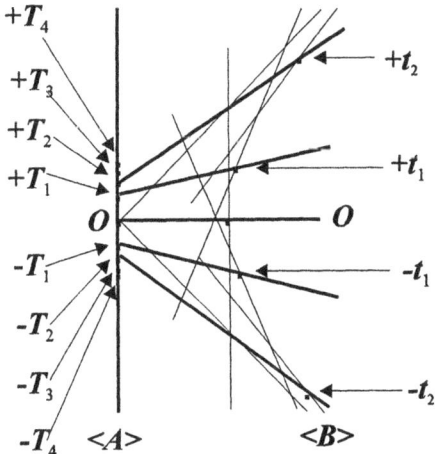

You can see that the five *ITL*s connect the Moments on both *(<A>'s and <B>'s) WL*s simultaneous from $<B_a>$'s point of view. Both the tangents and the corresponding *ITL*s form equal angles with Lightlines *(only one ITL is drawn to reach the O-Event)*. The limited number of Time markers and is sufficient to demonstrate the important relationships between $<A>$'s Time and $<B_a>$'s Time which are listed next:

(1) $<A>$'s Time markers are simultaneous with $<B_a>$'s Time markers, as taken from $<B_a>$'s point of view.

(2) $<A>$'s Time Segments between the *(<B_a>'s) ITL* crossings are *NOT equal* to those between $<B_a>$'s Time markers, something we should have guessed looking at the Figure 1 1.9 *(drawn from <A>'s Point of view)* first.

(3) The marked-off Time Segments on $<A>$'s *WL* are here shown true to scale *(because the Map shows <A>'s Home Frame)* these Time Segments are getting shorter and shorter the further they are removed from the *O*-Event while the corresponding marked-off Time Segments on $<B_a>$'s *WL* remain equal even as their Map line lengths *(drawn in <A>'s Other Frame)* get progressively longer. If this is confusing, just note that the *ST* Map above is drawn from $<A>$'s point of view but everything there is shown from $<B_a>$'s point of view. The markers on $<A>$'s *WL* and $<B>$'s *WL* are simultaneous only according to $<B>$! Again, Observer-specificity is the key to the confusion for you to consider, digest and accept *!*

(4) Extra interval markers *(indicated by extra arrows)* pointing at $<A>$'s *WL* show more clearly that $<A>$'s Time Segments become shorter and shorter *(Figure 11.10)* just as $<B_a>$'s Time Segments became shorter and shorter in duration on her hyperbolic *WL* *(Fig. 11.9, original roles assigned)*.

(5) If the same Virtual Experiment is shown entirely from $<B_a>$'s point of view, then her *WL* would be straight and vertical because $<B_a>$ would consider herself stationary and located in a gravitational field but she would now label $<A>$ to be $<A_a>$ and moving along a hyperbolic, constantly accelerated *WL !!!* Everything would then be the way $<A>$ observed $<B_a>$ except that the roles would be reversed. The reversed scenario is finally mapped the way $<B_a>$ observes it *(and is shown below without further details or labels)*:

F11.11

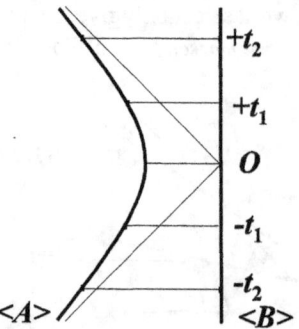

(6) Figure 11.9 showed everything in *<A>*'s Home Frame. In Figure 11.10, *<A>* shows *<B_a>* in *his* (*<A>*'s) Other Frame while Figure 11.11 shows everything in *<B_a>*'s Home Frame with *<A>*'s **WL** in *her* Other Frame. Does this summation clear up the fog?

Communication lines between <A> and <B>

First of all, let *<B_a>* keep sending a series of Signals toward *<A>* while approaching. Recall that communications can take place only though Signallines *(same as Lightlines)*. But when do you think *<A>* would be able to begin receiving those Signals? The surprising answer is shown graphically in the next illustration:

F11.12

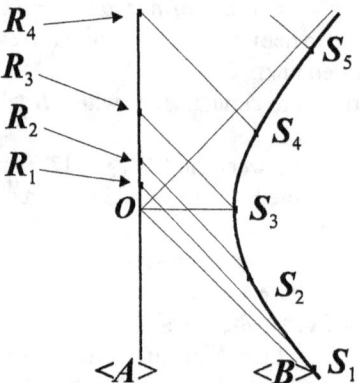

It is disappointing that ***the earliest Moment when <A> can receive Signals from <B_a> is only AFTER the O-Event, never before!!!*** *(Only a signal sent by <B> from infinite Distance would reach <A> exactly at the O-Event).* Can you imagine both of them trying to exchange information or discuss plans for a soft landing? Only by modifying the deceleration early on and cutting off thrust at the right *(but unpredictable)* Moment at the **O**-Event that is real close to *<A>* *(but unknowable as to WHEN)* can conditions be created for the necessary exchange of information. Without earlier contact, the entire sequence may well end up in a crash landing. If it is up to *<A>* to seek contact, he will have to send random, probing Signals repeatedly toward *<B_a>* while her spacecraft is still unobservable *(being Space-connected relative to <A>)* and about whose position in Space neither one has the faintest idea. If he succeeds making contact with *<B>*, the earliest response he can receive is, unfortunately, too late *(after the O-Moment!)* to be useful. So, soft landings are impossible unless the *(equivalence of)* Ground Velocity is manageably very-very slow. Otherwise, the entire trip will be a disaster in the making, our math impossibly complicated and totally useless, as well.

Constant Acceleration line = Equidistance Line

The hyperbolic *WL* of the *(astronaut)* Space Traveler at constant acceleration is identical by analogy with the Equidistance Line described in Chapter III. By definition, *any point on that curve is located at the same Distance from the Event, O* not according to *<A>* but *according to <B$_i$> traveling repeatedly through the O-Event at increasing, uniform Velocities (see Chapter III).* It was obtained by plotting the *ITLs* at a series of increasing *(uniform)* Velocities so that all the various *<B>*'s Distances from *<A>* duplicate the *(single)* value of *<A>*'s measured *X*-Distance. In other words, it belongs to the general world of all Inertial *(non-accelerating)* Systems moving at all the possible Velocities *(at v = from 0-to-c)* in 1-D Lineland and all of them passing the stationary *<A>* at Event, *O* *(somehow wisely avoiding what could be an inevitably disastrous traffic jam).* Each inertial-Travel Path is run at a different Velocity *(v varying from 0 to c)* but these can also be performed sequentially, the *O*-to-*Q* Distance at *v=0* kept equal at the different *<B>*'s velocities *(from v=0 all the way to v=c)* and the results plotted as if all the Passing Events had occurred simultaneously. Because all the astronauts in those runs are momentarily without relative Motion at the Passing Distance, they are also momentarily in the same Inertial System with *<A>* and agree with *<A>* about the *R*-Distance. This particular topic has interesting ramifications which, however, we are not prepared to pursue further in this book.

To repeat: *<B$_a$>*'s *WL* traces out the Equidistance Line in 1-D Space and the sesults are mapped on a hyperbolic curve. So *<B$_a$>*'s *R*-Distance from the *O*-Event is equal to all *<B$_i$>*'s Distances from the *O*-Event at Velocities possessed by *<B$_a$>* at such Moments on the *WL* mapped *(on the Equidistance Line)* by innumerable *<B$_i$>*s *(at velocities from v = 0 to c).* The *x*-markers in Fig. 11.13 below represent equal Distances of inertial *<B$_i$>*s *(and only by analogy <B$_a$>s !)* at Time Segments of the different *<B$_i$>*s and are marked by $t_{1, 2, 3, etc.}$ from the fixed *O*-Event at different *<B$_i$>*'s velocities on the Equitemporal Line. Only four such selected *(<B$_i$>'s)* t_i Time markers and *<B$_i$>*'s x_i-Distance markers are shown below to complete the limited visual argument:

F11.13

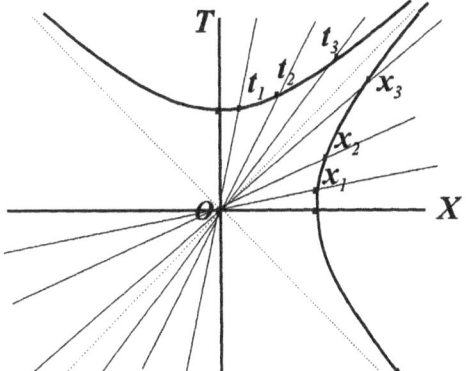

On the above Map, the *<B$_a$>*'s Distances from Event, *O* remain constant, unchanging from positions marked x_0, x_1, x_2, x_3, etc. and her Velocity, *w* previously indicated by Tangents at those Moments on her hyperbolic *WL* that correspond to Velocities shown by parallel lines to those Tangents drawn through the *O*-Events at equal Time Segments from that *O*-Event, marked as t_0, t_1, t_2, t_3, etc. on the Equitemporal Line.

There is more to the simultaneous relationships here. However, it is a topic worth dedicating an entire chapter to unravel its implications. Considering the elementary nature of the discussion in this book, it is best avoided altogether. But see what *YOU* can do with it *!!!*

Gravitational Acceleration = 1/ R

There is one more interesting question related to Constant Acceleration we can handle: What is the R-Distance that corresponds to the Gravitational Acceleration observed on Earth? In other words, in terms of the 2-D Motion with constant Ground Velocity, $v = c$, what would the Passing Distance, R have to be to provide a Gravitational Acceleration of 9.81 meters per second, the Acceleration observed *(and years later also measured)* by Galileo who once dropped a couple of objects of unequal weight from the leaning Tower of Pisa? Did he really get thrown out of the Academy for doing this and thereby disproving Aristotle?

Using a big packet of Light that does not hit us but contains enough extra photons to be scattered while in Motion *(Tyndall Effect)* to trace out its path for an Observer at a Distance, the Light path can be made observable. To get the R-Distance, we can use our newly-acquired expertise.

According to our familiar equation,

$$a = \frac{1}{R}, \text{ and the value of } R \text{ is: } R = \frac{1}{a}.$$

Recall that *(in the 2-D setting)* the *(Radial)* Velocity of Light relative to a ground-based Observer is always less than 1.0 if, and only if, the Light on its path does not hit the Observer directly. Let's use the figure, 9.81 *(m/sec²)* we specified for gravitational acceleration. Next we need to convert 9.81 meters *(per second per second)* to Lightseconds by dividing 9.81 by the Velocity of Light *(here in meters per second)*:

$$\frac{9.81m}{300,000,000m/sec} = \frac{3.27}{100,000,000} \text{ (in Lightseconds)} = \frac{1}{R}$$

To get R, we'll have to develop $\dfrac{1}{R} = \dfrac{3.27}{100,000,000}$ as follows:

$$R = \frac{100,000,000}{3.27} = 30,581,040 \ m = 30,581 \ km, \text{ the passing Distance.}$$

The Passing Distance seems rather far away but now we have at least some idea about its magnitude. In everyday terms, it is about 1/10 of the Moon's Distance from the Earth. Stronger gravitational attractions, of course, require shorter R-Distances. Which leads us to another interesting application of our acceleration equation. We'll introduce it in form of another question: How close from a Black Hole of 1.4 solar mass *(the minimum needed for it to form)* can an errant ray of Light come before it gets trapped and pulled into it. This Distance is known as the Event Horizon. Unfortunately, we cannot find the answer due to insufficient expertise. The question is posed only to point out that this and other applications of the equation, $a = 1/R$ belong to the domain of General Relativity and are, therefore, solvable but beyond our present reach, at least for now.

We have arrived at a stage in our exploration where we could begin to apply our newly acquired understanding to solve another very important problem of relativistic mechanics, that is $E = mc^2$. But before getting into this complex topic, it may be interesting to answer one more question related to Accelerated Motion. It is about how to calculate the Time flow observable in a constantly accelerating System, something we cannot measure directly? Don't miss this thought-provoking exercise. You'll find it in the next chapter. It is about Time as measured by a constantly accelerating Observer but correlated with the even progression of Time in the Home Frame of <A>, the stationary Observer. So the next topic is "Time In Acceleration."

XII - TIME IN ACCELERATION

Time slowing is progressive

In the task of extracting Constant Acceleration, a from 2-D Motion, we measured Time with a stationary clock in <A>'s Home Frame. With <A>'s Time (T) and Passing Distance (R), we had all we needed to calculate both Instantaneous Velocities and Acceleration. The rate of Time progression in the Accelerating System was not needed for anything. But ***Time clocked in an accelerating system with constantly increasing Velocity is OBSERVED going slower and slower** (and slower still).* So what? Can we use it for anything? Not really. We didn't need it for calculating Accelerating Velocities or Acceleration. Nevertheless, it is a subject of curiosity begging to be scrutinized. The path to wisdom here is much more complicated so be prepared for a lot more difficult lines of argument. Fortunately, our 2-D scenario has served us well so far and will be useful also here as a convenient starting point.

Let us review the data available to <A>, the stationary Observer at Moment, T_1 on our usual *ST* Map. The Accelerating Observer, <B_a> has moved from P to Q after decelerating to a momentary stop at the Passing Distance, R from <A>:

F12.1

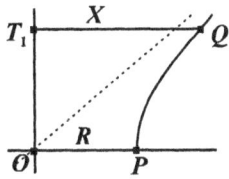

Visualizing slowing of Time

By drawing a Tangent to WL_a *(Worldline of Accelerating Traveler, <B_a>)* at Event, Q, we obtain a line that represents the Velocity attained by <B_a> at that particular Moment. Next, let's draw a line parallel to the Tangent so that it goes through the O-Event. This would give us the Worldline of an Inertial Traveler moving at a Constant Velocity equal to that represented by the Tangent at Q. By the additional Map line now available, a floodgate of new information is opened up. Study the next illustration and its labels in detail. Everything there should be self-explanatory:

F12.2

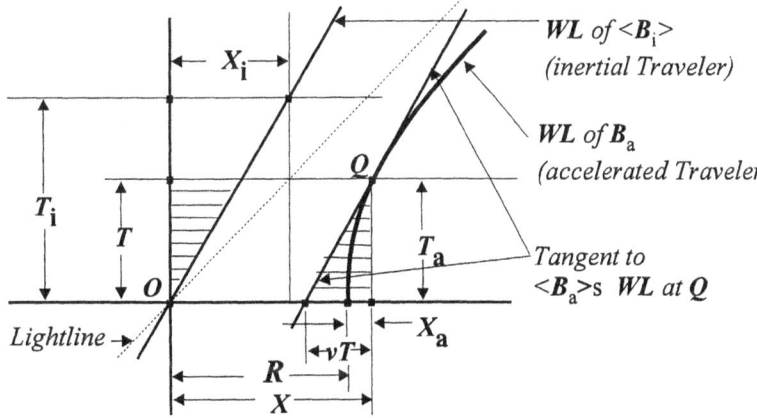

Let's list what we have obtained. Because all measurements are done by <A>, all his observations are given in upper-case letters:

    (1) <A> is our Stationary Observer,

    (2) <B_a> is the Accelerating Observer on her WL_a moving at a constantly accelerating Velocity, w.

(3) $<B_i>$ is an Inertial Observer at constant Velocity, v equal to w of $<B_a>$ at Q.

(4) T is the Time, OQ measured by $<A>$ as Proper Time, Oq in his Home Frame but Oq is also $<A>$'s Coordinate Time, $T_a=T$ of $<B_a>$'s Proper Time, OQ.

(5) R is the closest Distance from $<A>$ that $<B_a>$ reached on her decelerating-accelerating travel.

(6) The line, X_a is the Distance traveled by $<B_a>$ from Passing Event, P to Q, which amounts to $X_a = X - R$.

(7) The line, vT is the Distance an Inertial Traveler would have traveled during the Time Segment, T at the Constant Velocity indicated by the Tangent drawn to $<B_a>$'s *(accelerating)* WL as well as by the line drawn parallel to the Tangent, going through the O-Event and standing for the *WL* of an Inertial Observer *(moving at constant velocity)*.

(8) X_i is the Distance an Inertial Traveler *(moving at the Tangent Velocity)* would cover during the Time, T_i; T *(both X_i and T_i measured by $<A>$)*, with T_i being equal to X which, in turn, is equal to $R+X_a$.

If we compare the measured T and X *(which are $<A>$'s data about the Accelerating Observer, $<B_a>$)* with the T_i and X_i pertaining to the Inertial Observer, $<B_i>$, we should see that:

$$T = X_i \quad \text{and} \quad X = T_i \; !!! \quad \text{Therefore, as } v \text{ (of } <B_i>) = X/T, \text{ so } w \text{ (of } <B_a> \text{ at } Q) = T_i/X_i$$

and this is an additional nugget of information that makes reasonable the earlier *(in Chapter XI)* derivation of $w = T/X$, the Instantaneous Velocity of the accelerating $<B_a>$ at any Moment on $<A>$'s clock. The Moment in our example is shown and mapped in Fig. 12.2 as the Q-Moment.

In the discussion that follows, Observers who move are synonymous with *"Travelers."* The two kinds of Travelers, therefore, are the ***Inertial Traveler***, $<B_i>$ and the ***Accelerating Traveler***, $<B_a>$. The only Observer not moving is $<A>$.

Accelerating and Inertial Travelers

The ***ACCELERATING TRAVELER***, $<B_a>$ *at location*, Q and the ***INERTIAL TRAVELER***, $<B_i>$ *(WL parallel to Tangent at Q !!!)* have the same velocity but the Time, T is still measured by $<A>$ with his stationary clock. We, with $<A>$, cannot measure $<B>$'s Time directly and it makes no difference whether her moving clock is inertial or accelerating, it is still unknown and not available to $<A>$. All we can say is that at Moment, Q *(and ONLY at that Moment)*, $<B_a>$'s clock is *(momentarily)* ticking at a rate, t_a *(for a very-very short Time Segment)* $= T\sqrt{1-w^2}$

For $<B_i>$, the Time flow would be constant and always the *same*; but for $<B_a>$, the rate of Time flow would be ***constantly changing***. The change is dependent on T as well as on R, both accessible to measurement by $<A>$. Get it? *So $<A>$ has potentially all the information at his fingertips for calculating Cumulative Time measured by the accelerating clock in $<B>$'s Home Frame (from O to Q)*. The only problem is ***how to put it all together***, that is what kind of mathematical relationship is there between the two systems, $<A>$'s and $<B_a>$'s?

As stated before, It is easy to calculate t_i for any duration of T. Using only T and w in our estimation of t_a, we can only say that for very short segments of T, $t_a = T\sqrt{1-w^2}$. But even with knowing this much, we still don't have the tools to derive t_a from Event O to Q. So, we must digress a little and investigate a math tool new to us, called the ***FUNCTION*** *(also mentioned in the preceding chapter)* we could also call ***Relatedness***. We'll be doing it the slow and easy way. No shortcuts, though.

Visualizing Acceleration

If we draw a vertical axis for all values of t_i and a horizontal axis for all values of T, the slope *(of the straight line)* for a non-moving *(v = 0)* Traveler *(<B>)* is set at 45 degrees, showing that for all values of T, $t = T$:

F12.3

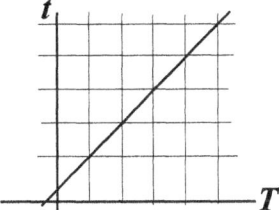

On the other hand, if the Traveler's Velocity is greater than zero, t_i would be less than T, or $t_i < T$. For repeat runs by *<B$_i$>*, each with increased but Constant Velocity, the slope of $t_i = f(T)$ would become flatter with greater and greater *(constant)* Velocities:

F12.4

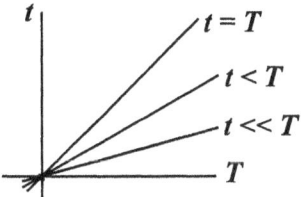

In the same fashion, we can easily plot instantaneous rates of t_i for a series of short Segments *(each at successively higher velocities)* of the hyperbolic, curved *<B$_a$>*'s Worldline *(t again shown as a function of T)*:

F12.5

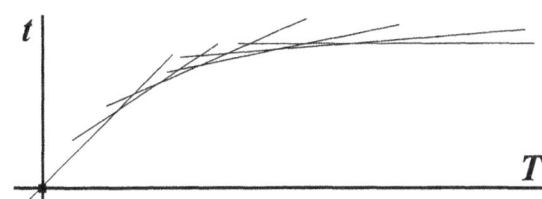

Next, we'll subdivide *<B$_a$>*'s **WL** into a number of very short Segments of t_a and calculate the average Time lapses for those. Then we'll add up these individual Segments and obtain an approximate total value of t_a. With small enough subdivisions of T, we can reduce our errors to as little as we wish. With a good calculator and plenty of patience, it'll be a cinch, clumsily mechanical but operationally simple. Our $t_i = f(T)$ curve can be made successively smoother and smoother until we get:

F12.6

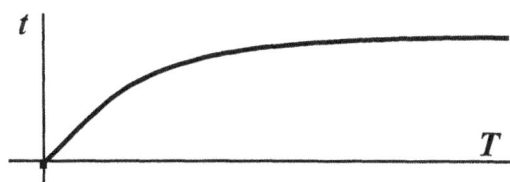

With accelerating velocity approaching *c*, the clock will slow down progressively until it is finally hard to tell *(provided we have access to the moving clock's data)* if the clock is ticking at all. The function curve starts out at 45 degree angle and becomes horizontal *(or almost so)* at the end. What we really need is an equation made up of information contained in just **TWO** measurements, *T*, the one and only variable, and *R*, the only constant. These are the only measurements needed by <.*A*>. Recall that *R* *(actually its inverse value)* defines the acceleration. The limits at both ends of the Function curve are as follows: **(1)** For a very short Segment of Time, at and very close to the Passing Moment, t_i is equal (or very close) to *T* $(t_i = T)$. **(2)** For velocities very close to *c*, the very short Segment of t_i is very close to **ZERO** $(t_i = 0)$!

Experimenting with different functions

Paying attention to the limits defined above is a step in the right direction. At least we know what kind of a curve we should get. Next we need to turn the task around and **(1)** calculate *a set of approximate numerical values* for several successive segments of t_a, **(2)** *add them up cumulatively* just to see what kind of approximate values we get, then **(3)** decide *what kind of a curve it represents* and, finally **(4)** *find an equation* that produces *(approximately)* the same calculated numbers for a series of *T* values thereby showing good enough a "fit" for the curve indicated. Doing it this way is *very much like conducting an actual laboratory (or field) experiment* that can provide us with numbers we can plot on a curve for which we, then, need to find the right equation that describes it, all in that order.

Let us look at some function curves just to find out if any of them have a shape similar to the curve in Fig. 12.6.

First, let us take two clocks, one connected to the other by a servomechanism that makes the other clock show exactly the same Time as seen on the first clock. The Time shown by the first, or **Master Clock** is the **INDEPENDENT VARIABLE** and the Time shown by the **Dependent Clock** is the **DEPENDENT VARIABLE** *(depending on the Independent variable)*. In a table, we enter *T* in minutes shown by the Master Clock and below that we'll show the corresponding *t* in minutes on the Dependent Clock. **Then,** let's draw the curve, actually a straight line in this case, showing the corresponding linear relationship:

F12.7

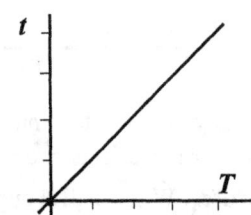

T 1 2 3 4 5 6.....
t 1 2 3 4 5 6.....

$t = f(T)$ and $t = T$

Now, suppose that the Dependent Clock has become stuck at 2 minutes while the Master Clock is moving at normal rate:

F12.8

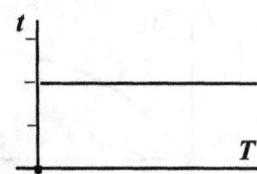

T 1 2 3 4 5 6.....
t 2 2 2 2 2 2.....

$t = f(T)$ and $t = 2$

Let us also look at some unusual curves that are no longer linear. Still, the relationships between the examples of T and t are fully contained in the equations given:

F12.9

T	0.1	0.5	1.0	2	5
t	10	2	1	0.5	0.25

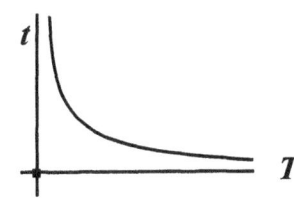

$t = f(T)$ and $t = 1/T$

F12.10

T	0.1	0.5	1.0	2	5
t	0.01	0.25	1.0	4	25

$t = f(T)$ and $t = T^2$

F12.11

T	0.01	0.25	1.0	4	9
t	0.1	0.5	1.0	2	3

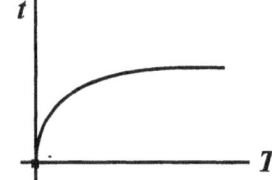

$t = f(T)$ and $t = \sqrt{T}$

F12.12

T	0.0	1	2	3	5	10
t	$-\infty$	0	0.3	0.5	0.7	1.0

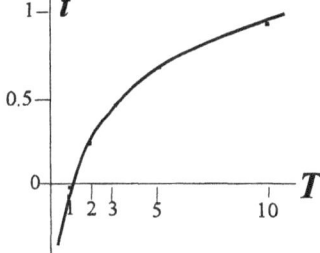

$t = f(\log T)$ and

$t = \log T$

Selecting a curve

The last two curves, if combined in some way, look promising. The general shape could be duplicated if these numbers could be only approximately obtained. Our final curve, however, should start out with $t = T$ *(Fig. 12.3)* with *(almost)* linear relationship between the two variables and only toward its end it should look more like $t = \sqrt{T}$. In some way, the curve should also follow a logarithmic pattern. But the linear pattern is actually deceptive. If the curve is plotted on a non-logarithmic, linear curve, we'll still approximate in part the parabolic curve of Fig. 12. 11. So, what should we do? Well, for Segments of T, let us calculate a number of approximate numerical values of t, enter them in a table, plot their cumulative sums on a curve and see if the results give us a hint. The purely math part of what follows may exceed the expertise of many readers of this book. But hold on for a while and try to follow at least the reasoning process.

Calculating approximate values

To repeat, what we need to do here is to first obtain a set of approximate numerical values of t_a calculated for a series of T segments and plot their cumulative sums on a curve. This will tell us (1) if the curve so obtained is what we expected. If so, then we'll have to (2) find an equation that is a good "fit" for the data we have calculated. To proceed, we need to go back to the last chapter *(XI)* and pick up the equations needed for calculating the average Time flow rates for various segments of WL_a. First the instantaneous Velocity is:

$$w = \frac{T}{X}$$

and t_a for very **short Segments** of T is: $t_a = T\sqrt{1-w^2}$.

But as we learned from Chapter XI, $X = \sqrt{T^2 + R^2}$ and $w = \frac{T}{X} = \frac{T}{\sqrt{T^2 + R^2}}$,

so *(trace out the derivation to make sure you understand it)*:

$$t = T\sqrt{1-w^2} = T\sqrt{1-\left(\frac{T}{\sqrt{T^2+R^2}}\right)^2} = T\sqrt{1-\frac{T^2}{T^2+R^2}} = T\sqrt{\frac{T^2+R^2-T^2}{T^2+R^2}} = T\frac{R}{\sqrt{T^2+R^2}}$$

Make sure you have followed this last "math blast." Next, let us fix the value of R *(a constant!)* at $R = 10$ and for T, let us select a series of $<A>$'s Time increment markers. Because **the slope of the WL_a curve changes most rapidly at smaller values of T**, we need to keep these T increments very short at first and increase them in successively larger steps. So, let's pick the following T- Markers: first 0, then 1.0, 10, 100, 1,000, 10,000, 100,000 and 1,000,000; and estimate the **average t** for these Time Segments by calculating t for each **half-segment** between the successive T- Markers.

Now, $t_{accelerating}$ at $T=0$ is also zero ($t=0$). The t for the T increment from 0 to 1.0 should be roughly the $t_{inertial}$ at halfway point between 0 and 1, that is at $T = 0.5$. Next we need to calculate the initial *(approximate)* t by setting up an equation according to what we learned in the previous chapter.

With R at the **constant** value of 10, let's find our first average increment of t_a for T from 0 to 1:

$$t_a = T\sqrt{1-w^2} = T_{(at\ 1.0,\ the\ marker\ selected)}\left(\frac{R_{(set\ at\ 10)}}{\sqrt{T^2_{(halfway)} + R^2}}\right)$$

$$t_a (from\ 0\ to\ 1) = 1\frac{10}{\sqrt{0.5^2 + 10^2}} = \frac{10}{\sqrt{0.25 + 100}} = \frac{10}{\sqrt{100.25}} = \frac{10}{10.0124922}$$

$$= 0.99875233$$

So the initial t_a increment for the T increment from 0 to 1 is, therefore, 0.99875233.

The t_a increment for the next T increment would then be: $t_a = 10\frac{10}{\sqrt{4.5^2 + 10^2}}$ *(but*

calculate it out to get the numbers).

First, let's tabulate all the T increments:

T	T borders	T segments	Half of T segments
1	0 - 1	1	0.5
10	1 - 10	9	4.5
100	10 - 100	90	45
1,000	100 - 1000	900	450
10,000	1,000 - 10,000	9000	4,500
100,000	10,000 - 100,000	90000	45,000
1,000,000	100,000 - 1,000,000	900000	450,000

Next let's calculate average t_a increments for the increasingly larger increments of T and tabulate the t_a increments for each T increment.

The following calculations below were done with a hand-held calculator. The table first shows the T- column then the $\frac{1}{2}T$- column *(0.5, 4.5, 45, 450, 4,500, 45,000 and 450,000)* after which the resulting t-values provided for the entire range of average T-values (from *0 to 1,000,000*) in increments. By cumulatively adding each t increment to the sum of all previous t increments, we'll get the cumulative total of all approximate t - increments:

T_i	$\frac{1}{2}T_i$	Approx. t_a increments	Approx. t_a cumulative
0	0	0	0
1	0.5	0.99875233	0.99875233
10	4.5	9.11921505	10.01796736
100	45	21.69304578	31.71101317
1,000	450	22.21673729	53.92774046
10,000	5,000	22.22216735	76.14990781
100.000	45,000	22.22222167	100.59429683
1,000,000	450,000	22.22222222	122.81651905

Next, we need to plot the cumulative sums on the curve to see it reveals a pattern that matches that of Fig. 12.6. At low values of T, t is initially very close to T but for higher T values, the increases become progressively less and less until the increments become constant for each successive ten-fold *(logarithmic)* values of T. Next, we'll plot the entire curve on separate charts, each at progressively reduced scale and labeled only with last two T markers.

The second chart in Fig. 12.13 deserves extra comment. The displayed T-Segment there is split into extra five *(smaller)* increments *(which were not entered in the table above)* to allow us see the most pronounced changes in the slope *(of the curve where these changes happen to be the most pronounced)* in better detail. In addition, the curve is plotted on a non-logarithmic scale in order to show more clearly the progressively reduced t-increments where *(on the second chart)* the more rapid change in the slope of the curve is seen after which the typically parabolic progression becomes the rule. The shape of the curve in each successive chart becomes increasingly flattened because of progressively reduced scale in the charts. The total elapsed T shown in those charts is multiplied successively by ten in five steps from 10 to 1,000,000 so that the reduction in scale of these individual charts also makes the height of the curves progressively flatter, a price paid for displaying everything within a minimum of space.

F12.13

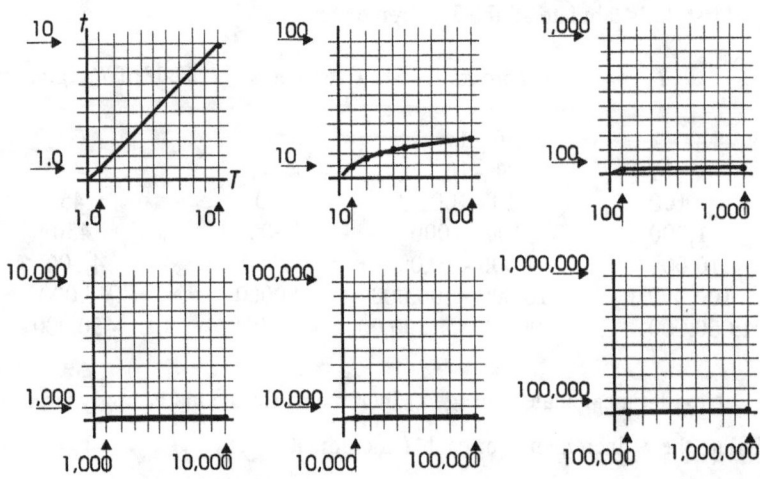

Deciding about characteristics

What do we see? Initially, at low T values, the curve starts out very close to $T = t_a$ just as we expected and best shown on Fig. 12.6. With each ten-fold increase in the total value of T, the incremental t_a additions become proportionally diminished, soon reaching a constant t-value for successive ten-fold increases in the T-values. Because of a progressive reduction of scale, this change is best seen in the second graph where additional values of T from 10 to 50 of were calculated. But except for a short, linear start, the rest of the curve is essentially *LOGARITHMIC and parabolic!* The Function that gives us the same incremental t_a values for each tenfold increase in T increments must then be $t = f\,(log\ T)$ in which the T is combined in some ways with unchanging, constant R.

The logarithms we need here are those used for "natural growth" processes that utilize ϵ *(Greek letter epsilon = 2.71828)* for base, not the number 10 most folks associate with logarithms. The equation must, therefore, incorporate the Independent Variable, T and the constant, R both obtainable by <A>. The actual units *(in meters, inches, Lightseconds, etc)* should not matter: only the comparison measure expressed by *the ratio of T to R* or *R to T* would provide the progressive increase in the cumulative total t_a values. How the equation is finally derived is not within the scope of this book because our mathematical expertise was not presumed to extend beyond elementary algebra. But ultimately, the solution should rest only on T and R and indirectly also X, all of which are the only data <A> can obtain firsthand.

At this point, we should also recall that **(1)** the entire *T & R* information along with X is contained in a condensed form in the *Doppler constant* which can be derived *for ANY instantaneous velocity* for *any value of T* and that **(2)** R always remains fixed *(constant)* throughout. Therefore, we have a choice either basing our final equation **(1)** on the ratio of T to R or **(2)** simply on the value of k that contains the same *T&R* information.

The solution that fits the data

For the final result, we'll select *the simplest and the most elegant equation,* one that contains the least number of elements and also "fits" the data obtained. All this is achieved either by **(1)** using a trick of the mathematical trade or **(2)** cowardly *(perhaps cleverly)* picking it from a handbook of mathematics *(more specifically from its table of integrals)* or **(3)** finding the solution from an exceptionally good undergraduate physics textbook. ***So, here it finally is, tah-taah!!!:***

$$t_a = R \, log_\varepsilon \, k$$

What a beauty of an equation, a supreme example of elegant simplicity*!* The Doppler constant incorporates all the information that Observer, *<A>* has or can determine firsthand.

Converting *k* to directly obtainable data : *T & R*

But it is more insightful to work with the familiar parameters, *T* and *R*. Their meaning is eminently more palpable. So, by using the known relationship between *k* and *w* *(w in place of v)*, we can work out *k* in terms of *w*:

$$k = \sqrt{\frac{1+v}{1-v}} = \sqrt{\frac{1+w}{1-w}} = \sqrt{\frac{1+\dfrac{T}{X}}{1-\dfrac{T}{X}}} = \sqrt{\frac{X+T}{X-T}}$$

Next, if we multiply both Numerator and Denominator by $\sqrt{X+T}$, we get:

$$k = \frac{(\sqrt{X+T})(\sqrt{X+T})}{(\sqrt{X-T})(\sqrt{X+T})} \quad \text{which comes out as:} \quad k = \frac{X+T}{\sqrt{X^2-T^2}}$$

But we know that as $X^2 = T^2 + R^2$ so $R^2 = X^2 - T^2$, and $\sqrt{X^2-T^2} = R$ thus conveniently simplifying the Denominator in the above equation for *k:*

$$k = \frac{X+T}{R}$$

The last equation can be split into $\dfrac{X}{R} + \dfrac{T}{R}$ and rearranged as $\dfrac{T}{R} + \dfrac{X}{R}$.

Now the difficult-to-measure *X* in *X/R* can be substituted with its derivative *(known from earlier exercises)* $X = \sqrt{T^2 + R^2}$, so that we can eliminate the *X* entirely:

$$k = \frac{T}{R} + \frac{X}{R} = \frac{T}{R} + \frac{\sqrt{T^2+R^2}}{R} \quad \& \quad k = \frac{T}{R} + \sqrt{\frac{T^2+R^2}{R^2}}$$

$$= \frac{T}{R} + \sqrt{\frac{T^2}{R^2} + \frac{R^2}{R^2}}$$

And the solution is . . .

In the last derivation, the Numerator under the square root sign is $T^2 + R^2$ is divided by R^2 which resolves that part of the equation into $(T/R)^2 + 1$. This will give us the final value of k in terms of just T and R both combined into a *ratio* that is insensitive to, and un-influenced by, specific measurement units:

$$k = \frac{T}{R} + \sqrt{\left(\frac{T}{R}\right)^2 + 1}$$

At last, by substituting the derived value of k into our initial equation, we obtain the Cumulative Time, t_a clocked by the accelerating Traveler, $<B_a>$ and using only the measured values *(obtained by $<A>$)* of T and R:

$$t_a = R \; log_\epsilon \left(\frac{T}{R} + \sqrt{\left(\frac{T}{R}\right)^2 + 1} \right)$$

Comparing results

To show that the approximate total t_a values and the curves are essentially identical to those obtained with our final equation, let's set them side by side in the next table. Because our final equation gives us only total t_a values, t_a segmental values are calculated as differences between the consecutive total t_a values:

t_a SEGMENTS		CUMULATIVE *(TOTAL)* t_a	
Approximate	Calculated	Approximate	Calculated
0.998752338	0.8998340788	0.998752338	0.998340788
9.11921505	7.815395082	10.01796739	8.81373587
21.693045578	21.16849363	31.71101317	29.9822295
22.21673729	23.0011941	53.92774046	52.98342366
22.22216735	23.0256034	76.14990781	76.0090271
22.22222167	23.0258484	100.59429683	99.03487555
22.22222222	23.0258509	122.81651905	122.0607265

You can see that both the approximate and the accurate *(calculated)* t_a segments have exactly the same pattern of progression and the totals of both series are surprisingly close. If we had wanted our approximate results to be more accurate, we could have subdivided T from 0 to 1,000,000 into much smaller increments to reduce errors to as small as we wished. But the above exercise is sufficient to illustrate our general approach and at the same time utilize the least amount of data. Let us rejoice that with this equation we can finally get an accurate t_a for any value of T, a really nice reward for our efforts.

How about T from t_a

A new question does come up: can we work our way back from t_a to T? Of course, this should be possible. But there is the usual complication: So far, we considered all events as being simultaneous to Observer, $<A>$. Now, $<B_a>$ would have to consider everything from her point of view, that is according to what is simultaneous to her. And we know that this does not agree with $<A>$'s viewpoint. Relativity of Simultaneity and Observer-Specificity again*!!!* That's why we cannot use the same equation turned around to get back to our original T values

To illustrate these different points of view, let us compare them visually:

F12.14

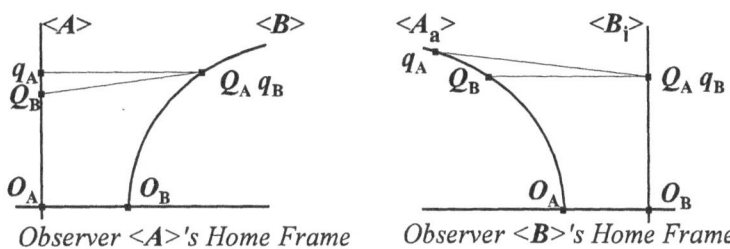

Observer <A>'s Home Frame   *Observer <B>'s Home Frame*

If <B_a> would calculate <A>'s Time (*T*) using the same equation but with *T* and *t* switched, she would get figures different from what we listed for <A> **BECAUSE** doing so would make us *descend the Spacetime Ladder by one rung!* That's what we learned in a previous chapter *(Ch. VII)*. Remember: what is simultaneous to <B> is *NOT* simultaneous to <A>*!!!* Again, *Relativity of Simultaneity* (*along with Observer Specificity*) is always ready to raise its ugly head*!* To get back to <A>'s original numbers, we'll have to solve our logarithmic equation backward from our hard-won *T*-to-*t* equation.

Back to <A>'s *T* from *t*$_a$

To assist those who want to get back from *t*$_a$ to *T* *(without descending the ST Ladder)*, the following worked-out derivation should be easy enough to follow. The complex equation of *t*$_a$ can be simplified by letting *k* stand for the entire bracketed part of the expression. This gives us our original equation that is much simpler and easier to use as a road sign:

$$t = R\ log_\epsilon k \quad \text{and} \quad k = \left(\frac{T}{R} + \sqrt{\left(\frac{T}{R}\right)^2 + 1} \right)$$

Dividing both sides of the initial equation by *R,* we get:

$$\frac{t}{R} = log_\epsilon k$$

The next step requires some knowledge of logarithms. We can orient ourselves by working with the more familiar base-10 logarithms first. We can agree that 10 to the power of 2 equals 100 and this allows us to state that:

$$log_{10}\ 100 = 2, \quad \text{thus} \quad 2 = log_{10}100 \quad \text{and} \quad 10^2 = 100$$

With natural logarithms, we use the quantity, ϵ instead of the familiar 10 as base. Therefore, by the rules of the above example, ϵ to the power of *t /R* must equal *k:*

$$\epsilon^{\frac{t}{R}} = k, \quad \text{same as} \quad k = \epsilon^{\frac{t}{R}}$$

and, therefore:

$$\epsilon^{\frac{t}{R}} = \frac{T}{R} + \sqrt{\left(\frac{T}{R}\right)^2 + 1}$$

Multiplying both sides of the equation with *R,* we get *(some intermediary steps not written out):*

$$R\epsilon^{\frac{t}{R}} = T + \sqrt{T^2 + R^2}$$

From this easily follows that:

$$T = R\epsilon^{\frac{t}{R}} - \sqrt{T^2 + R^2}$$

We are finally going in the right direction. We must isolate the T^2 under the square root radical and bring it to join the other T on the left side of the = mark. But first, let us keep the radical on the right side of the = sign while bringing all the other quantities to the left:

$$R\epsilon^{\frac{t}{R}} - T = \sqrt{T^2 + R^2}$$

then square both sides of the equation:

$$R^2\left(\epsilon^{\frac{t}{R}}\right)^2 - 2R\epsilon^{\frac{t}{R}}T + T^2 = T^2 + R^2$$

In the above equation, there are two T^2 expressions on opposite sides of the = sign. These we can (1) cancel out *(by subtracting them from both sides of the equation)*. Then (2) let's keep the expression with the remaining T in it on the left side of the equation, all others placed to the right and finally multiply both sides with -1:

(1) $R^2\left(\epsilon^{\frac{t}{R}}\right) - 2R\epsilon^{\frac{t}{R}}T = R^2$ Then (2) $2R\epsilon^{\frac{t}{R}}T = R^2\left(\epsilon^{\frac{t}{R}}\right)^2 - R^2$

Now, we can solve for T by dividing both sides by the Multiplier of T, that is by $2R\epsilon^{\frac{1}{R}}$, which step gives us:

$$T = \frac{R^2\left(\epsilon^{\frac{t}{R}}\right)^2 - R^2}{2R\epsilon^{\frac{t}{R}}}$$

Bringing R^2 forward in the Numerator gives us:

$$T = \frac{R^2\left[\left(\epsilon^{\frac{t}{R}}\right)^2 - 1\right]}{2R\epsilon^{\frac{t}{R}}}$$

In this last equation, one of the R in the R^2 in the Numerator and the single R in the Denominator cancel out giving us:

$$T = \frac{R^2\left(\epsilon^{\frac{t}{R}}\right) - 1}{2R\left(\epsilon^{\frac{t}{R}}\right)} = \frac{R}{2}\left(\frac{\left(\epsilon^{\frac{t}{R}}\right)^2 - 1}{\epsilon^{\frac{t}{R}}}\right)$$

The bracketed part of the equation can be further developed and the **R/2** expressed as ½ of **R**, giving us the final derivation of **T** from t_a:

$$T = \frac{1}{2}R\left(\epsilon^{\frac{t}{R}} - \frac{1}{\epsilon^{\frac{t}{R}}}\right)$$ which is our final "reverse" conversion result.

By the way, **R** was **<B>**'s Distance from **<A>** *(and vice versa!)* at the Moment when her (**<B>**'s) deceleration brought her spaceship to a full stop and before she started to move away from **<A>**. At that Moment, the Distance between them was the same to both (**<A>** and to **<B>**). That's because at that very Moment, both were in the same Inertial System *(without Motion between them)* and could totally agree about Time and Distance values. If, however, **<B>** would subsequently want to calculate cumulative Time on **<A>**'s clock using her *(<B>'s)* own clock data and using the same equation we worked out for **<A>**, she would not get back the original **<A>**'s **T**. Once **<B>** gets moving, they would never again agree about anything as to Space or Time measurements, and about Simultaneity. If **<B>** would calculate **<A>**'s cumulative Time *(with our original equation, **T** and **t** switched)*, she (**<B>**) would simply move down the "**ST** Ladder" *(see Chapter VII for review)*. That's Relativity for you, case closed.

This ends the above side trip to Time In Constant Acceleration. But wait! There is one more interesting side trip we must take.

A side trip with Kepler

Do you recall Kepler's Second Law of Planetary Motion? It says that the line *(secant)* connecting the Sun with the planet in orbit sweeps out equal areas of the elliptical orbit *(actually a near-circle)* in equal times!

F12.14

How does it apply to accelerated motion? Where is the planet? Well, here **<B>** is the "planet." But first, let's go back to **Ground Map** of 2-D Motion. The Path #2 here *(of a passing car or spaceship)* is placed horizontally for better utilization of this page. As the ground velocity is constant, the line connecting the stationary Observer with the moving object sweeps out equal area triangles in equal Times:

F12.15

Stationary Observer

Direction of motion

As the Area of each triangle equals half of the product of base and height, you can see that all the triangles are equal and none are "more equal" than others.

Now, let us take the same setting but featuring uniformly accelerating motion in 1-D Space. That the hyperbolic curve is part of a **ST**- warped "circle" or "cloverleaf ellipse" *(see*

Chapter III) was brought to your attention previously. Here the line that connects Event, **O** with the WL_a of the Accelerating Observer also sweeps out equal Map areas in equal Times (t_a) according to the clock on board with the Accelerating Observer, all this from the point of view of <A>. Again, very interesting!

F12.16

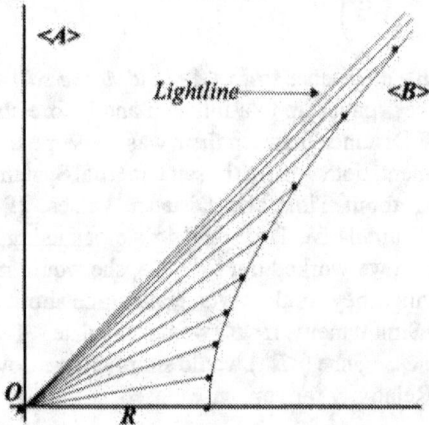

That's all about this topic. Time to move on to $E = mc^2$.

XIII - RELATIVISTIC MECHANICS

A glance at the universe

The universe at large seems eternally tranquil, undisturbed in celestial splendor. Yet, on closer scrutiny, it is engaged in relentless expansion, violent birth and death of stars, collision of galaxies, stellar cannibalism by hungry, black holes. And more action is contained in the sub-microscopic world of atoms. We ourselves both contain, and are surrounded by all kinds of *CHANGE.* Hardly anything remains the same forever, much of it *powered by,* and resulting in, all kinds of *MOTION.* But what powers Motion, what's behind it*?*

Elementary mechanics

To understand the driving force in nature, *we need more than Space, Time and (Uniform) Motion* so far at the focus of our attention. It is fortunate that *Classical Mechanics* still provides many of the answers we seek. Reviewing it briefly would, therefore, be the best introduction to our limited exploration of Relativistic Mechanics presented here as an extension of the topic, Acceleration From 2-D Motion *(Chapter XI)*.

Aristotle taught that *to KEEP things in Motion, you had to KEEP PUSHING or PULLING them*. It is quite descriptive of the ancient and not-so-ancient worlds of chariots, ox carts and sailing ships. Newton's laws of Motion revised all that, brought it up-to-date, set it straight. So, let us restate his most important Law of Motion: *An object will stay at rest or in Uniform Motion until (and only until) it is acted upon by a Force.* Now, that *FORCE, F* will cause a body with *MASS, m* to undergo *ACCELERATION, a,* while *(and only while)* that Force is being applied. What a neat package of insight*!* Let's put it into symbolic, mathematical language:

$$F = ma$$

Here we have an equation with three components. There is a description for F but the other two elements have separate identities of their own, expressed as follows:

$$m = \frac{F}{a} \qquad\qquad a = \frac{F}{m}$$

The last equation tells us that Acceleration is proportional to Force and inversely proportional to Mass. This agrees well with our everyday experience as things respond by Acceleration *more willingly to greater Force* and *more stubbornly if there is more Mass.*

In Classical Mechanics, Mass was found to be *conserved,* that is, not gained or lost no matter what was done to it. What Newton considered to be the "Quantity of Motion" is known as *MOMENTUM, p* and it was also *conserved.* The mathematical relationships here are also simple and linear:

$$p = mv \qquad\qquad m = \frac{p}{v} \qquad\qquad v = \frac{p}{m}$$

Accelerating an object for a duration, T will produce a Velocity, v or:

$$v = aT \quad \text{and} \quad a = \frac{v}{T}, \quad T = \frac{v}{a}$$

If we lift up a heavy bowling ball from the floor and place it on a table, we perform *WORK, W* that is equal to the lifting Force applied times the *DISTANCE, D* lifted, or:

$$W = FD, \text{ also } F = \frac{W}{D} \text{ and } D = \frac{W}{F}$$

Now, the bowling ball, elevated to table height, can do Work if the force of its weight is allowed to drive some machinery on its way back down to the floor. So, the Work performed in lifting was converted into **potential ENERGY** that could be retrieved as work and that would be equal to the Work performed by lifting it upon the table, or:

$$W_{done} = E_{potential} = W_{recovered}$$

What we have shown here is a small number of very **simple equations that specify important relationships between a small number of mechanical concepts.** It is possible to **combine** some of these to **obtain new** relationships by taking an equation and substituting one of the elements with its derivative from another equation. Let us see how we can combine **Force** and **Velocity** through **Acceleration** and obtain a **new expression** *(equation)* for **Momentum:**

$$F = ma, \text{ but } v = aT \text{ and } a = \frac{v}{T}, \text{ thus } F = m\frac{v}{T} = \frac{mv}{T}$$

Restated: $F = \dfrac{mv}{T}$

But $mv = p$, thus $F = \dfrac{p}{T}$ and $FT = p$, restated: $p = FT$

this time meaning: Force, F applied for Time, T producing Momentum, p.

This is just one of the many ways we can combine information contained in different equations to obtain new information.

We'll limit ourselves to just these concepts of Classical Mechanics. It is all very elementary, indeed. We'll return next to Constant Acceleration in the 1-D world, also inherent in Uniform Motion observed in 2-D Space *(Chapter XI),* and see how we can extend our understanding gained there to matters of **Force and Energy** in the fast-moving world of Special Relativity.

Reviewing Acceleration math

To review the Constant Acceleration *(Chapter X)* pertinent to mechanics, let's look again at the familiar Ground Map of the 2-D Uniform Motion, Path #2 and the **ST** Map of 1-D, Uniformly Accelerated Motion:

F13.1

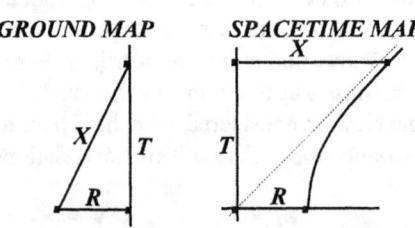

GROUND MAP   *SPACETIME MAP*

As the only data *(measured quantities)* available to work with are X, R and T, the basic relationships we need to bring over from Chapter XI, are:

$$X^2 = R^2 + T^2, \qquad R^2 = X^2 - T^2, \qquad T^2 = X^2 - R^2 \text{ and}$$

$$X = \sqrt{R^2 + T^2}, \qquad R = \sqrt{X^2 - T^2}, \qquad T = \sqrt{X^2 - R^2}$$

At **O**-Event, $X = R$ and $T = 0$

At Event, Q Motion has covered a Distance, $D = X - R$.

Because of constantly changing relative Velocity *(increasing from zero at O)*, the Velocity, w at any moment on the Path of Accelerating Motion is *NOT (X-R)/T* but:

$$w = \frac{T}{X}$$ from which we also get $T = wX$ *(a Velocity-times-Distance product known to us from the Spacetime Ladder, Chapter VII)*.

And the relationships between a and R are:

$$a = \frac{1}{R} , \quad R = \frac{1}{a}$$

Because on the Ground Map, T equals the path traveled at Constant Velocity, the Velocity, v is actually strictly proportional, or completely equal, to vT, therefore:

$$T^2 = X^2 - R^2 \quad \text{or} \quad (vT)^2 = X^2 - R^2$$

and the Ground Velocity, v is *(haven't we seen it a few times already?)*:

$$v = \sqrt{\frac{X^2 - R^2}{T^2}}$$

If the maximum possible Velocity, w can be no more than the Ground Velocity, v then the accelerated Velocity, w at any Moment, T is:

$$w = vM_{(multiplier)} = v\frac{vT}{X} = \frac{v^2}{X}T$$

If the limiting Velocity in a Constant-Acceleration scenario is c *(which equals 1.0)*, then our Instantaneous *(accelerating $<B_a>$'s)* Velocity at any Moment measured according to $<A>$'s clock *(as shown in Chapter XI)* is:

$$w = \frac{T}{X}$$

Now we have all the needed mathematics brought into the present Chapter.

Acceleration in Spacetime

What we would like to see is *how* (1) *accelerating bodies put on weight (gaining Mass)* by "eating" *(or accelerating by consuming)* Energy and *how* (2) *Mass is, therefore, equivalent to Energy.* In practically all undergraduate textbooks, the gain in Mass is simply shown in print, derivation kept top secret!? It is up to us to dig it out by combining different equations and developing the resulting combinations further. But first, let's recall that *Force is needed to obtain Acceleration, F = ma.*

And recall also that: $a = \frac{1}{R} , \quad R = \frac{1}{a}$

Taking R from the equation that relates the variables R, X and T:

$$R^2 = X^2 - T^2$$

and substituting R in it with its *(derived)* equality, $1/a$ we get:

$$\frac{1}{a^2} = X^2 - T^2$$

Now we'll develop it into:

$$1 = a^2 (X^2 - T^2)$$

But as $w = T/X$ and $T = wX$, the T^2 can be replaced with $w^2 X^2$

$$1 = a^2 (X^2 - w^2 X^2)$$

Bringing X^2 out of the brackets:

$$1 = a^2 X^2 (1-w^2)$$

and using the last equation above to solve for $(aX)^2$:

$$a^2 X^2 = \frac{1}{1-w^2} \quad , \quad \text{restated:} \quad (aX)^2 = \frac{1}{1-w^2}$$

From this we get a very-very interesting result:

$$aX = \frac{1}{\sqrt{1-w^2}}$$

Take a very good look at it *!!!* In the second half of the equation, there is the inverse form of our β, the Accelerating-Velocity version of $\sqrt{1-v^2}$. This is the tool we really need.

Next, let us go back to $T = wX$ *(derived from $w=T/X$)* and multiply both sides of the equation with Acceleration, a and obtain:

$$aT = waX$$

We saw previously that $a = \frac{F}{m}$ *(from classical equation, $F = ma$)*, so multiplying both sides of this classical equation shown above *(a = F/m)* with T, we get:

$$aT = \frac{FT}{m}$$

Now we have two derivations for $aT :$ $aT = waX$ and $= \frac{FT}{m}$.

But as both derivations are equal, so $aT = waX = FT/m$ and we can restate the preceding equalities to be: $\frac{FT}{m} = waX$ and because $aX = \frac{1}{\sqrt{1-w^2}}$ *(see above)*, we can say that

$$waX = w\frac{1}{\sqrt{1-w^2}}, \text{ so}$$

Recall that $\dfrac{FT}{m} = w\dfrac{1}{\sqrt{1-w^2}}$ and $FT = \dfrac{mw}{\sqrt{1-v^2}}$

Recall also that *FT* is none other than ***Classical Momentum, mv*** but here featuring *w* instead of *v*. So this part of the equation,

$$\frac{mw}{\sqrt{1-w^2}}$$

is actually ***the MOMENTUM obtained as a result of ACCELERATION !!! You can see that this Momentum amounts to a quantity greater than the original Classical Momentum (p = mv).***

$$p = \frac{mw}{\sqrt{1-w^2}}$$

What a remarkable revelation, clearly different from what Newton would have said *!!!* But the resultant Momentum, *(FT =)* $\dfrac{mw}{\sqrt{1-w^2}}$ can be written as either

$\dfrac{m}{\sqrt{1-w^2}}w$ or $\dfrac{w}{\sqrt{1-w^2}}m$. Both would be mathematically correct.

However, in the first equation, the Mass, *m* increases infinitely as Velocity increases to *c*. In the other equation, the Velocity, *w* increases infinitely. But *we know (theoretically and from actual experiments) that no Velocity can exceed c, so w cannot exceed c.* The Velocity, *w* can only reach *c* but not go beyond. So which choice would be the correct one *?*

Making the right choice

As Velocity, *w* cannot really equal or exceed *c*, not to mention infinity, it must be the Mass that increases without limit, even to infinity:

$$m \rightarrow \frac{m}{\sqrt{1-w^2}}$$

Force applied adds to Mass

The result we obtained is not only interesting but it is *NOT* what we know from Classical Mechanics according to which Mass is something that is *conserved*, always the same, even when pushed hard, accelerated.

What we have worked out here is *a NEW LAW OF NATURE*: *as Energy is applied in form of FORCE over a period of Time, we get less and less additional VELOCITY and more and more MASS as a reward for our efforts. It also says that Mass is not what is conserved. Instead, it is the combined ENERGY + MASS that is CONSERVED.* Moreover, *along with Mass, it is also Momentum that is increased at increasing Velocities!!!* This non-classical result is so important that it is worth repeating it along with its corollary:

At higher velocities, ENERGY is CONVERTED into MASS. Therefore, ENERGY is EQUIVALENT to MASS. Also, *if ENERGY contains (or is) MASS, so MASS contains (or is "concentrated" or "packaged") ENERGY !!!*

Actually, it works both ways: we have converted *(kinetic)* Energy of Motion into Mass and we can do it also in reverse.

Exploring Force <=> Mass Transformation

By all accounts, the extra Mass obtained from Energy should be again convertible into Energy. That certainly must be so. If we slam a high-velocity object *(missile)* into a solid, brick wall, it should do more damage than obtainable from a simple $p = mv$ combination, m here taken as Rest Mass.

Let's continue thinking along classical lines: Work: (1) W equals FD, (2) $D = X-R$, same as $D = X-(1/a)$ and (3) $F = ma$.

By multiplying both sides of the classical equation, $F=ma$ *(that connects F, m and a)* with D, we get:

$$FD = maD.$$

But $D = X - R$, so

$$FD = ma(X-R)$$

and as $R = 1/a$, so

$$FD = ma\left(X - \frac{1}{a}\right)$$

and furthermore, the above equation develops into:

$$FD = maX - \frac{ma}{a}$$
$$= maX - m$$

Again, as aX (in maX) $= \dfrac{1}{\sqrt{1-w^2}}$ and $FD = W = E_{\text{potential}}$, therefore:

$$W = E_{potential} = FD = \frac{m}{\sqrt{1-w^2}} - m$$

This last equation gives us the new total for our increased Mass minus Rest Mass which represents *the amount of increase (or gain) over and above the "Rest Mass" due to Motion.* It is what we get by application of Force, F upon Mass, m over Distance, D less "Rest Mass."

Now, at slow speeds, $\sqrt{1-w^2}$ equals almost exactly 1.0 so *the increase in Mass added to the Rest Mass* at slow speeds can be calculated:

$$\text{as } \sqrt{1-w^2} = \sqrt{1-0^2} = 1.0, \quad \text{thus } \frac{m}{1.0} - m = 0$$

What the last derivation means is this: There is *NO* observable *(measurable)* increase in Mass at slow Velocities, exactly the way we are used to and have always believed. Settled! Anyone disagree?

Let's look again at Work or Energy at very-very high Velocities, starting out with:

$$W (= E) = FD$$

Let us remind ourselves that at Velocities close to that of Light, *w* itself does not budge, does hardly increase any more and the accelerated Velocity becomes *(almost)* a constant. So, *we can use the symbol, v for a constant velocity* instead of *w* in our next argument.

Now, the Distance, *D (as on our 2-D Ground Map, math identical with that of 1-D Acceleration)* is also *vT (D = vT)* and so *E = W = FD* changes into:

$$W \;=\; E \;=\; FD \;=\; FvT \;=\; FTv$$

But *FT = p* which is Momentum, and is also equal to *mv*, therefore *FT = mv* and:

$$FTv = pv = mvv, \quad \text{so} \quad E = FTv = mvv = mv^2$$

Restated: $E \;=\; mv^2$

And, finally....

At velocities of *v = c*, the above equation becomes:

$$E \;=\; mc^2$$

Now, this, finally, is the famous equation that says *ENERGY equals MASS times VELOCITY of LIGHT squared !!!*

The above statement depends on v = c !!!

But wait a minute! Don't we know that nothing made up of Mass can go that fast. This is an impossibility! Therefore, we must qualify the above statement. Two different but qualified conclusions can be drawn from the equation:

Conditional, qualified conclusions

(1) *LIGHT* itself, which is *ENERGY*, has *MASS!!! All ENERGY must, therefore, have MASS. Moreover, LIGHT must also have MOMENTUM*, closely related to both Energy and Mass. And the Mass of Light is:

$$m \;=\; \frac{E}{c^2}$$

which is another way of saying that *ENERGY is EQUIVALENT to MASS, or is SAME as MASS or acts like Mass !!!*

(2) *All MASS has ENERGY, or is ENERGY (at least POTENTIALLY), therefore, Mass is CONVERTIBLE to ENERGY !!!*

Both conclusions contain an *intuitive leap, a CONJECTURE based on parallels to, and extrapolations from, logically founded and mathematically justified derivations*. Both conclusions have since been supported by a great mass of experimental data so that we can consider the theory about *ENERGY = MASS* as proven beyond doubt.

The classical laws of conservation of Mass and of Energy are combined here into a law of *conservation of Energy+Mass.*

Final comments

Incidentally, there are a number of alternative derivations of the equation, $E = mc^2$ all proceeding from different premises. The one shown here resulted from combining the mathematics of Uniform Acceleration developed in Chapter XI with the elements of standard Classical Mechanics and extracting from the mix of different equations the bold, relativistic conclusion.

The mathematical argument is basically identical with the one used by Sam Lilley in his book for the intelligent lay person *(see Preface, page A-4, line B-2)* but the presentation here has the advantage of proceeding from the more transparent, 2-dimensional geometric model *(of Chapter XI, Acceleration From 2-D Motion)* that is clearly more intuitive.

The results of Special Relativity at low Velocities are reduced to those obtainable by Classical Mechanics and these agree comfortably with our mundane sensibilities. On the other hand, at very high velocities, Classical Mechanics needs replaced with Relativity to remain true to Mother Nature. ***Classical Mechanics is, therefore, a low-speed approximation of Special Relativity and Special Relativity is a much more general statement*** *(or a revision)* ***of Classical Mechanics.*** All this is accomplished without pushing the older version of Mechanics off the exalted pedestal it has occupied over three centuries.

XIV - ABERRATION OF LIGHT

What is it?

When viewing objects through complex optical devices, we no longer notice colored halos around them. It used to be a rather annoying problem caused by the off-center prism-like action of lenses and was appropriately called **CHROMATIC ABERRATION**. Ingenuity in the design of compound lenses and other optical components has practically eliminated it in cameras, microscopes and refracting astronomical equipment. The introduction of concave *(Newtonian)* mirrors avoided it in telescopes early in the history of modern astronomy. But a subtle shift in the observed location of stars, now known as **GEOMETRIC ABERRATION,** came to light gradually as increasingly accurate instruments were put into use by the early 18th century. A British astronomer, Bradley *(1727)* was the first to point out that *a part of the seasonably variable star locations could be distinguished from parallax* and argued that *the difference was due to Motion of the telescope along with its platform, the Earth, through Space.* That this was less than an arc second is why it was not noticed earlier.

First evidence to support Copernicus

Bradley's discovery provided the first observational evidence in support of the new Copernican heliocentric view of the world. Bradley used the carefully worked-out planetary orbits resulting from successful application of Kepler's laws of planetary Motion *(about 1618-28)* and, later, of Newton's laws *(1687)* to celestial mechanics. Additionally, he had available the first accurate estimate of the Velocity of Light by Roemer *(1676)* based on calculated Earth-Jupiter distances along with carefully timed occultations of a Jupiter's moon. Despite many theoretical successes in astronomy, observational evidence of Earth's Motion was still missing until Bradley's discovery. So, it could always be argued that the Earth was standing still and, true to Ptolemy, it was not only Jupiter but everything else in the sky that moved. *Bradley's conclusions were strictly circumstantial but heavily weighted in favor of Earth's Motion. That's because it was impossible that all the stars in the firmament first* moved in one direction *then in the other, causing geometric aberration. So it had to be the Earth that was moving and not the stars.*

Before Bradley, the Copernican view was officially just a convenient, *mathematical device* useful for making the erratic movement of planets rational in the abstract, geometric sense. Some even took the politically risky step and openly pronounced it a description of reality at a time when direct observational evidence for it was still missing. Just imagine Galileo's confident cockiness reduced to stammers when sternly pressed by the Vatican council of learned cardinals for irrefutable, solid evidence that the Earth was actually moving around the Sun.

A model from common experience

To develop an intuitive feel for the Bradley's effect, let's start out with a specific, familiar experience on a rainy day and see how it could be used to explain geometric aberration.

Let's *imagine sitting in an automobile and seeing water droplets streaking down the side windows*. The droplet direction would be straight down on a windless day but down and toward the rear with the wind blowing against the front window or with the car moving forward. A *similar slant in the path of Light rays is seen if the telescope (with its mount, the Earth) happens to be moving at some angle to the incident Light.* Because of the very high Velocity of Light compared to that of the Earth in its orbit *(0.000,1c)*, the rearward slant of Light entering the telescope is actually very slight, noticeable only with very accurate mounts and evident only after numerous measurements taken over a long time were compared.

What did Bradley find?

It is not possible to determine exactly where within the aperture the starlight actually enters the telescope. But the small image of the star is usually adjusted to strike the center cross hairs of the telescopic field in order to determine its location in the sky. If the observed star is directly overhead and the Motion of the Earth happens to be directly toward or away from the star, the actual position of the star would be exactly where it is observed *(Fig. 14.1A)*. If, however, the Light enters the telescope through the exact middle of the aperture and the telescope is moving sideways, it would strike the mirror, and be viewed, slightly off center *(Fig. 14.1B)*:

F14.1

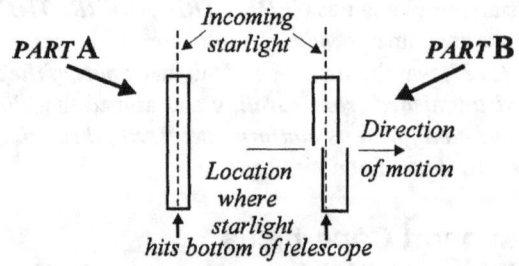

To obtain the star's image at the exact center of the telescopic field, *the instrument has to be tilted slightly toward the Earth's Motion* in order to correct for its movement while the Light is speeding down the telescope barrel. This tilt is greatly exaggerated in the illustrations to make the geometric aberration appear graphically more obvious:

F14.2

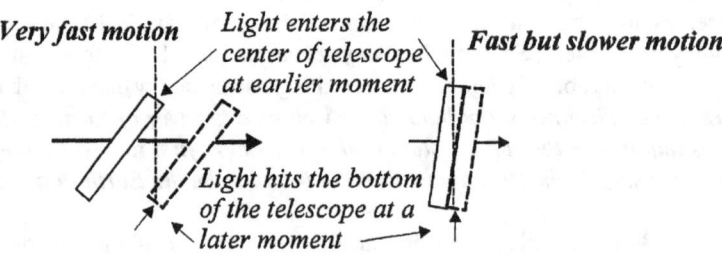

As noted by Bradley, the telescope had to be kept ever so slightly tilted in the direction of the Earth's orbital Motion throughout all the seasons:

F14.3

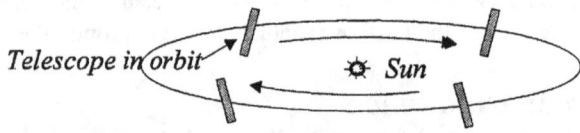

Geometric representation

The path of telescope Motion and the direction of Light can be represented as follows, the angle of tilt again exaggerated in the illustration for emphasis:

F14.4

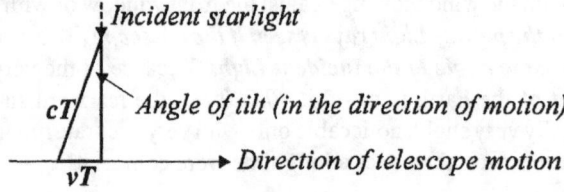

Definitions:

(1) vertical line - direction of incident Light, in this case perpendicular to the movement of the telescope to provide us with a right-angle triangle.

(2) cT - Length of Light path through the telescope.

(3) vT - Distance of telescope travel while the Light is traversing the length of the telescope.

(4) α - The angle of the telescope tilt toward the direction of motion proportional to the Velocity of the Earth through Space.

The trigonometric function applied to the right-angle triangle *(here true to scale)* shown above can be given as follows:

$$sin\ \alpha\ =\ \frac{vT}{cT}\ =\ \frac{v}{c}\ =\ v\ ,\quad v\ \text{expressed as a fraction of } c.$$

Restated: $sin\ \alpha\ = v$

You should see that $sin\,\alpha$ is Earth's velocity, v through Space given as a fraction of c. The above description follows the established convention of giving velocities in the language of trigonometry, something we have generally avoided. But because of the lesser importance of this particular topic in our study, we'll try to acquire enough understanding of Light aberration without the otherwise mandatory math in order to prepare us for our last chapter about the most intriguing although a somewhat peripheral feature of Special Relativity: Sighting *(of)* Blue Streaks.

Note that the telescope tilt, α cannot be measured directly. But it is always there, **_hidden_** in all measurements. *As the tilt of the telescope is in the direction of orbital Motion, it is oriented in opposite directions on any two dates 6 months apart. The deviation from actual direction of incident starlight is, therefore, only half of the difference between two such measurements.* This last statement ignores the considerable difficulty determining the actual direction of the telescope movement in relation to the star while not only orbiting the Sun but also orbiting the axis of the Earth. The contribution of the solar system's orbit around the galactic center is, fortunately, too small to be of significance.

But the result of this deviation is important: *The TRUE POSITION of the star is not exactly where it is observed. Its VIRTUAL POSITION, the position where it is actually observed, is always shifted slightly toward the forward direction of the telescope movement.* If we know the *Velocity, the direction of the telescope movement* and *the angle of tilt*, we can calculate the *true (but unobservable) position of the star.* In the illustration below, the telescope is shown first in positions occupied 6 months apart then placed to form an angle that is twice that of the shift in the apparent star's location from its true *(but unobservable)* place in the sky:

F14.5

Half of the angle between telescope directions 6 months apart

Direction of telescope pointing

Now that Bradley's effect has become clear *(or has it?)*, we should also ask: what kind of a shift is actually observed if the telescope is moving away from the star at an angle and is pointed obliquely in the backward direction of telescope movement? Let's see:

F14.6

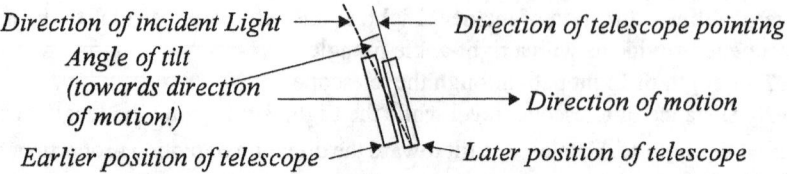

A surprise*!* The tilt is still toward the direction of motion. But what if the telescope is pointed at the star located directly ahead or directly behind, let's see:

F14.7

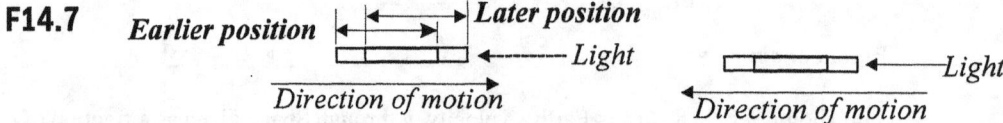

As you may have guessed: no shift.

It is interesting that this shift in image position from unobserved *(real)* to observed *(virtual)* is the source of *two closely related phenomena both due to the SAME SHIFT of the image (or images) toward the direction of telescope movement but produced by two different directions of telescope pointing.*

Bradley's Effect

First, there is this **BRADLEY'S EFFECT, *produced by pointing the telescope at th*e *Star's actually observed VIRTUAL IMAGE*** as shown in the two illustrations below, first with Light coming in at perpendicular direction then at a slant with regard to telescope Motion:

F14.8

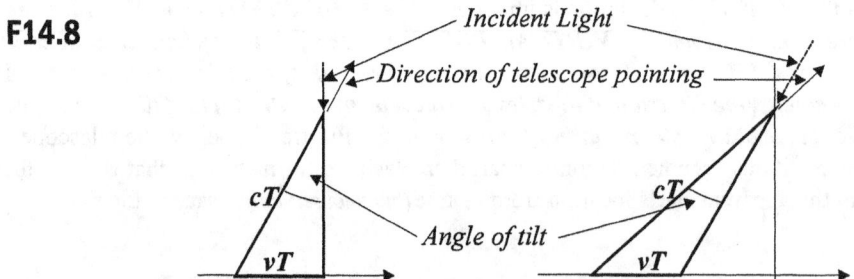

On the left side in the above illustration *(Fig. 1.4)*, the angle, α_1 is located between the presumed true but unobserved direction of the incident Light *(coming from the unobserved star position)* and the direction that is perpendicular to the telescope Motion. The angle, α_2 is between the *(actually)* observed direction of Light *(the virtual location of the star)* and the direction perpendicular to the telescope Motion. The angle between the actually observed direction of the *(virtual)* image and the true but unobservable direction of the star would then be equal to $\alpha_2 - \alpha_1$, or simply α *(without the subscript)* the way as it is usually given in most textbooks.

F14.9

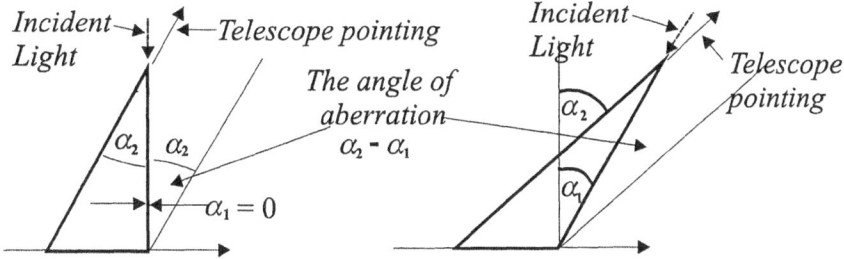

The Headlight Effect

Secondly, there is this "***HEADLIGHT EFFECT***" which is obtained by ***pointing the telescope in the direction of*** *(TELESCOPE)* ***MOVEMENT in Space.*** Here the angle, ϕ_2 of the star's virtual *(that is observed)* position and ϕ_1 of the star's true *(but unobserved)* position in the sky are both measured from the telescope pointing direction that coincides with the direction of telescope movement. Two different cases of the Headlight Effect are next shown, Fig. 14.10:

F14.10

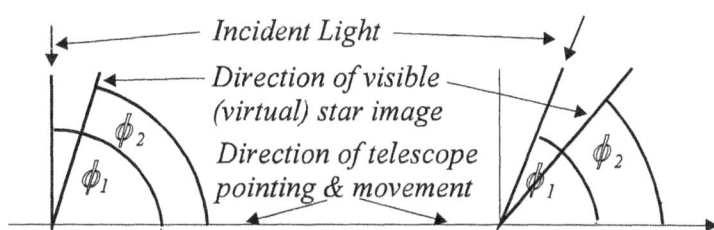

It may be confusing that for the description of the Bradley's Effect, only a single angle, α is used while the Headlight Effect is usually specified by two angles, ϕ_1 and ϕ_2.

If the ***REAL*** *(unobserved)* position of the star is perpendicular to the direction of motion *(see Bradley's Effect, on the left, Fig. 14.9)*, the angle between the ***REAL*** *(unobserved)* position and the perpendicular direction is ***ZERO*** and there is only one angle, our $\alpha (= \alpha_2 - \alpha_1 = \alpha_2 - 0)$ we can use. In case the Light coming in at a slant, the single α is meant to represent the angle of shift from the true to the virtual direction. We can always use, unconventionally, two angles, α_1 and α_2 in all cases of Bradley's Effect to underline its basic similarity with the Headlight Effect. Once you understand the difference between them, you should have no problem understanding how these two Effects are related.

The Headlight Effect can benefit from additional visual demonstration. If we look at the starry sky through the telescope, we always see a multitude of stars. With more powerful optics, more of the dim stars between the brighter ones become visible. The visual field of the telescope attends only to a relatively small angle far less than the almost 180 degrees ideally available to maximal wide-angle view. It is like looking at just a very small circle in the sky. So, to demonstrate the Headlight effect, we can point the telescope straight ahead to a star in the direction of the movement and note, let us say, five prominent stars within the field of vision. Also, let us include a few stars outside this narrow visual field and tag them by numbers:

F14.11

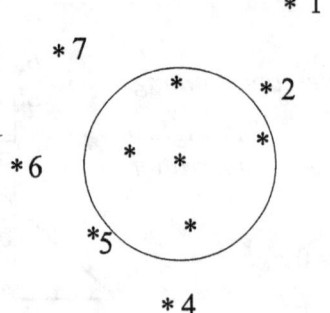

Next, let the telescope *(now placed in a space vehicle)* move at a very high Velocity towards the star in the middle. The high Velocity amplifies the usually subtle effect of slower Motion making it more obvious. All the stars would now appear closer together so that some of the stars previously outside the visual field would be seen *(within the visual field):*

F14.12

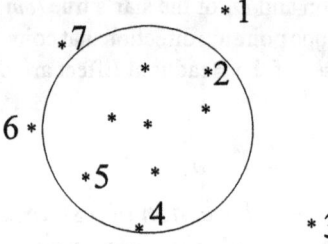

Comparing Bradley's and Headlight Effects

We can now unify the Bradley's Effect with the Headlight Effect by means of the following equalities:

$$\alpha_1 + \phi_1 = 90^\circ, \quad \text{and} \quad \alpha_2 + \phi_2 = 90^\circ$$
$$sin\,\alpha_1 = cos\,\phi_1 \qquad\qquad sin\,\alpha_2 = cos\,\phi_2$$

and show them both in the following illustration:

F14.13

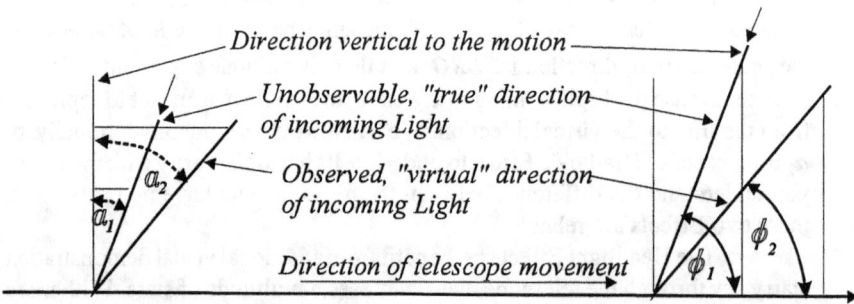

Note that the Hypotenuses of the two triangles bordering α_1 and α_2 are formed by the *directions of* (1) *the unobserved (true, α_1 yet to be calculated) position* and (2) the *actually observed (virtual α_2) position of the star.* Similarly, the same lines forming the Hypotenuses bordering ϕ_1 and ϕ_2 are also defined by *the same direction of the true but unobserved position*

of the star and the direction of the virtual but observed position of the star. While two triangles with the angles, α_1 and ϕ_1 & α_2 and ϕ_2 always share the same Hypotenuse, the telescope is pointed in the direction of the observed *(but virtual)* position of the star for Bradley's Effect and in the direction of telescope movement for the Headlight Effect.

The direction of Motion referred to in this chapter needs additionally restated as to *who* or *what* is considered to be moving. In Special Relativity, the inertial Observer always considers him*(her)*self stationary. But in Bradley's Effect, it is considered impossible that celestial objects can first move in one direction then six months later move in the opposite direction. *It is,* therefore, *the Observer who, according to the new Copernican thinking (though contrary to our relativistic orientation), is considered to be moving.* In addition to star clusters, we may need to consider also artificial, man-made objects moving in Space in our direction and visible in the telescope or in plain sight. In all cases, it is the *direction of the "Effective Motion" of the Observer relative to the observed objects* that matters. This exceptional stipulation of Motion to the Observer allows us to state that in all cases, be it Bradley's or Headlight Effect, *the positions of all observed objects become moved closer to the Observer's direction of actual or "Effective" Motion.* To keep everything consistent with Special Relativity, we can say that approaching *(multiple)* objects always become positioned closer together and receding objects spread out while inertial Observers can be considered as stationary, non-moving. Is it clear Special Relativity does not need to be bent just to please Bradley's ghost.

Conclusions

Bradley's and Headlight Effects are basically one and the same phenomenon but the results obtained differ not only because the telescope is pointed differently in both but also because more than one object is needed for the production of the Headlight Effect. *Despite the different telescope orientations, the shift of visible images is always the same: always towards the direction of "Effective Motion" of the Observer which is equivalent to the visible objects approaching. The faster the Motion, the greater the shift. In Bradley's Effect, the telescope is aimed directly at the virtual (visible) image while in Headlight Effect, the telescope is aimed in the direction of its "Effective Motion" through Space.* If the observation is carried out on a group of approaching or receding space vehicles, the Observer is still considered to be moving in relation to them. The determination of actual *("effective")* direction of Motion of the telescope *(mounted on a platform, the Earth)* in relation to celestial objects is a technical matter related to astronomy and beyond our capabilities at this time. And, once more, in summary:

Bradley's Effect is most appropriately applied to a single object under observation. This is how Bradley first discovered it. The telescope was pointed at a single star and knowing the direction of Earth's movement on its orbit around the Sun was all that was needed. The pointing of the telescope and the direction of movement did not have to coincide. With a telescope *(or binoculars),* many stars are usually brought into the field of vision, so the telescope needs pointed in the actual forward direction of Motion to obtain the Headlight Effect. Just keep in mind that as any single star within 180 degrees of forward Motion is shifted toward the center of the field of vision so all stars *(or objects)* would be seen crowding toward the center of the field. In case of receding Motion, the shift *(of all receding objects)* would be directed toward the periphery of the telescopic *(or binocular)* field. In Startrek movies when the Spaceship Enterprise disappeared from sight into distance at "warp" speed, the crew *(and the movie audience)* looking in the direction of travel, all forward images of stars were seen converged to a single, bright spot. In the less spectacular rearward view, on the other hand, everything in the sky would have been seen spread out, individual star images markedly enlarged, faint and, perhaps, even barely visible.

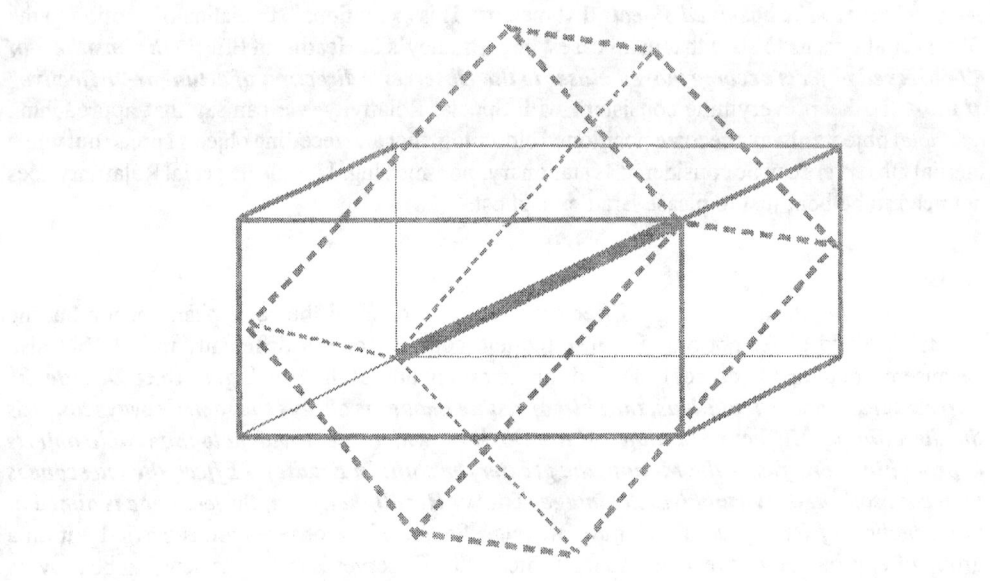

XV - SIGHTING BLUE STREAKS

What are those Blue Streaks?

Sighting Blue Streaks is all about *VISUAL KINEMATICS, or how objects moving real fast would appear to us*. The problem is not only about their looks at high Velocities but also how to catch a glimpse of those elusive birds in flight.

The *time limitations* are severe. For example, only 0.1 second is needed for a near-Light-velocity object to approach us from a Distance of 15,000 km, pass us at close range and disappear behind another 15,000 km giving us hardly enough time for a picture opportunity. Using a wide-angle camera, all we would get is a big blur with outlines of the moving object smeared beyond recognition. With a fast-action camera, the image caught on film or magnetic memory would be shaped by *Lorentz-contraction, Doppler Effects (on image and color)*, *Bradley and Headlight Effects* and still *another Transformation* we need to add to our list. What all this adds up to is a new territory in Spacetime we need to explore by the step-by-step method that has served us well with other topics. We'll revisit a few adventure trails in Lineland *(4-D "compressed" into 2-D)* already familiar to us as we finally charge head-on into the full 4-D Spacetime for a closer look at a few less known applications of Special Relativity.

Who has seen them?

Blue Streaks share one important distinction with UFOs: *NO AUTHENTICATED SIGHTINGS !!!* If they exist, then where are they? If not here now, future technology, no doubt, will produce them. Then, like ornithologists good at stalking birds, we should know what to look for, where to spot them and how to snap their pictures. The most important equipment we have to find, or invent, is an ultra-fast camera to "freeze" them in flight. A picosecond shutter speed is exactly what is called for. It would give us 0.3 mm resolution, more than adequate for even small objects. *(Light travels 0.3 mm in one picosecond defining the resolution possible at that speed!)*. Once we have the pictures, we can take our time to decipher the Transformation of their original shapes into Blue Streaks.

As to the make-up of these Blue Streaks, all we can do for now is to speculate pre-empt-ively about them because nothing known to exist has ever come close to answering the criteria of Blue Streaks. To prepare for future sightings, we'll have to put together a presumptive list of all the possible relativistic disguises. This challenging topic has already received some expert attention in the literature but the last word has not been written. We may not get everything right this time but, then, why shouldn't we have fun trying.

What is Perspective?

First, lets's review briefly *what ordinary matters like DISTANCE and POSITION* can do to the looks of stationary or slowly-moving objects. The entire subject was extensively developed during renaissance first in Italy, then also in other countries and is known as *Linear Perspective* or, simply, *PERSPECTIVE*. It has been a favorite tool of artists and architects for giving depicted structures their realistic, 3-dimensional appearance. It is a necessary starting point because relativistic effects of fast Motion always become grafted upon those of Linear Perspective. Considerations of brightness, shadows, intensity of colors, sharpness of lines and other matters important to artists do not concern us. *It is only the outline of objects that is imp-ortant* and will be at the focus of our attention.

To study image transformations, let's select a simple, geometric figure of a *CUBE* to assist us. It has a shape that duplicates the Cartesian coordinate system in the 3-dimensional Space familiar to us from everyday life and has been a test object in visual kinematics before. So, let's start out with the picture of a Cube:

F15.1

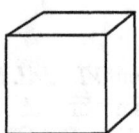

What a beautiful, simple shape*!* Actually, this particular position presents to our eyes *all the four edges of the exposed three Sides*, the maximum possible number and *two Edges of each of the other three hidden surfaces.* But as you see, no Side remains totally unnoticed.

Looking at identical Cubes in a row at eye level from extreme left to extreme right with generous amount of space between them, we can make a number of observations:

(1) All parallel lines appear to converge to points at the extreme left and right on the horizon where the Cubes appear so small and point-like that their visible detail is lost completely.

F15.2

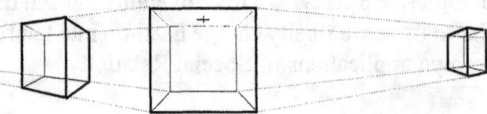

(2) With our line of sight shifted upward a little above the row of Cubes, an additional visual point of convergence is created at the center of the horizon. Also, the straight, transverse lines in the closest part of the row now appear just a little bowed:

F15.3

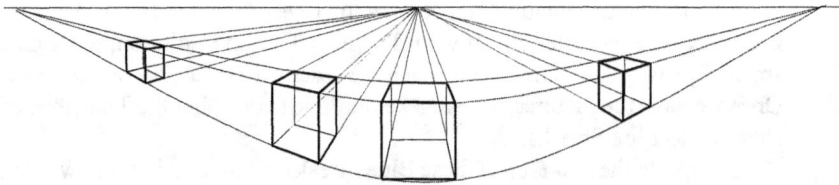

(3) By zooming in next on a row of Cubes moving in a L-to-R direction, what we'll see is mostly the Right Sides of the distant Left Cubes and mostly the Left Sides of the distant Right Cubes. To clearly identify all the Sides, let's label them as follows*:* *F*- Front Side facing our side on the Transverse line-up of Cubes, *L* - the Trailing Left Side, *R* - the Leading Right Side:

F15.4

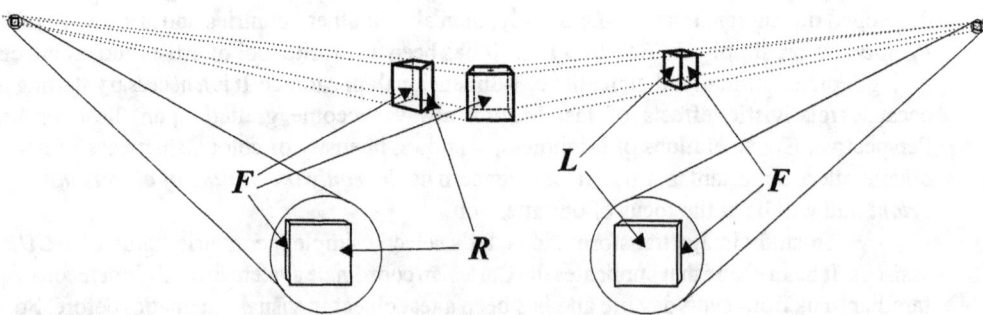

Considering the visual results of a single Cube moving from the extreme left to the extreme right along a straight, transverse line, we can say that **(1)** The Cube in such a Linear Motion undergoes an apparent *(near)* 180 degree **ROTATION** while also displaying **(2)** a continuous **Distance-dependent change in image SIZE.** In addition, whatever part of the Cube is hidden behind any Side of the Cube cannot be seen because **(3)** *Light does not travel around corners.*

Splitting Transverse Motion into its components

Observing a Cube moving from left to right along a straight line at some Distance, what we see is **TRANSVERSE Motion** in 2-D Space. With the moving object still very far, the Motion observed approximates pure Approaching Motion, that is *Linear Motion* in the nearly one-dimensional Lineland, the kind of **Relative Motion** in our direction we have been studying all along. Let's call it the **RADIAL COMPONENT of Transverse Motion** or, simply, **RADIAL Motion.** Next, note that the Cube at the Moment of Passing, exhibits a strictly Sideways Motion, perpendicular to our line of sight : the **ANGULAR COMPONENT of Transverse Motion,** or **ANGULAR Motion.** These two components can be shown geometrically by marking off equal Segments on the Transverse Line corresponding to equal Distances traveled in equal Times *(shown here without the effects of Perspective).* By radial arrows *(or vectors)* toward the Observer's position, we can indicate the magnitude of the Radial Components of Transverse Motion and by arrows *(or vectors)* perpendicular to the radial arrows the magnitude of Angular Components of Transverse Motion:

F15.5

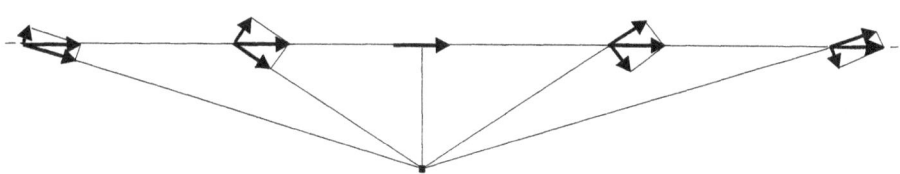

This breakdown of Transverse Motion into two components is very important. It reinforces our previous observations: **(1)** *At extreme Distances on both sides, the Motion observed is almost pure Radial while the Angular Component is minimal, or negligible.* **(2)** *At the Moment of Passing, we have almost pure Angular Motion while the Radial Component is minimal, or negligible.* **(3)** At intermediate Distances, there is a mix of both in a proportion that always keeps changing.

To keep everything as simple as possible, let us pay attention only to Motion at three locations: close to the extreme left and right and one in the middle. Cubes at the intermediate Distances would show a mix of Radial and Angular Motion. Idealization of these *(Radial and Angular)* components helps us make their contribution to the visible shape of the Cube a great deal easier.

Each component of Motion goes with a different Effect

To be fussy, we must admit that the Cube at the exact Passing Position presents a problem. The Motion there is all Angular but, strictly speaking, only the midpoint of the Cube would be at the exact Position of Passing. The Leading Edge of the Cube would already be receding while the Trailing Edge is still approaching. Of course, the problem arises only with very large objects. With smaller ones, the Radial Components of the Leading and Trailing Edges would be very small and, therefore, negligible. Additionally, at very high velocities, Lorentz Contraction would shrink the visible width of the Front Side *(facing us)* thereby further diminishing not only its horizontal width but also the very small Angular Components at extreme Distances.

Let's digress and recall that with a single *(Light or Radar)* Signal Source we can observe the Doppler Effect on the Time intervals between successive, regularly paced Signal bursts and also on the wave length of a steady Signal. This happens only *with **Radial Components of Transverse Motion***. With Reflecting Targets attached to the *(two)* Ends of a Stick and the entire Stick viewed in Radial Motion, *the Radial Component produces a Doppler Effect on visible Sightlengths.* This in approaching mode is *Stretch* and in receding mode is *Compression.*

If you are rusty about the Visual Doppler Effect, please, review the last pages of Doppler Optics where considerable care was taken to develop the *Sightlength concept* analogous to the *Red and Blue Shifts* that tint the coloration of all fast-moving objects. *Visual Doppler Effect,* however, *is notably absent in Angular motion.* As the Transverse Motion at the Moment of Passing is Angular, all we have there is *pure Lorentz Contraction,* no Doppler.

Although *Bradley's and Headlight Effects* produce shifts in image location, *Radial Motion produces only Headlight Effect and Angular Motion only Bradley's Effect.* The magnitude of these depends on the angle between the line of sight and the direction of the Cube's Motion or, as we can say, the "Effective Direction" of the Observer's quasi-Motion *(see Chapter XIV for review).* By direction of Motion in case of Bradley's and Headlight Effects, we did not consider *(in Chapter XIV)* the Motion of the stars but the "Effective *(quasi-)*Motion" of the Observer which under consideration in this Chapter *(XV)* would be the opposite in direction to that of the moving Cube *!!!*

The Headlight Effect of Radial Motion toward the Observer reduces the visible size of the Cube in all dimensions *(perpendicular to the Radial Motion).* *Radial Motion away from the Observer produces apparent enlargement of the Cube. Bradley's Effect is observable only in Angular Component* of Motion and *shifts its image in the direction opposite to that of the Cube's actual Motion (or in the direction of the "Effective Motion" of the Observer).* So the resulting visible *LAG* of the virtual image behind its *(unseen)* "real" location of the Cube is proportional not only to the Velocity of the Cube but also to the Observer's Distance from the Position of Passing.

Unequal Distance Effect

What we have described so far is not as surprising or complicated as the next one, the *UNEQUAL DISTANCE EFFECT.* Despite of its unfamiliar name, it is not entirely new to us. We have already described it in the Doppler Optics chapter but we applied it only to Sticks in 1-D Space. Here we need to extend it to the second dimension.

The Unequal *(Space)* Distance Effect means exactly what is says. As the Cube occupies more than one dimension *(more than with the Stick)*, each part of it is located at a different Distance and in all 3 Dimensions from the Observer. Light Signals from parts of the Cube that are closer to us take less Time to reach us than Signals from the more distant parts. Those originating from all those Distances, therefore, reach our picosecond shutter speed cameras with varying amounts of delay. The sum total of these *Signals (of non-simultaneous origin)* brings us visual information that *transforms (or "distorts") the images* of these moving objects. This visual distortion *(or Transformation)* poses a considerable burden to our intuition. For that reason, we need to study this Effect in detail.

It is inherent in the nature of our vision *(and of our picosecond shutter-speed camera)* that *we do NOT see all parts of the moving Cube at their simultaneous positions. What we see instead is (1) what is REVEALED to our EYES* *(or cameras)* *by the SIMULTANEOUS*

ARRIVAL of information-bearing (Light) Signals!!! And we see only (2) WHAT happens to be on our SIGHTLINE !!!. All this is produced by the same mechanism that produced Doppler Visual Effect in the Lineland (1-D) setting (see Chapters XI and XIV about observing in 2-D) We have already seen that *in RADIAL Component, it resolves into Doppler STRETCH and Doppler COMPRESSION giving us Visual Lengths or SIGHTLENGTHS (by now an "old hat" with us) but in ANGULAR Component, this visual effect is really ODD.* It was mentioned only briefly at the end of the Doppler Optics chapter. So let's explore what it does to the image of the Cube at the exact Moment of Passing when our eyes are aimed straight at its middle and slightly above its Upper Rim *(to see at least some of its Top Side).*

The Rhomboid Shear

In previous chapters, we dealt exclusively with one-dimensional Distances and Lengths, with the Observer also occupying the same, single dimension and everything located in our Line of Sight and in pure Radial Motion. What is different here is this: *the Cube we are looking at is an object occupying three dimensions of Space so that our observations must be conceptualized in at least two dimensions of Space.* At the Moment of Passing, the direction of Motion is exactly Angular, the only position where pure Lorentz Contraction *(and no Doppler)* is observed. The Front and Back Sides of the Cube *(as well as the Top and Bottom Sides)* are parallel to the direction of Motion and parallel to each other. As these Sides are pointed in the direction of Motion and move at the same Velocity, all sustain the *same amount of Lorentz Contraction.* The Leading and Trailing Sides, on the other hand, are perpendicular to the direction of Motion and do not undergo Lorentz Contraction. These Sides are subject only to Linear Perspective and to the just-mentioned Unequal Distance Effect important enough to deserve an appropriately discriptive name of its own. And this is *RHOMBOID SHEAR.*

To investigate this newcomer to our group of Transformations, let's start out with a Lorentz-contracted Cube placed within the "shadow" of its non-contracted outline:

F15.6

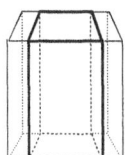

What Rhomboid Shear does is re-arrange the visible positions of ALL six Sides of the Cube, most obviously the Sides that do not undergo Lorentz Contraction. Let's find out what we can see with our picosecond shutter-speed camera.

During the *(picosecond)* shutter opening, Signals arriving from the more *DISTANT* parts of the moving Cube originated at an *EARLIER MOMENT* compared with the *simultaneously arriving* Signals from the nearer parts. Applying the *Sightline* concept to the Cube in Motion at the Passing Moment, we can visualize the Upper Rim of the Back Side *LAGGING BEHIND* the Front Side facing us but with both Sides still remaining parallel to each other, pointing in the direction of Motion and undergoing the same amount of Lorentz Contraction. *The amount of this lag depends on (1) the Distance between the closer Front and the more distant Back Sides (along the Sightline that does not undergo Lorentz Contraction) and on (2) the Velocity of Motion.* The Light that originates from the more distant Upper Rim takes a certain amount of Time before it reaches the Upper Rim of the closer Front Side where it combines with the Light originating there. Signals from the Front and Back Sides then proceed toward the Observer. The resulting *VISUAL LAG* of the more distant part of the Cube *changes the visible Top Side from a perceived rectangular shape to a Rhomboid (further modified by the Perspective Effect).* The resulting Transformation is what is ***the RHOMBOID SHEAR, essentially the visual distortion of the Cube's image.***

In the next illustration *(Fig. 15.7)*, we'll develop this Transformation of the Cube's image by first drawing: **(A)** the stationary Cube with only the Perspective Effect. Below its front view, there is its "Floor Plan" or "bottom print" as if displayed on a vertical surface without the Perspective Effect. **(B)** Next we'll see the Lorentz-contracted *(moving)* Cube *(within its non-contracted shadow)* and its Floor Plan; then **(C)** the Lorentz-contracted Cube shown together with its superimposed "earlier" position *(shadow)* which is also shown in the Floor Plan. **(D)** The spacial shift between "earlier" and "later" allows the development of the Rhomboid Shear indicated by dotted Diagonals connecting the distant corners of the "Earlier" Cube with the nearer corners of the "Later" Cube. The resulting Rhomboid Shear is further modified by the addition of the Perspective Effect. Finally (***E***), the Cube is drawn as it would actually be visible with Perspective, Lorentz Contraction and Rhomboid Shear Effects combined. Both the Front View and the Floor Plan show these three Effects:

F15.7 Five Cubes with Floor Plans undergoing three transformations

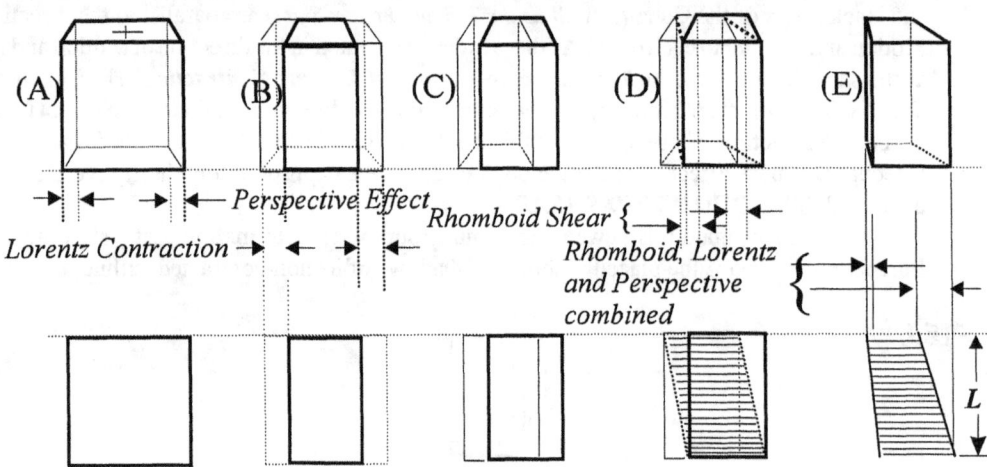

Note, again, that the Near and the Far Rims *(of the Front and Back Sides)* remain parallel to each other and *(because their velocities are equal)* they undergo equal amounts of Lorentz Contraction. Also note that the Front Side facing us does not show Perspective Effect. But the more distant Back Side is seen as smaller compared to the closer Front Side. The Leading Right Side and Trailing Left Side are perpendicular to the Direction of Motion and show a slight "tilt" to our Line of Sight because of the Rhomboid Shear. This Effect makes the ***earlier*** position of the Back Side and the later position of the Front Side visible at the same Moment. The Leading Right Side facing toward the Direction of Motion *(although concealed from our view)* is attached to the Leading Vertical Edges of the Front and Back Sides thereby sustaining a visible tilt in their positions relative to our line of sight. In addition to the Rhomboid Shear, the observed position of the Back Side is further influenced by the Perspective Effect that brings the Vertical Edges of the Back Side closer together. As you can see from the Fig. 15.7 and 8, the combined result is, therefore, asymmetrical. To the Leading Edge of the Back Side, the Rhomboid and Perspective Effects are ***ADDED***, at the Trailing Edge, the Perspective Effect is ***SUBTRACTED*** from the Rhomboid Shear.

To illustrate the Rhomboid Shear in Angular Motion in better detail, the Lorentz-contracted **(D)** and **(E)** are redrawn in Fig. 15.8 *(next page)* this time without the Floor Plan

F15.8

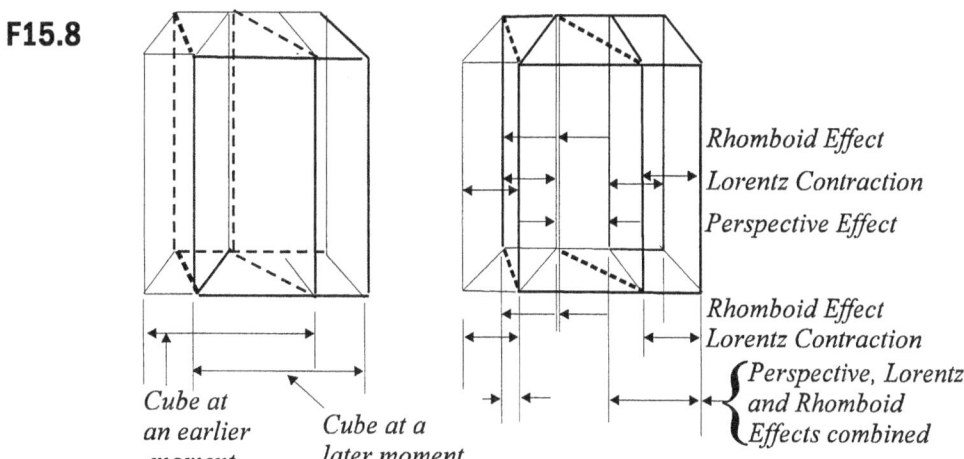

Rhomboid Effect
Lorentz Contraction
Perspective Effect

Rhomboid Effect
Lorentz Contraction
⎰ Perspective, Lorentz
⎨ and Rhomboid
⎱ Effects combined

Cube at an earlier moment

Cube at a later moment

Rhomboid Shear makes Light go around corners? !!!

There is one peculiarity of the fast-moving Cube that you have to see on the *ST* Map to really believe. Rhomboid Shear is capable of exposing the hidden Side of the Cube to the Observer earlier than would otherwise be possible. This was shown in Fig. 15.7E & 15.8 and is further detailed in the Floor Plan, Fig. 15.9.

The above Floor Plan *(Bottom Print)* shows how the identical Top Side would look *without the Perspective Effect*. The more distant *VERTICAL Edge of the Trailing LEFT SIDE,* usually hidden behind the *Trailing Vertical Edge of the FRONT SIDE (facing us),* can become visible to the stationary Observer *just a little before the Trailing Side actually reaches the exact Passing Position*. This unbelievable statement can be justified by mathematical reasoning.

Let's now go over what the Rhomboid Shear does to the Floor Plan as indicated by symbols *(L, L', T' and v)* used to label the key triangle in the Floor Plan, Fig. 15.9:

F15.9

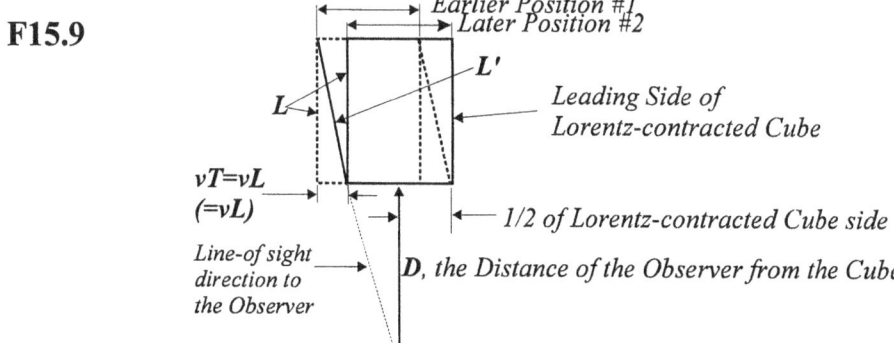

Earlier Position #1
Later Position #2

L'

L

Leading Side of Lorentz-contracted Cube

$vT = vL$
$(=vL)$

1/2 of Lorentz-contracted Cube side

Line-of sight direction to the Observer

D, the Distance of the Observer from the Cube

Location of the Observer

The Rhomboid Shear moves the entire closer-positioned Front Side with its Trailing Vertical Edge forward and *OUT OF THE WAY for the Light* coming from the more distant

Vertical Edge of the *(otherwise hidden)* Trailing Left Side **BEFORE** the Trailing Vertical Edge of the Front Side *(joined to the closer Vertical Edge of the Trailing Left Side)* would reach the exact P-osition of Passing. The result, therefore, **clears the visual obstruction** from the path of Light Signals coming from the more **distant** Vertical Edge of the Trailing Left Side thereby allowing the non-simultaneously originating Signals from the two unequally-distant Trailing Edges to reach the Observer simultaneously *(that is at the same Moment)*. The Light on its way, therefore, describes a **DIAGONAL Path, L'** *(see Floor Plan, Fig. 15.9)* toward the stationary Observer. This diagonal path appears longer than the **width of the Trailing** *(not Lorentz-contracted)* **Left Side** of the Cube.

Let's identify the components shown on the **Floor Plan,** Fig. 15.9:

(1) Earlier Position - refers to the earlier position of the Cube when the Light from the Trailing Vertical Edge of the Back Side started to move toward the Observer.

(2) Later Position - refers to the position of the Cube when the Light from the more distant Trailing Vertical Edge of the Left Side reached the closer Trailing Vertical Edge of the Left Side attached to the Front Side of the Cube seen located at a Later Moment when the earlier Light Signal *(from the more distant Vertical Edge)* has already arrived at the nearer Vertical Edge and from where both Light Signals then traveled together to the Observer, arriving there simult-aneously.

(3) L - the **NOT-** Lorentz-contracted Trailing Left Side of the Cube placed perpendicular to the direction of motion.

(4) L' - The Rhomboid-Sheared Diagonal Length of the **slanted Light Path** from the more Distant Vertical Edge to the Closer Vertical Edge of the Trailing Side as it would appear to the Observer. To another Observer riding along with the Cube, however, this Light path would be traversing completely parallel to the Trailing Side and perpendicular to the direction of Motion.

(5) vT' - The Length of travel by the Cube during the Time, T' needed for Light to travel the longer diagonal *(Light)* path *(as seen by the stationary Observer)* and across the entire width of the Rhomboid-Sheared Trailing Left Side. The Front and Back Sides *(attached to the Near and Far Edges of the here visible Bottom Print)* of the Cube *(positioned in the direction of Motion)* having undergone Lorentz Contraction and additionally having participated in the production of Rhomboid Shear. The Bottom Print with attachments to the Front and Back Sides shows the extent of the Lag but without the Perspective Effect.

Calculating Rhomboid Shear

The Time taken by Light to travel the longer **diagonal** path, L' on Figure 15.9, is T' because Light would take longer to travel parallel to the Trailing *(NON-contracted!!!)* Left Side of the moving Cube. This longer-lasting Travel Time *(according to <A>)* needs expressed next in terms of L' :

$$vT' = (L'/c)(v/c) = vL'/c^2, \text{ As } c^2 = 1.0, \text{ so } vT' = vL' \text{ and } T' = L'$$

The triangle in Fig. 15.9, formed by L', vT' and L, can be used to calculate vL' :

$$(L')^2 = (vL')^2 + L^2. \text{ From this, we get: } L^2 = (L')^2 - v^2(L')^2 = (L')^2(1-v^2)$$

Restated: $L^2 = (L')^2(1-v^2)$ giving us $L = L'\sqrt{1-v^2}$, and so $L' = \dfrac{L}{\sqrt{1-v^2}}$.

So the Rhomboid Lag, vL' is: $vL' = \dfrac{vL}{\sqrt{1-v^2}}$

But *realizing that vL', the Distance traveled during T' is pointing in the direction of Motion (L remaining transverse to the direction of Motion)* necessitates that *vL'* be *subjected to Lorentz Contraction.* This transforms the last equation into:

$$vL' \ = \ vL \ \frac{1}{\sqrt{1-v^2}} \ \sqrt{1-v^2}$$

which gives us the *Lorentz-contracted Rhomboid Lag, or Shear.* Canceling out the $\sqrt{1-v^2}$ in the Numerator and in the Denominator gives us *vL AFTER Lorentz Contraction:*

$$vL' \ \rightarrow \ vL \ \text{ which at Velocity, } c \text{ is equal to } L$$

What is seen at the Velocity of Light

The *L* now is the Length of the Lorentz-contracted Travel Path of the Cube at Light Velocity. It is equal to the non-contracted Trailing Left Side that happens to be visibly equal to the Front Side of the Cube without Motion. Note, however, that the Front Side width at Light Velocity has contracted to zero*!!!* Is it clear*?*

This unexpected result allows us *re-calculate the diagonal Path of Light (the Rhomboid Shear)* in the triangle and obtain an actual numerical value *(in the Floor Plan, Fig, 15.9)* for a revealing new equation:

$$(L')^2 \ = \ L^2 + (vL)^2 \ = \ L^2(1+v^2) \ \text{ and } \ L' = L\sqrt{1+v^2}$$

But at the Velocity of Light, $v = c = 1. = 1.0$:

$$L' = L\sqrt{1+v^2} \ = \ L\sqrt{1+1} \ = \ L\sqrt{2} \ = 1.4L \ !!!$$

Now that we have a numerical result for the Diagonal Path, let's visualize it.

Maximal Shear at Velocity of Light illustrated

If the numerical result is puzzling, let's clarify it by posing a blunt question: *What would the Cube look like at the Passing Position and at (exactly) the Velocity of Light?* Never mind the impossibility of achieving such a Velocity in the real world, let's just answer the question. But we have already done the math*!* Hmm*! The Front and the Back Sides would be Lorentz-contracted to a vanishing, negligible thinness. The diagonal path of Light, L' of the Rhomboid-sheared Cube would transform the originally almost-line-of-sight Trailing (Left) Side into a silhouette as wide as L, the non-Lorentz-contracted Trailing Left Side which is now seemingly rotated toward us !!!* Note that 1.4*L*, in front-view silhouette, appears to be only 1.0*L, the same as calculated above for the Rhomboid Shear* after Lorentz Contraction at the velocity of Light. This result is shown again but now more graphically in the following illustration (Fig. 15.10):

F15.10

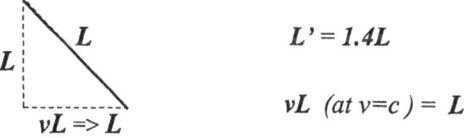

$$L' = 1.4L$$
$$vL \ (at \ v=c \ ) = L$$

Once *v=1.0* equals *c*, then the Lorentz-contracted Rhomboid Shear Lag, *vL'* equals *L*. While Light moves toward us along the Trailing Side, the Cube also moves forward by a Distance that equals the Trailing Left Side of the Cube*!!!* The ***Floor Plan*** of the Rhomboid-sheared Cube shown below most accurately explains the final visual image of the Cube at the exact Moment *(and Position)* of Passing. The thickness of the horizontally contracted Cube is grossly exaggerated in the illustration to show that the Front and the Back Sides still remain parallel to the direction of Motion. At the velocity of c, the thickness of the Lorentz-contracted Cube actually becomes zero and the Cube's Front and Back Sides vanish completely from sight *!!!*

Here now is the Floor Plan:

F15.11

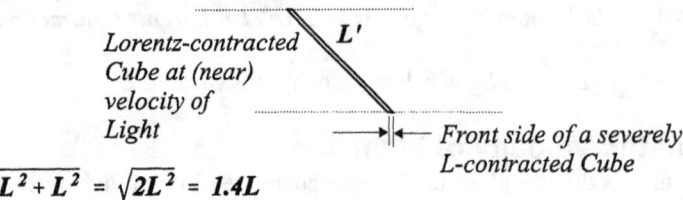

Lorentz-contracted
Cube at (near)
velocity of
Light

Front side of a severely
L-contracted Cube

$$L' = \sqrt{L^2 + L^2} = \sqrt{2L^2} = 1.4L$$

How would all this look *(visually)* to the Observer whose line of sight lets him see $L' = (\sqrt{2})L$ and also the Top Side? Well, take a look:

F15.12

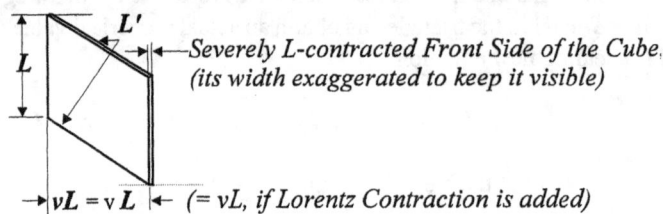

Severely L-contracted Front Side of the Cube,
(its width exaggerated to keep it visible)

$vL = vL$ (= vL, if Lorentz Contraction is added)

Summing up Rhomboid Shear

Without Motion, the Trailing Side of the Cube has to be just slightly past the Passing Position to become barely visible. The Unequal Distance Effect gives us Rhomboid Shear. *So with Motion (see Fig. 15.9 again)*, the forward movement of the ***closer Vertical Edge*** of the Trailing Left Side would shift away the visual obstruction to the otherwise still hidden, ***more distant Vertical Edge*** of the Trailing Left Side thereby exposing the hidden Far Edge, and the entire Trailing Side at a slant, to our view slightly ***before*** it *(the Trailing Side)* actually reaches the exact Position of Passing. In this way it becomes possible for the ***Light from the Distant Edge to join the Light from the Closer Edge to reach the Observer at the same Moment BEFORE the Trailing Side actually reaches the exact Position of Passing***. Well, ***here the Light finally seems to travel AROUND the "FAST CORNER" after all !!!***

Terrell's account

Among distinguished physicists, the Unequal Distance Effect was first described in 1959 by N.J. Terrell but with a twist slightly different from our conclusions. He came up with a ***ROTATION*** of the Cube, the Relativistic Effect of a very-very fast Motion. A close look at his explanation allows us to bring it into harmony with Rhomboid Shear.

To set the stage, let's remind ourselves again that **(1)** Lorentz Contraction causes narrowing of the visible image of the facing Front Side to $L\sqrt{1-v^2}$ and that **(2)** the amount of the "sheared" positioning of the Front and Back Sides *(or the Rhomboid Shear)*, is $L(v/c)$ or vL after Lorentz Contraction. Next, let us set up in a table the sizes of Lorentz-contracted Front and Back Sides and the amounts of Rhomboid-Shear lags *(of the far Edge)* for a sample of Velocities from zero to *c*. Because all measurements are given as multiples of *L,* only the calculated Multipliers need be shown. At zero Velocity, there is no Lorentz Contraction and the amount of Front-to-Back Side displacement *(Rhomboid Shear)* is zero. At *(almost)* the Velocity of Light, the Facing Side is Lorentz-contracted to *(almost)* zero and the apparent, visually presented width *(silhouette)* of the Trailing Side *(at the exact passing position!)* is horizontally *(almost)* equal to the entire Trailing Side of the non-moving Cube *turned,* that is *ROTATED directly to face us (stationary Observers)*. At Velocity, *v* = 0.71, however, the Lorentz-contracted Front Side and the visually presented, Rhomboid-sheared Trailing Side *(projected on transverse line)* seem equal in horizontal view:

Velocities:	$v = 0$	$v = 0.1$	$v = 0.5$	$v = 0.71$	$v = 0.9$	$v = 1.0$
Lorentz-Contracted $L = L\sqrt{1-v^2}$	1.0	0.995	0.866	0.71 0	0.045	0.000
Rhomboid Sh'd $L = vL$:	0.000	0.1	0.5	0.71	0.9	1.000

The corresponding Front-view measurements are next illustrated by actual true-to-scale drawings but with *our eyes positioned NOT above the cube but at its mid height* so that we *cannot see the Top Side.* The progressively contracting Facing Side and the progressively exposed Trailing Side are shown as seen by the Observer *(without Classical Perspective)* at the Moment of Passing at the above sample Velocities using a picosecond shutter-speed camera. Incidentally at *v*=0.71, the Lorentz-contracted Front Side and the horizontally presented visual Lag caused by the Rhomboid Shift reduced by Lorentz Contraction are equal*!!!:*

F15.13

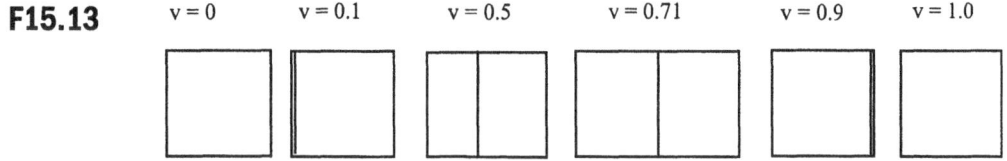

Terrell's Rotation argument is quite persuasive. So let us look again at the horizontal visual images *(Fig. 15.13)* of the Cube each taken at the exact Moment of Passing and at the same selection of Velocities but this time the non-moving Cubes are placed at various degrees of "Rotation" directly above those horizontal images, Fig. 15.14. The angle of Rotation is given as *α.* The amount of the Trailing Side exposed to our view by this "Rotation" corresponds to the *LAG (vL)* produced by the Rhomboid Shear. The amount of Rotation is given both visually and mathematically, the latter by *L sin α* = *vL (note: v = sin α).* The angle on the *ST* Map standing for velocity, *v* is exactly the angle of Terrell's "Rotation:"

F15.14

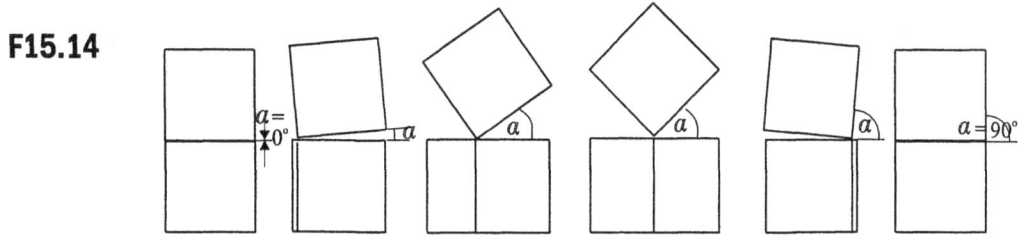

Terrell's Rotation vs, or the same as, Rhomboid Shear

The interesting thing is this: *Terrell's "Rotation"* and *Rhomboid Shear* are *visually identical but only as long as we keep our line of sight fixed at midlevel of the cube so that we do NOT see the TOP of the CUBE!* The horizontal, eye-level views of the Rhomboid-Sheared Cubes match exactly those of the *NON-MOVING, ROTATED CUBES*. So $\sin\alpha$ equals exactly the Velocity of the Cube ($\sin\alpha = v$) and the Lorentz-contracted Facing Side, $L\sqrt{1-v^2}$ equals exactly L times $\cos\alpha$, or $L\cos\alpha$. If, on the other hand, we expose the Top Side of the cube to our view, we would be immediately presented with a more complete visual arrangement of all Sides, especially the Top Side, and the image Transformation is revealed to be Rhomboid Shear, Fig. 15.15:

F15.15

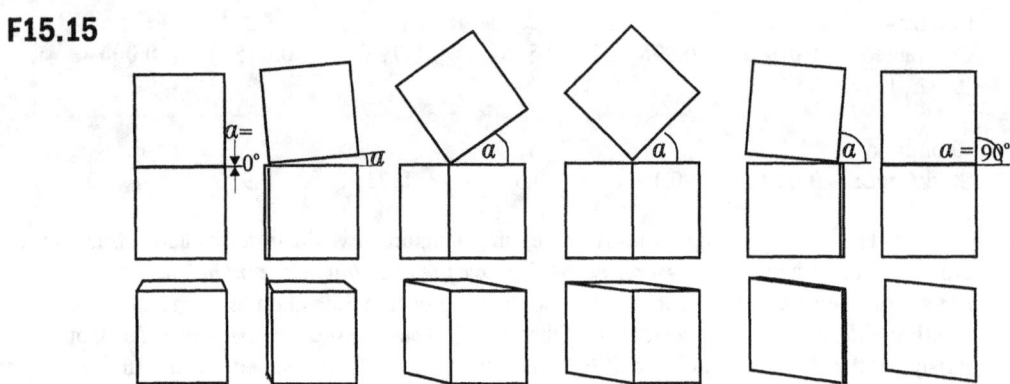

Prof. Terrell is correct by not insisting that the Cube itself becomes rotated. He points out that *it only LOOKS that way.* This we have seen. The horizontal image at $v=0.71$ happens *(very-very instantaneously)* to appear the way expected from a half-way ($45°$) rotated stationary Cube. At Velocities above $v = 0.71$, there is further *APPARENT ROTATION* that finally becomes full 90 degrees at the velocity of Light. This would also be expected according to Rhomboid Shear if we do not see any part of the Top Side. He does not insist that the Leading and Trailing Sides actually *(physically!)* turn away from the direction of Motion and he allows us to understand that it may only be *CONCLUDED* to be so according to *(deceptive!)* visual appearances. But again, seeing the Top of the Cube would dispel any impression of "Rotation" by revealing a more complete image of the Cube at the Moment of Passing. *Rhomboid Shear is, therefore, a better way to visualize and understand this particular image Transformation due to very-very fast Motion.*

One more Unequal Distance Effect

Keep in mind that all vertical Edges whether in Motion or not are never undergoing Lorentz Contraction because these are not pointed in the direction of Motion. If, however, the Cube is large enough and is passing us close enough, *the midpoint of the closest vertical Edge is momentarily closer to us than its upper and lower parts.* The Unequal Distance Effect can, therefore, be applied to all these visible vertical lines *BUT* with a result that is different from all the other Unequal Distance Effects *!!!* *The upper and lower parts of all vertical lines LAG behind the mid parts that are closest to us.* This *BENDING EFFECT* was pointed out by D. E. Mook and T. Vargish.

F15.16

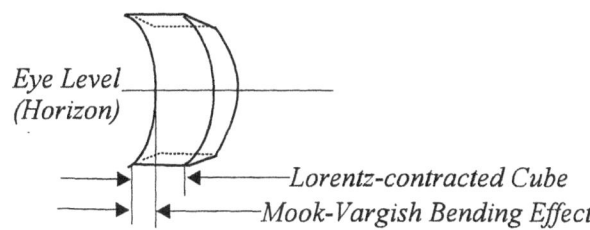

*Eye Level
(Horizon)*

————Lorentz-contracted Cube
————Mook-Vargish Bending Effect

Because it is very unlikely that we'll ever see anything that big passing us at that dangerously close a range, we can safely assume that this ***Bending Effect*** *(bending of all vertical lines)* at reasonably close range will usually be minimal *(if the Observer along with the Blue Streak are not literally incinerated by the searing heat from friction with the atmosphere)* and, therefore, most likely negligible. The Mook-Vargish Bending Effect had to be mentioned in order to be reasonably complete in discussing all Unequal Distance Effects. And some day we may just see the Blue Streak in a picture taken with a fancy pico-second shutter-speed camera.

A naive alternative

Let us now try out *a common-sense argument against the Rhomboid Shear/Rotation* of solid Cubes. For this, let's first look at two identical Cubes placed side-by-side in front of us with their adjacent vertical surfaces touching each other and located exactly in our line of sight. Seeing some of the Top Side would not make any difference here:

F15.17

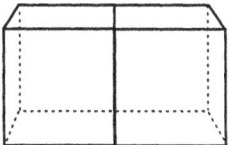

Let's remove the Cube on the left and consider the Right Cube without Motion. What we see of the Cube is the entire Facing Side. Of course, we cannot see its Side on the right because it is hidden behind the Front Side. The Left Side *(with or wthout Motion to the right)* is not visible either because it happens to be in the exact line of our sight, perpendicular to the frontal plane and not yet exposed to our view. If the Cube were just a little more to the right, we would just about begin to see the Trailing *(Left)* Side. But because it is not more to the right, we don't see the Left Side.

The key question is this: ***Can the hidden Trailing Side, still out of sight, ACTUALLY become slanted in very-very fast Motion and thereby become visible in that position to a pico-second camera?*** Note the word *"become"* rather than *"appear."* Let's consider the answer in left-to-right Motion.

The Light coming from the surface of the ***HIDDEN*** Trailing Side can travel only rearward and within 180 degrees sideways direction. The most forward direction a Light Signal can take is still restricted to one that is ***PARALLEL to the instantaneous position of the Trailing Side*** and perpendicular to the direction of Motion at any position of the Cube at any particular Moment. This Side becomes visible only ***AFTER*** the Trailing Side will have arrived ever-so-slightly beyond the exact Position of Passing, ***NOT BEFORE***. The Rhomboid Shear ***CANNOT*** help out because it is a ***VIRTUAL Effect***, one of ***IMAGE PRODUCTION*** and, therefore, ***does NOT result in a DEFORMATION of the actual physical shape of the cube.*** So, exposing the

Trailing Side toward the Observer cannot happen **BEFORE** it is **ACTUALLY** *(ever-so-slightly)* beyond the exact Position of Passing. Only those parts of the Cube that point ever-so-slightly in the direction of the Observer, whether with or without Motion, can appear to be positioned at such *(ever-so-slight)* a slant:

F15.18

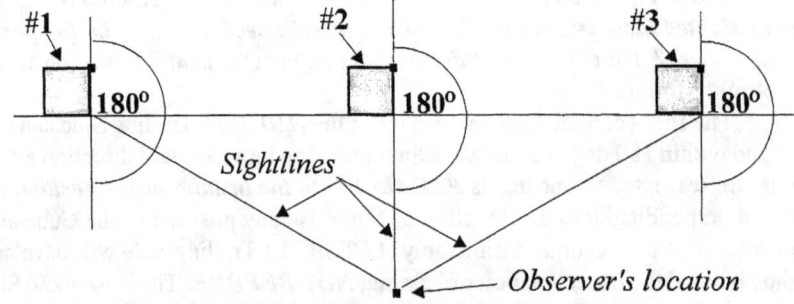

Problem location from where Light has trouble reaching Observer

The Cube shown at three locations

#1   #2   #3

180°   180°   180°

Sightlines

Observer's location

Velocity itself does not deform the Cube which *in its own Home Frame is not deformed.* So nothing can expose the hidden Left Side toward the Observer **BEFORE** its exact Position of Passing is reached. Light could come from the totally hidden Trailing Side before the Passing Moment and reach the Observer *only if the Velocity of the Cube were FASTER than that of Light!!!* The Near Edge of the Trailing Side presents an essentially stationary, non-moving obstruction to Light emitted *(or reflected)* from its Far Edge. Anything made of matter is as good as always standing still to Light. To repeat: *In its own Home Frame, the Cube is stationary* or non-moving. In real world setting, the Trailing Side must reach a position very slightly past our line of sight before Signals from the Far Edge *(of its Trailing Side)* can travel to us. Thus, there is **NO deformation of the Cube itself, only of its image.** And *it is the Cube, not its image that sends us Light Signals by emission or by reflection.*

What about the Leading Side? Before Passing, its disappearance from our view is in the same quandary. By the same argument, it can send Light in any direction but only forward within 180 degrees perpendicular to the direction of motion, never backward, not ever-so-slightly around the corner. Rhomboid Shear, being only *a change in the (virtual) IMAGE, not of the actual SHAPE of the Cube,* cannot have any modifying effect upon what can be seen. The nearer vertical Leading Edge chokes off Light from the far Vertical Edge exactly at the Moment of Passing when both of them no longer remain in our direct line of sight.

F15.1

#1   #2   #3

180°   180°   180°

Sightlines

Observer's location

It is impossible that even the Bradley/Headlight Effect can come to rescue by helpfully "shifting" the image of the still **HIDDEN** Side, somehow "compensating" for the failure of the

Light to travel around the corner. Any way you want to try, it doesn't work. Remember:*Visual Transformation does not mold or deform the Cube!!! All it can do is to transform its virtual image!!! And images do not emit or reflect Light, actual objects do!!!* This common-sense argument does not consider Blue Streak Transformations to be anything other than *changes in images*. The image of the Trailing Left Side cannot be seen before its Source actually points in our direction. That can happen only *AFTER* the Moment of Passing.

Now you have been exposed to different descriptions of the Unequal Distance Effect at the exact Moment of Passing. Just mull them over, try to find fault with any of those and pick one best suited to your taste. But beware of intellectual indigestion if you swallow any argument without mentally chewing it well.

Illumination is magic

If you refer to Doppler Optics, page VI-19 summing up Single-Signal Radar Probe, you'll be reminded that you can calculate Lorentz-contracted Lengths, nothing more. In the same chapter, pages Vl-20-24, the rationale for the Sightline Effectis explained. But our results also depend on the type of illumination used, *Constant Illumination vs. Single Flash*. It is best to point out that a *single picosecond-duration Flash of Light* would give us a result that is the *exact reverse* of what can be seen with *Constant illumination*. With a Single Flash from the direction of the Observer, a Doppler Compression is obtained in approaching mode and Doppler Stretch in receding mode. Moreover, *the appearance of Rhomboid Shear is also reversed.* Remember: Lorentz Contraction is not observable to the eye or camera in pure Radial Motion, only in Angular Motion.

What was explained is this: A *single, 1-picosecond FLASH with OPEN-LENS camera* is a scenario different from *Constant Illumination and 1-picosecond shutter-speed camera*. The Distance information in the first case is returned and recorded by the camera not at a single Moment but in two separate Moments occurring sequentially. As the nearer reflection is happening first, while the more distant part of the Cube is still traveling toward the oncoming Flash, the recorded result is *Doppler Compression in approaching Motion*. The opposite result is found with the picosecond shutter-speed camera. And you may have also guessed: a *Single, 1-picosecond Flash and open-lens camera in receding mode gives us Doppler Stretch !!!*

Other conclusions

After having worked through these virtual exercises, we can also conclude that *what we see is NOT WHAT IS (or was) there, true to the actual shape of the Cube (in its own Home Frame). The IMAGE does NOT represent TRUE REALITY which is ALWAYS an INVARIANT under any and all Transformations. And again, we can SEE the Invariant only under certain conditions (see Postscript)!!!* Even stationary objects do not present to our view all their parts existing at exactly the same, simultaneous, Moment. The question of *Image versus Invariant* has interesting and far-reaching implications. Some comments about it will follow here next but can also be found in the Postscript. Don't miss them.

What we see and record with camera is not what ultimately exists. In other words, *what exists is not always what we see.* Shutter speed is important but the *method of illumination* also counts. (1) A *constant, omnidirectional Light source* illuminating an approaching object and a fast shutter gives us *nearer parts ahead* and the more distant parts lagging. (2) A *fast flash (obtainable with Light or Radar Probe),* an approaching object and open shutter gives us nearer parts lagging. With a passing Cube, an impression of *ROTATION* is created if we do not see the Top Side. Unequal Distance Effect always produces an obvious Rhomboid Shear which can be correctly perceived if some of the Top Side is made visible. Another form of **apparent rotation** is produced by Linear Perspective. But after all those details, we can ask: **WHAT is it that we see** and record *(in a photograph)?* Is it actually how it is, **WHAT ACTUALLY IS THERE** or is it merely how it **LOOKS**, that is a matter of its **IMAGE?** As you can see, the image is shaped by a **number of Transformations due to very-very fast motion**. And **LIGHT** in some specified circumstances **may literally succeed traveling around corners !!! Or does it ???**

What a mess getting bogged down in arguments about the Unequal Distance Effect in Angular Motion and we haven't even discussed the combination of all these in positions intermediate between the most distant and the nearest. The effects of the Radial and Angular Components of Transverse Motion combine there in a proportion that always keeps changing with the position of all moving objects relative to the Observer.

Summation

We better summarize everything again at those three different locations and at a Velocity of almost that of Light so that the various effects we have studied are maximal. Visual transformations at intermediate locations and slower speeds should suggest themselves reasonably loud and clear. Note that illumination in a standard situation is constant and omnidirectional, shutter speed 1.0 picosecond. The extremely fast shutter speed guarantees good resolution. With open-lens camera, just like with our open eyes, all we would see is the proverbial Blue Streak chosen for the title of this book.

IN APPROACHING MODE

Lorentz - Lorentz Contraction cannot be seen in Radial Motion because we cannot see what is simultaneous to our *NOW*-Moment. Lorentz Transformation cannot, therefore, reduce Doppler Stretch either. Lorentz Contraction can be seen most clearly at the Passing Moment when both the Leading and Trailing Edges happen to be equidistant from us

Doppler Stretch - The Unequal Distance Effect in the direction of Motion is observed as Doppler Stretch.. At intermediate positions, we'll see some of both: Lorentz and Doppler. The Sides of the Cube positioned in the direction of Motion appear equal to kL, with k being the Doppler Constant and L being the Proper Length. The k at the speed of Light equals infinity but Blue Streaks are never expected to go that fast. The resulting image is Doppler-stretched. The Tail End would be hard to make out seeming to be so far away, maybe Lighthours or more behind its Nose End. Who knows? All depends on the actual velocity of Motion.

Headlight - The approaching *(part of the)* Cube becomes pinched almost to a point which is then Doppler-stretched to a thread*!*. This makes its Tail End *(if still very far)* difficult to see even with zoom.

Bradley's - Nothing that we can clearly notice. What we get is Headlight *(pinch)* Effect instead, see above description .

Doppler Shift of color - Bright purplish blue.

IN RECEDING MODE

Lorentz - Contraction not visible as explained above and not influencing Doppler Compression.

Doppler Compression - to $(1/k)L$ causing flattening of the Cube's image in the direction of motion.

Headlight - Widening of the image to not quite 180 degrees of visual angle, marked fading of brightness and of nearly of all contrast, thereby obscuring detail.

Bradley's - None noticeable. Headlight *(flare)* Effect instead.

Doppler Shift of color - Hard to see anything but everything made out visually looks pale red.

AT PASSING MOMENT

Lorentz - Shrinking of the Front Side *(facing the Observer)* to a very narrow wafer. That's because, as already explained, the Leading and Trailing Vertical Edges are momentarily equidistant from the Observer and there is no Unequal Distance Effect of those Edges individually in the direction of Motion.

Doppler - No Doppler Stretch or Compression. A close relative of Doppler Effect, the Rhomboid Shear is at work here, see below. It, combined with Lorentz Contraction, exposes the trailing Left Side at a slant that horizontally appears to be of the same full width as the Trailing *(Left)* Side turned toward us as if having undergone a *(near)* 90 degree Rotation and as if it is seen momentarily without any Motion at all. *Doppler Shift of color* - None. Maybe just a little reddish tint at its Advancing Edge and a hint of *(perhaps greenish)* blue at its Trailing Edge.

F15.20

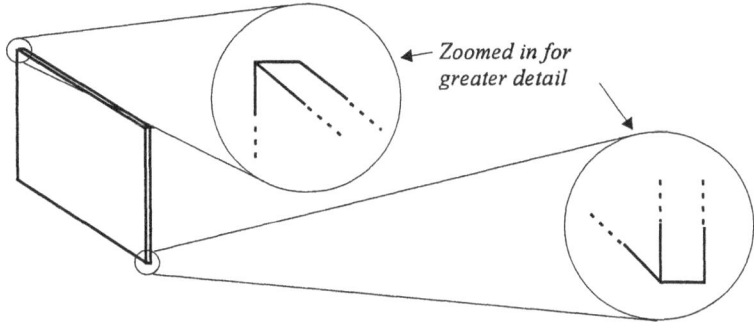

← *Zoomed in for greater detail*

Bending Effect - negligible unless the Cube is huge and passing very close and very-very fast. The upper and lower ends of all vertical Edges are bowed backward in proportion of proximity of Passage and Velocity of Motion.

Headlight - none.

Bradley's - What is actually seen of the Cube at the Passing Moment is nothing but Bradley's Effect*!!!* Its visible image is like a shadow trailing the true actual location *(which we actually cannot see)*. When the image is seen at the Passing location, the Cube has already passed us by a Distance that *(almost)* equals our Distance from the passing location.

Rhomboid Shear - acting in the direction perpendicular to the direction of Motion.

Last word has it

Now, what kind of image would we get of the three-dimensional Cube masquerading as a Blue Streak? A picture obtained by a wide-angle, open-lens camera would show a big, long blur. **(1) Approaching part:** A Doppler-stretched and very strongly Headlight-pinched bright blue tapered tail-like object; **(2) Passing part:** A horizontally Lorentz-contracted, wafer-thin Front Side joined to the Rhomboid-sheared Trailing Back Side with its Vertical Size unchanged. However, the Vertical Edges appear bent backwards in case the Cube is very tall and its visible horizontal extension just about equal to that of a stationary Trailing Side, color changing from blueish to reddish at its middle, and **(3) Receding part:** Doppler-compressed, Headlight-flared and markedly dimmed, indistinct, reddish blunt blob of a Nose Endpointing in the direction of the wide open yonder from whence it came and to where it is now heading, all in a brief Moment before you are even ready to say "wow."

The description given above incorporates all the Relativistic Transformations without over-emphasizing the Bending Effect. A rough but dramatic, open-eyed impression is provided on the cover of this book. Seeing these at this close range, however, is an impossibility because staging the experiment on Earth is likely to prove fatal to any Observer standing without substantial protective gear. The shock wave accompanying such a speeding object would also destroy any kind of equipment intended to record this unique experience. Moreover, the Blue Streak is likely to disintegrate and burn up soon after hitting the atmosphere. To think of it, there is something missing in the summary description of all Blue Streaks at the end of this chapter. It is the all-pervasive Visual Perspective Effect that is missing*!!!* Adding it to the existing summary description is left as an assignment to those readers who have managed to work their way through the seemingly endless, exacerbating details to the very end to inspire and spur them on to complete the description of the unavoidably elusive Blue Streaks.

This brings us to the end of the last chapter and also to the end of the book. Everyone must by now be *enlightened* about the nature and looks of these elusive Blue Streaks and also about the polemics surrounding them. The most important effects of fast Motion have been laid out qualitatively *(and to some extent semi-quantitatively)* for all to see. The *intuition-oriented description of Blue Streaks* has, hopefully, made us all *ready for sighting them* wherever and whenever future technology makes them happen. The math, had we attempted it in greater detail, would have taken us into a domain far beyond our expertise and perhaps into debates with those who already know everything about them. After some real fun with imaginative guessing games, we can now lay back and wait for the *empirical answers* to roll in, something only Future can provide. On second thought, an actual Blue Streak experiment may not be that difficult after all and, in addition, could be the cheapest scientific adventure in our high-tech Space Age. Sooo, *let's wait and see it done* or, if we are impatient, *do it ourselves.*

C - POSTSCRIPT

The Queen of Understanding

After considerable detail about Relativity, the time has finally come to reflect upon our learning experience. A realization that *understanding is profoundly intuitive* is, hopefully, of lasting benefit. Actually, understanding is possible only by *INTUITION*, so intuition is the ultimate faculty of comprehension, the *QUEEN of UNDERSTANDING*. And it should now be clear that *RELATIVITY is NOT ANTI-INTUITIVE*. It is more *ANTI-COMMONSENSE* than anti-intuitive. But does it mean that we must distrust common sense? Far from it. It took a good deal of intuition to develop it in the first place. We can improve upon it by additional doses of intuition until it becomes *"UNCOMMON SENSE."* Once acquired, we are not forced to abandon common sense. This is not an either-or choice. Common sense and uncommon sense can live in peaceful coexistence in our minds. An open mind can accommodate both.

The Rosetta Stone of Measurements

Another grain of wisdom comes from recognition that *observations can be unreliable.* Two Observers in different inertial systems measuring Distances *(or Lengths, both being Space Segments)* and Time intervals *(or Segments)* cannot get identical results. But armed with *Lorentz Transformation,* they can determine if their divergent measurements actually agree even though their numerical values differ. *Lorentz Transformation is, therefore, the ROSETTA STONE OF MEASUREMENTS*. So the phrase: *"invariant under Lorentz transformation"* does have a clear and unambiguous meaning. Because *Lorentz rescued turn-of-the-century science from going down in a quicksand of unexplained measurement chaos,* his Transformation Theory was once called *A THEORY OF MEASUREMENT,* a designation still appropriate today.

The New Absolute

If Lorentz Transformation is the Rosetta Stone of measurements in Special Relativity then *INVARIANT is the new ABSOLUTE in nature*. The *Invariant* is *the 4-dimensional measure yielding the same value to all Observers in differing Inertial Systems.* It can be directly observed and measured if, and only if, it can be specified exclusively in one dimension in which case it is also located in the Observer's Home Frame. These measurements give us *Proper or Rest values (Proper Time, Rest Length)*. *Rest Mass*, however, is a dynamic concept outside the narrowly defined realm of inertial systems. It is the resistance to acceleration offered to an applied force and is, paradoxically, measurable as *weight* if the Space coordinates remain equal to zero throughout the measuring process, that is, *if it does not move.*

What's in the name of the theory?

As *all Invariants are Absolutes,* no wonder Einstein's generalization of this new physical theory was referred to as the *Theory of Invariance*. An important part of the theory was the role of the Observer as the one and only reference point to which all Motion and all measurements had to be related *(Observer Dependence!!!)*. So, all measurements were, well, *"relative"* *(to the Observer!)*. That's why the Theory of Invariance was nick-named the *Theory of RELATIVITY* and the name stuck. But as far-reaching as it was, it had applicability only to *Uniform, Linear Motion* that was clearly an abstraction, *a special case*, something *not found in the universe at large except as an approximation.* That's why the Theory of Relativity became known as the *Theory of SPECIAL Relativity*. And now you know what's in the name of the theory we have been studying.

The two additional names not actually used in this book need pointed out. These are contained in the title of Einstein's 1905 paper: "On the Electrodynamics of Moving Bodies." The *"moving bodies"* we already know. These are none other than the Observers and objects in Motion at various, different velocities. We have also learned a great deal about *"electrodynamics,"* the Light and Radar Signals, the only messengers capable of bringing us information along Sightlines across Space in our external world. Our observations are influenced by the nature of the Signals we receive as well as by the Relative Motion of the objects under observation. These influences are responsible, as we have seen, for the many Effects and Transformations we have explored at great length. But in the final analysis, *it is the finite velocity of the electromagnetic Signals that ultimately shapes the results of our observations.* With certain qualifications, we can say that if these Signals could reach us instantaneously, without needing any Time at all for travel, then all simultaneous Events in the Space-connected Present would be revealed to us instantaneously, all at the same Moment without delay and our observations obtained this way would be totally free of those Transformations that always change *(the images of)* very-very fast-moving objects into *(the images of)* Blue Streaks.

Reality

To get *from Absolute to REALITY* is a small step. Practically all quantitative measurements, are **OBSERVER-SPECIFIC** providing different values for all Space and Time Segments, a conflict imbedded in all observations many of which can be reconciled only by Lorentz Transformation. The Invariant, therefore, rescued **REALITY** from the chaos of "relativism" ultimately threatening to give scientific theories and facts a sociological interpretation. The Invariant, the new Absolute, has proven solidly trustworthy and a clear choice for something that must really exist at the bottom of things. In other words, it represents Reality, as closely as it has been achieved at this time in history. It correlates well with a wide variety of concpts and initial observations that on closer scrutiny have proven merely partial or totally deceptive and spurring us on to an honest quest for the *"HIDDEN VARIABLES"* or the *"TRUE NATURE"* behind observations. But Reality is a word that means different things to different people. In the context of Relativity, *Reality is still something to be discovered behind disguises or APPEARANCES, also called Phenomena,* the immediate result of all observations. When carefully done, observations *(which include also measurements)* possess the virtues of *accuracy (minimal errors)*, *precision (reproducibility)* and *actuality (measured or calculated from measurements, not estimated or conjectured)*. Still, what we ultimately observe is *REALITY TRANSFORMED to IMAGES or PHENOMENA.* And we have seen that high velocities produce a number of *Visual Transformations* that are added to the linear-perspective images of stationary objects. Thus the Invariant, our ultimate Reality is often *a HIDDEN variable* to be be revealed only through *logical-mathematical reasoning, our* "mind's eye," a process we routinely use to find deeper relationships in nature.

Once we go beyond the domain of grossly observable, macroscopic world, the word "Reality" acquires even a less palpable meaning. It figured heavily in the 30-year debate between Einstein and Bohr. The issue was **OBJECTIVE REALITY versus PHENOMENALISM.** <u>**Objective Reality**</u> means that a *"REAL" WORLD* does exist even if no-one is aware of it so that the act of observation is not needed for it to exist. It is there even when we are in sleep. It was there for our great-grandparents and it was there before our world was populated by homo sapiens. *Physical entities* have always existed before they were discovered. As more of them are currently being discovered, they do not simply materialize out of nothing at the moment of discovery. *Clues* to their existence did exist even before they were discovered. There is consistency, regularity, order, pattern, a persisting organization and interaction by various forces in the universe that we can is perceive only by our "mind's eye," our intelligence, and all this is also a part of Reality. *__Phenominalism__*, usually dressed up as *Positivism,* concludes that all we ever observe are *Phenomena*. There is no possibility of finding *hidden Reality* behind Phenomena, only other Phenomena. Thus Phenomena can always be explained by other Phenomena. *Reality, therefore, becomes a(n almost) meaningless word* and we can never know what it is if at all there is such a thing. Questions about Reality at this level do not belong to Special Relativity and we have stayed wisely clear of them in all these 15 chapters.

Knowledge

And finally, we could ask what *KNOWLEDGE* is? What is it that we know? How do we know it? Do we have more of it after reading this volume? These probing questions belong to *Epistemology*, the study of what constitutes Knowledge. It includes the optics of image formation, the broad field of neurophysiology of sensory perception, the workings of our native and culturally inherited *(learned)* *intelligence*, a partially hidden, conceptual *fabric* connecting all pertinent but linguistically imprecise statements about Observations. All the epistemological subtleties hinted to pertain to the meaning of knowledge we have failed to define.

As you can see, the nature of knowledge remains hard to pin down. A few additional *QUESTIONS point out the difficulty:* **(1)** Do simple creatures who have learned to avoid danger and find food have knowledge or must knowledge be always expressed verbally, mathematically or in some other form as information before we can be considered to possess knowledge? **(2)** Can knowledge be equated with information? If so, can knowledge be coded in genes, stored in a non-verbal, kinetic memory, stored in a computer memory, contained in a book, an encyclopedia, a whole library, or found in an undeciphered writing on a clay tablet? **(3)** What kind of correspondence is there between knowledge and its subject to which it pertains? **(4)** Do abstract logical and mathematical rules of inference also constitute knowledge? **(5)** What is the difference *(if any)* between knowledge and understanding, is it possible to know something without understanding it? And so on and on. *Let these unanswered questions remain simmering in our minds to awaken curiosity that leads to a search for more wisdom in knowledge.* Moreover, is there wisdom in knowledge? What is wisdom? There is no end to questions we can ask and ponder.

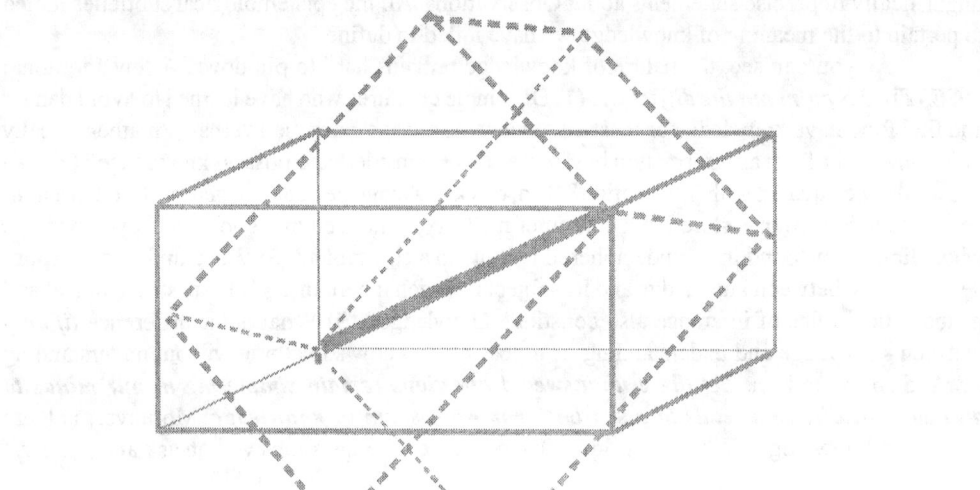

D - INDEX

ABBREVIATIONS

<A>, <B>, <C>, etc. - Observers, each identified by a capital letter enclosed in an angular bracket and in various states of Motion at Velocities from zero to c, the Velocity of Light.

Absolute (see also below): anything possessing the same magnitude in all Frames of Reference.

Absolute Velocity *(see also below)*: the Velocity generally considered the fastest possible Velocity that allows all Space locations on the path of Motion to be passed through at the same Moment, or simultaneously according to another Observer *(and as further qualified by Relativity of Simultaneity).*

c - the Velocity of Light, Radar, any electromagnetic Signal.

ETL - **Equitemporal Line** *(see also below)*: a Spacetime Map line that connects the **O**-Event to the **Q**-Events on the Worldlines of all the possible inertial objects/Observers moving at Velocities ranging from zero to **c** so that all measured **OQ**-Proper Times are equal.

ITL - **Isotemporal Line** *(see also below)*: a Spacetime Map line connecting all Events simultaneous to the Observer at the crossing of that **ITL** Map line with the Observer's Worldline.

MME - **Michaelson-Morley Experiment** *(see also below)*

O - **Event** *(see also below)*: a Spacetime Event of two or more Observers in relative, near-colliding, passing Motion fleetingly sharing the Passing Moment so as to allow synchronization of clocks by all participants for any shared experiment or observation.

Q - **Event** *(see also below)*: the Spacetime location where a sent Radar, Light or any other electromagnetic Probe Signal hits its target and is reflected back to the station that sent it.

R - **Event** *(see also below)*: the Event of receiving back a reflected Signal sent.

S - **Event** *(see also below)*: the Event of sending an electromagnetic *(Light or Radar)* Signal for the purpose of probing the target's Distance from the Observer's Spacetime location or of its Proper Time.

SL *(see also below)*: **(a) Sightlength** - the Doppler-influenced, Radar *(or Light)* Probe-imaged Distance between the nearest and the farthest Ends of a single-signal-probed object in Motion; or **(b) Sightline** - the Lightline between the Present and the Past realms which is the only line-of-sight direction or path capable of forwarding information in form of Light or other electromagnetic Signals from a variety of distant objects unreachable by other means.

UFOs *(see also below)* - Unidentified Flying Objects.

v - Velocity *(see also below)*.

w - **(a)** sum of two or more Velocities added relativistically, or **(b)** instantaneous Velocity of an accelerating object or Observer.

WB - Worldbundle

WL - Worldline

TECHNICAL TERMS & NAMES

A

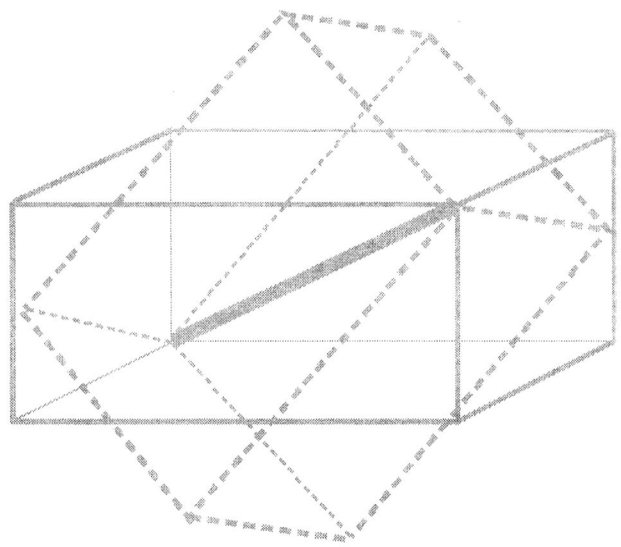